# Introduction to
# work study

# Preface

## to the first Indian edition

This highly successful international textbook on work study by the International Labour Office has been adapted by us to suit the needs of readers in India. While attempting this, we have tried to maintain the lucidity inherent in the ILO edition, and as such have tried to maintain the same framework and accordingly have reproduced verbatim from the ILO edition wherever required.

Some of the chapters have been modified or rearranged so as to make their presentation more suitable for Indian readers. A few examples have been changed. The major changes have been effected in Chapters 1, 9 and 24. In chapter 1, the focus has been changed from "productivity and standard of living" to "productivity" per se. In chapter 9, various methods of determining space requirements have been added. Chapter 24 has been redesignated as Chapter 25; in its place a new chapter 24 has been added. This chapter contains 2 integrated exercises.

Although this adaptation is based on the original ILO work the responsibility for the revision rests with us.

M.N. Pal
A.K. Chatterjee
S.K. Mukherjee

Calcutta

# Contents

Preface to the first Indian edition  v

## PART C. WORK MEASUREMENT

# TABLES

# Part A

## Productivity
## and work study

# Chapter 1

# Productivity concept and definitions

## 1.1  Introduction

Throughout the history of civilisation, there has been a constant effort to improve the utilisation of resources. A necessary requirement in this respect has been to define appropriate measures of performance for systems. The concept of productivity provides us with one such measure. Operationalised as the ratio of output to input, the productivity measure aims at identifying how efficiently the resources in a system are used in producing the desired output. The measure is applicable for an enterprise, an industry as also the whole economy. The expressions for outputs and inputs, however, vary depending on the system for which productivity is measured. Variations of the traditional definition of output-input ratio have evolved mainly due to the need to highlight the status of the system in part or in full, and to indicate the future directions of improvement for the system.

The objective of this chapter is to put the topic of productivity in perspective. Accordingly, the different definitions and the measurement problems at different levels are presented in the next two sections, and the benefits of higher productivity are discussed in the final section.

## 1.2  Definitions of productivity

As already noted in the introduction, productivity is defined as:

> **The ratio of output produced to the input resources utilised in the production.**

The input resources in any productive system are one or more among these four: land, material, machines and men. Over and above the overall measure of productivity, where all the inputs are taken together, there is a need to have partial measures relating output to individual input, in order to pin-point the areas of deficiency.

Two basic variations around the above definition have thus evolved:

> **Total productivity is the ratio of the aggregate output to aggregate input. Partial productivity is the ratio of the aggregate output to any single input.**

The definitions apply for an enterprise, an industry or an economy as a whole.

Partial productivity measures, such as, labour productivity and capital productivity, are to be used together with the total productivity measure in order to make any meaningful interpretation. The aggregate output of a system, for example, may increase because of replacement of old machineries at a certain cost; if the labour input remains unchanged, the labour productivity, defined as output per man-hour, will show an increase, while the capital productivity, expressed as the ratio of output to capital assets employed, will register a decrease. The change in total productivity will depend on the relative proportion of increase in output and the proportion of increase in input costs due to new machineries. It will be erroneous here to act based on the labour productivity figure *per se*. Both types of measures are to be used for proper interpretation of performance and subsequent action.

Within the above broad definitions, different expressions for outputs and inputs have been used by different researchers. The measurement of output at national level poses quite a different problem *vis-a-vis* the measurement at industry or enterprise level. Similarly, depending on whether the objective of measurement of productivity is to highlight the performance of a system over time, or to highlight the relative performance of a system with respect to similar systems, the expressions for outputs and inputs may vary. For example, if we are interested in comparing the labour productivity of two enterprises marketing the same product, with one enterprise engaged in both manufacturing and assembly, and the other doing only the assembly, obviously the measure of output per man-hour will be a wrong one to use. If the expression of output is changed to 'value-added output', whereby we deduct all purchases made by the organisation from the total sales, we get a better picture through the resulting productivity measure.

Thus, depending on what we want to highlight, and the type of system for which productivity measure is sought, different indices are found in use. In the next section, we take up the issue of productivity measurement at different levels. The discussion on value-added concept and other indices are deferred to the second chapter where we will examine productivity at the enterprise level.

## 1.3  Productivity measurement at national, industrial and enterprise level

The basic objectives behind productivity measurement are:

a) to study performance of a system over time;

b) to attain a relative comparison of different systems for a given level; and

c) to compare the actual productivity of a system with it's planned productivity.

At the macro levels of a nation or an industry, as also at the micro level of an enterprise, the development of productivity measures have been guided by the above objectives.

Though the concept of productivity is a fairly straightforward one, it's measurement poses a serious problem. Given the different types of output a system often produces, it is necessary to have a common unit of measurement to arrive at the aggregate output. Similarly, whenever different kinds of input like capital and labour are to be added to arrive at the aggregate input figure, we again need a common unit of measurement for these inputs. Even for the partial productivity measures involving only one kind of input, problem exists in terms of aggregating over different kinds of the same input. For example, labour as an input may have different categories like skilled and unskilled, that need to be aggregated.

The above difficulty in arriving at the aggregate output and input is commonly known as the Aggregation problem. This is a well-researched problem and forms a subject of study in itself. For the purpose of this book, we will be dealing with this topic in brief and in very general terms.

The most common way out of the Aggregation problem is to express both the outputs and the inputs in monetary terms. This can be achieved by multiplying the outputs and the inputs by their respective prices. To eliminate the effect of price variations over time, it is customary to inflate or deflate the price figures. However, given a situation, the imputing of prices of different kinds of outputs and inputs raises a host of problems, hence the resultant measure would normally entail a number of assumptions. This is further compounded by the problem of absence of data at macro levels.

Both at the national and the industry level, labour productivity is found to be the most common measure. This may be due to the fact that advances in any country have taken place by the displacement of labour by capital. Thus, the labour productivity measure of output per worker in an economy, is taken as a better indicator of the level of prosperity of the people in the economy compared to any other partial measures, with a higher labour productivity figure being interpreted as an advancement of the economy.

At the national level, the measure of output stems from 'national product', which connotes the market value of the output of 'final' goods and services produced by the nation's economy. The word 'final' is used to connote goods and services which are not resold. Variations of this, like gross national product (GNP) and gross domestic product (GDP), are used as measures of output for an economy. At the industry level, value-added output is a common measure of the numerator of the productivity measure. The input of labour is normally operationalised at the macro level by taking the total number of workers in the system.

The above gives only a broad idea on productivity measurement at the economy and the industry level. For a more detailed discussion on the topic, the reader is referred to the readings at the end of this chapter. For estimates of productivity of the Indian

Economy (1980–1989), one may refer to the study conducted by the National Productivity Council, cited in the references [1].

At the enterprise level, the problems of aggregation and the absence of data are much less pronounced compared to the above. However, the need to come out with meaningful measures, in line with the objectives narrated at the beginning of this section, has compelled many researchers and practitioners to work on the area. Here again, we find the dominance of the labour productivity measure, mainly because of easy availability of data and the relatively easier treatment involved in the approach *vis-a-vis* the other partial measures. However, the type of measure to be used depends on the nature of the enterprise. An enterprise with a high capital investment would naturally be more interested in the capital productivity, whereas, one using very costly raw materials would be concerned about the material productivity. Our major concern in this book being productivity improvement at the enterprise level, we take this up separately in the next section.

## 1.4   Benefits of higher productivity

From the definition of productivity, it is apparent that a higher total productivity may result, if:

a) more output is produced with the same or lesser input,

b) the same output is produced with lesser input, and

c) more output is produced with more input; the proportional increase in output being more than the increase in input.

In the first case, it is obvious that more goods and services are made available to the community as a whole, and as such, higher productivity contributes to a greater material well-being, hence better standard of living.

In the second case, we find the productivity gain occurring due to the reduction of input. The common fear that higher productivity leads to unemployment has its roots in such a situation. Given the different types of input like labour, capital and material reduction in input may actually be in terms of any one or a combination of these different types. Thus, labour may always be the one that is reduced to achieve the productivity gain in a given situation. The unemployment that is created in such cases may be obviated by using the excess labour for production of other goods and services. If this is achieved, then unemployment is only a temporary phenomenon, and in the long-term society gains in terms of a better standard of living.

Finally, when both the output and the input increase, with the proportional increase in output more than the proportional increase in input, it implies that the productivity gain has occurred by the displacement of one input by another. Historically, productivity gain normally occurs in this fashion. Displacement of labour by capital, resulting in higher capital input and lower labour input, with the overall input increasing and the output increasing in greater proportions, has quite often been the story of advancement of nations. Once again, in such cases, the drive for higher productivity has to be accompanied by corresponding plans to generate employment.

So far in our discussion, higher productivity has been treated synonymous to increase in total productivity. This is true in general. As is obvious from the last paragraph, as also the basic definitions, higher total productivity may be accompanied by a decrease in some of the partial productivity measures. The importance to be given to any particular partial measure depends on the objectives of the nation, the industry or the enterprise for which the total productivity gain will accrue. The effect of the differential treatment to different resources is normally taken care of in the long run by proper planning resulting in higher productivity in all spheres, and hence a better standard of living for all.

Our concern in this book is to look at the problem of raising productivity at the enterprise level. As such, we are interested in examining the ways in which one can increase the productivity of different kinds of resources individually and together. In the present chapter, we have introduced the concept of productivity. We now take up the issues of measuring and raising productivity at the enterprise level.

## REFERENCES

1. **Productivity in the Indian Economy: Some Tentative Estimates**, NPC Research Section, *Productivity* **29(4)**:372.

# Chapter 2
# Productivity in the individual enterprise

## 2.1 Introduction

Enterprise forms the smallest unit of an economy. Planning for higher productivity at an enterprise level, though predominantly guided by it's own profit-making and related objectives, has to be undertaken under the broad framework of the national plan. Taxation policy, labour laws, general availability of different resources, general level of demand of different goods and services in the country, are but some areas where an individual enterprise does not have much control. These areas are normally covered at the national level, and they define the constraints under which the individual enterprise is to operate. Productivity improvement at the enterprise level is thus initiated by identifying the resources, the improvement of the utilisation of which will contribute positively to the organisational objectives. The subsequent steps of improving and implementing are guided by the constraints as discussed above.

Operationalisation of productivity improvement is achieved mainly through partial measures. If raw material forms the major cost component of the unit cost of a product of a company, it is perhaps natural for the company to examine it's material productivity figure. While the analysis of the total productivity figures over time reveals how the company is doing in overall terms, a similar analysis of partial measures suggests the direction for future improvement. Techniques for productivity improvement are thus based on the directions suggested by analysis of partial measures. While any technique may be chosen initially with a view to raising the productivity of any particular resource, it is almost always accompanied by changes in productivity of some other resources. Measurement of total productivity assumes importance at this stage, to ascertain whether a technique goes to improve the overall productivity.

The objective of this chapter is to examine the process of raising productivity of an enterprise. Approaches to obtain the overall productivity measure of a company is first taken up. This is followed by a discussion of the partial measures, with a view to explore the process of raising productivity through these measures. In the final section, we provide a broad outline of the various factors that contribute to productivity improvement.

## 2.2 Productivity measurement approaches at the enterprise level

Different approaches to measurement of total productivity have evolved over time. The basis of such approaches has rested on the issue of what constitutes a valid overall performance measure for an enterprise. Three approaches to measurement of overall productivity are presented in this section. The approaches are in no way exhaustive, and one may refer to the readings at the end of this chapter for other approaches.

As already stated in the first chapter, total productivity is expressed as the ratio of the aggregate output to the aggregate input. That the overall performance is captured in this ratio, becomes apparent, if we examine the relationship between this ratio and the age-old performance measure of profit. If the outputs and the inputs for the period for which productivity is measured, are expressed in rupees, then under certain restrictive assumptions one can write:

Aggregate output = Gross sales = $G$ (say); and
Aggregate input  = Cost        = $C$ (say).

Thus, total productivity $= P$ (say) $= \dfrac{G}{C}$ .......... (1)

From the broad definition of profit, we have:

Profit $= \pi = G - C$ .......... (2)

From (1) and (2), $P = 1 + \dfrac{\pi}{C}$, which implies that zero profit will give a productivity value of 1, while a loss will give a value less than 1. The profit to cost ratio will determine the increase in productivity.

The above relationship that demonstrates that increased profit to cost ratio will lead to increased overall productivity, is consistent with our expectation on how an overall performance measure should behave. However, it suffers from a number of drawbacks, some of which are listed here:

a) Given that our objective in productivity measurement is to capture the efficiency of utilisation of resources, the effect of price variations over time need to be corrected. Thus, aggregate output should be equal to gross sales suitably inflated or deflated with respect to a base year.

b) Equating output to sales implies, whatever is produced in the particular period is sold. Possibility of inventory, material manufactured for own use, etc., are not taken into consideration.

c) Equating aggregate input to cost raises a host of problems and involves several restrictive assumptions. How to account for the fixed investment and working capital, whether to take the fringe benefits into account etc., are but some of the problems.

The different approaches to measurement have arisen mainly in the context of correcting the above drawbacks. The approaches are now discussed in the following paragraphs.

## Approach 1: Total productivity

The index of total productivity has been the most common approach for measuring the overall productivity of an enterprise. While the basic definition of aggregate output to aggregate input has remained unchanged, measurement of these has generated interest in many. We present here a crude version of the approach developed by Kendrick and Creamer, and subsequently modified by Craig and Harris, and Sumanth [1, 2].

For aggregate output calculation, we proceed as follows:

Step 1. For the period under consideration, find the production figure for each of the products. Let, $X_i$ = number of units of product $i$ produced during the period.

Step 2. For each of the products, obtain the data on the base period price. Base period may be taken as an earlier period with average production. Let $Y_i$ = the base period price for product $i$ (in rupees).

Step 3. For each product, multiply $X_i$ by $Y_i$ and then find the summation of $(X_i \cdot Y_i)$ for all the products to obtain, $Z = \Sigma X_i Y_i$ (in rupees). Then $Z$ is the aggregate output.

Aggregate input is arrived at in monetary terms by converting each of the input resources separately from physical units to rupee value. As outlined earlier, all the input resources in a productive system can be classified under the four broad heads of material, man, machine and land. The steps for aggregation are given here:

Step 1. Find the value of the material input $M$ in rupees. Raw materials for different products form the major material input. Other such inputs may be energy, consumables etc. Let, $M_i$ = quantity of material $i$ used during the period for producing the products, and $C_i$ = cost per unit of raw material $i$ in the base year: then $M = \Sigma M_i C_i$.

Step 2. Find the labour input $L$ in rupees. If $l_i$ = number of man hours put in by labour category $i$ during the period under consideration, and $h_i$ = base year's hourly wage rate; then $L = \Sigma l_i h_i$.

Step 3. Find the value of the capital and the land input in rupees. The capital input can be divided into two parts: the fixed assets, like machineries and equipments, and the current assets which include cash, inventory etc.

Let, $D_i$ = Depreciation (in rupees) for the fixed asset $i$.
$I$ = cost of capital in base period (percentage).
$A$ = total working capital in the period (in rupees).
$R$ = rental or equivalent of the value of land in base period (in rupees).

Then, capital and land input = $K = \Sigma D_i + AI + R$.

Step 4. Find the aggregate input $I = M + L + K$.

From the above, the total productivity can be found.

## Approach 2: Total factor productivity

Total factor productivity has been defined in the literature as the ratio of the total value-added output to the total capital and labour input. The major difference between this approach and the one described above lies in the treatment of the material input. The rationale behind using the value-added output has been discussed in brief in the first chapter. Here, we elaborate on it further.

Value-added output is obtained by subtracting the material input from the aggregate output, with both expressed in rupees. The resultant output figure then represents the amount that has been added by the company by it's own production process. Raw materials, parts etc., that are purchased by the company are the output of some other organisation, hence it's inclusion in the productivity measure can only lead to a diffused picture of overall performance. The steps involved in the measurement of Total Factor Productivity (TFP) are as follows:

Step 1. Find the value-added output VA.

Let, $S$ = Sales (in rupees) in base period prices.
$I_1$ = Inventory (in rupees) beginning of the period.
$I_2$ = Inventory (in rupees) end of the period.
$N$ = Items manufactured for own use (in rupees).
$E$ = Exclusions (material input in rupees + rentals).

Then, $VA = S + I_2 - I_1 + N - E$

Step 2. Find the labour and capital input. This can be done as in approach 1.

Step 3. Calculate $TFP = VA/$(Labour + Capital).

The value-added approach has been used in many different situations. We have given here one possible way of measuring TFP.

Depending on the situation, the exclusion E may include factors other than material input. One may refer to the TFP measure developed by Taylor and Davis for further details. The measure is summarised in brief in Sumanth's book [2]. Finally, one may notice that we have not brought in the land input in the above. In manufacturing enterprises, the change in productivity of land is much less frequent *vis-a-vis* the changes in productivity of other resources. Thus, it is felt that the land input be taken out from both the numerator and the denominator of the productivity measure.

## Approach 3: Return on investment

This approach has its basis in the relationship between profit and productivity that we have already derived at the beginning of this section. Return on investment (ROI) is defined as the ratio of profit to total investment. Both the figures being available in the financial accounts of an enterprise, the measurement process is simplified. ROI is well recognised as an overall performance measure of an enterprise, however, it has a number of drawbacks. One of the major drawbacks is that the price variations are not corrected. This would mean that the effect of changing prices over years will affect the productivity figures. An increase in productivity may thus be registered without any actual increase in the efficiency of any of the resource utilisations. Further details on this approach may be found in the reference at the end of this chapter.

As our major objective in productivity measurement will be to initiate the process of improved resource utilisation, we will be adopting views similar to the first two approaches while looking for a performance measure.

## 2.3   Productivity of materials

As noted in the introduction of this chapter, improvement of productivity is operationalised through partial measures. A productivity improvement technique is primarily
applied with the objective of increasing the utilisation of some particular resource, though
it may often be accompanied by better utilisation of some other resources as well. The
relative importance of each of the resources varies according to the nature of the enterprise, the country in which it is operating, the availability and cost of each type of
resource and the type of product and process. In this section and in the next one, we
discuss the broad way of increasing the utilisation of different resources.

There are many industries in which the cost of raw material represents 60%
or more of the cost of the finished product, the balance of 40% being divided between
labour and overhead costs. Many countries have to import a very large proportion of their
basic raw materials and pay for them in scarce foreign currencies. Under either of these
conditions the productivity of materials becomes a key factor in economic production or
operation; it is likely to be far more important than the productivity of land or labour,
or even plant and machinery. At the design stage, material saving may be effected by the
proper choice of product design and economical selection of plant and equipment. At the
operations stage too, proper training and motivating of workers may lead to considerable
material savings, by way of better handling of materials and reduction of rejections, to
mention a few. The techniques for improving material productivity may thus require a
separate volume. Work Study, our major concern in the current volume, contributes to
material productivity mainly in an indirect way. As such, in this book, we will be primarily
dealing with productivity improvement of plant and labour.

## 2.4   Productivity of land, buildings, machines and manpower

In the industrial setting, better utilisation of land and buildings can normally be effected
through space saving techniques. Work Study, as a scientific method for investigations
of different kinds of movements in a workplace, provides us with tools for identifying the
potential areas for space savings. Land and building productivity will be discussed in this
context in two later chapters of this book.

Finally, we come to the discussion on raising productivity for the two important
resources of men and machines. This assumes importance, as the thrust of Work Study
lies in improving productivity of these resources. Improvement here can be effected both
at the design and at the operations stage. At the design stage, both product and process
design can be chosen so as to ensure the most economical use of labour and capital
input. Similarly, at the operations stage too, inefficiencies in the use of these inputs can
be removed.

In order to see how the above can be achieved, let us take a fresh look at the
two partial measures of labour and capital productivity. As per our earlier definition on
partial measures:

$$\text{Labour productivity} = \frac{\text{Aggregate Output}}{\text{Labour Input}}, \text{ and}$$

$$\text{Capital productivity} = \frac{\text{Aggregate Output}}{\text{Capital Input}}.$$

Since aggregation of inputs is not a problem in this case, it has been customary to use physical units like man-hour and machine-hour for labour and capital inputs respectively. The aggregation of different skill categories for labour, or the different types of machines for capital can be obviated by proper weighing, to obtain an 'equivalent man' or 'equivalent machine' to represent labour or capital. Time is implicit in any productivity measure, as both the output and the input are to be measured for the period for which the measure is sought. Thus, for any output, the corresponding number of men (equivalent) employed multiplied by the number of hours put in by each will give the total labour input in man-hour. If the aggregate output is also converted into some equivalent output, then the labour productivity can be expressed in terms of output per man-hour. Productivity of machines can be obtained in a similar way.

From the above discussion, it can be seen that a way to effect improvement in partial measures would be to reduce the time required to produce one unit of an output. For example, if $Y$ denotes the output produced (number of units) during a period, and $X$ denotes the man-hour required to produce one unit (implying $X$ men, each working for one hour to produce one unit), then the total man-hours required for the production of $Y$ units will be equal to $(XY)$.

Taking labour productivity $(L)$ as the ratio of output to man-hour, we have $L = \dfrac{Y}{XY} = \dfrac{1}{X}$; which implies that if $X$ is reduced, then the productivity can be increased. The above rests on two fundamental assumptions; the first being that it is possible to reduce $X$, and the second is that there is no difference between the required man-hour and the actual man-hour. The first assumption raises the issue of level of technology and the like, while the second assumption, in case it is violated in practice, points towards another direction in which productivity improvement may be achieved.

It is thus felt necessary to have a working definition of the time (man-hour required to produce one unit), in order that we can look into the ways of reducing it and hence improving labour productivity. Finally, we realise that there could be a gap between the required man-hour and the actual man-hour, because of various kinds of inefficiencies in the operations. Thus, apart from reducing $X$, reduction of the above gap will help us in increasing the labour productivity. These issues are dealt with separately in the next chapter under work content and ineffective time. In the above discussion, we have used labour productivity as a vehicle for explanation, as an extension of the concepts to capital productivity is obvious.

In the final section of this chapter, we will try to summarise the factors contributing to higher productivity.

## 2.5  Factors contributing to productivity improvement

From the foregoing discussion, it is apparent that the following two issues assume importance in the context of raising productivity of labour and capital:

a) the time required to produce one unit of an output; and

b) the difference between the actual input that has gone into the production process, and the required input as calculated from the data on actual output and the unit time.

Both (a) and (b) above are expressed in either man-hour or machine-hour. Any factor effecting reduction in either (a) or (b) basically contributes to productivity improvement. Reduction in the former would involve factors related to the design of the product or the process, while reduction in the latter could be achieved by removing operational inefficiencies.

Some of the design aspects that are important for raising productivity in the above context are presented below:

a) Factors related to product design

- Standardisation of components
- Quality Standards
- Provision for use of high production machineries wherever possible

b) Factors related to process design

- Specifications of the operating conditions
- Interactions among men and machines
- Working Methods of the operatives

Thus, proper standardisation, fixing proper quality standard, removing wasted movements of operatives that may be inherent in the process design etc., go for contributing to higher productivity.

At the operations stage, inefficiencies often arise by way of improper planning of work, improper maintenance of machines, carelessness and negligence on the part of the operatives, etc. All these eventually go to increase the gap between the actual input used and the input that is otherwise required technologically. The factors that are important for raising productivity in such cases, can be summarised under two broad heads:

a) Managerial Efficiency, and
b) Workers' Efficiency.

Management, its tasks, responsibilities etc., are beyond our scope of discussion. We just note here, that for any enterprise, the management is primarily responsible for making proper utilisation of the resources. A broad outline of the managerial role is given in Figure 1. Proper planning of operations improve the operational efficiency and hence raises productivity. The workers' efficiency, on the other hand, is predominantly guided by the managerial efficiency. However, there are certain factors like maintaining the clock time and working with certain targets that are more under the control of the workers. As human beings, the labour resource has an unique position vis-a-vis other resources. Unlike other resources, the workers themselves can act to improve their own efficiency as also the efficiency of any other resources.

The above gives a summary of the different factors for raising productivity at the enterprise level. This will be taken up in more details when we discuss the productivity improvement techniques in the next chapter.

*Figure 1. Role of the management in coordinating the resources of an enterprise*

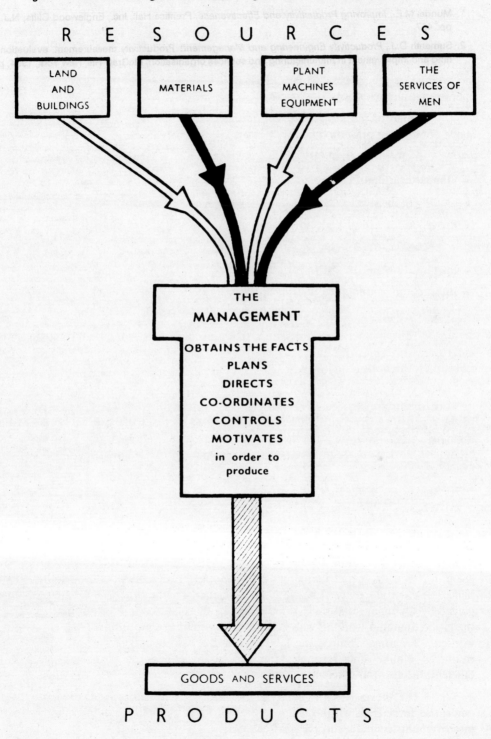

## REFERENCES

1. Mundel M.E., *Improving Productivity and Effectiveness*, Prentice-Hall, Inc., Englewood Cliffs, N.J., 1983, pp. 467

2. Sumanth D.J., *Productivity Engineering and Management*: Productivity measurement, evaluation, planning and improvement in manufacturing and services organisation, McGraw-Hill, New York, 1984, pp. 547

# Chapter 3

# Techniques for productivity improvement

## 3.1 Introduction

Higher productivity in organisations leads to national prosperity and better standard of living for the whole community. This has motivated several works on productivity improvement at different levels. Consequently, today we have a significant body of literature on the subject. The objective of this chapter is to discuss some productivity improvement techniques at the enterprise level.

While different frameworks can be adopted for discussing the above, for our purpose, we will develop it based on the issues raised in the last two sections of the last chapter. The importance of unit time and the difference between actual and required input for raising productivity have been noted there. Subsequently the factors that can contribute to improvement by reducing the above two have been outlined. In this chapter, these concepts are first formalised through the definitions of 'Work Content' and 'Ineffective time'. The improvement techniques are then presented in the context of reducing Work Content and ineffective time. In the final section, the management of productivity is discussed.

## 3.2 Work content and ineffective time

The objective of this section is to formalise the concept of 'time to produce one unit'. Given a level of technology, one would expect this time to be fairly constant for any given product. In reality, an improper design of the product or process, as also inefficiencies in operations, lead to different figures. If we assume that it is possible to define a theoretical minimum of such times, to which some excess and ineffective time gets added, then raising productivity can be looked at as a drive for approaching the theoretically minimum. A further classification of excess and ineffective time helps us in identifying the types of techniques that may be suitable for the purpose. The following definitions are presented in this context.

> **Basic Work Content is the irreducible minimum time theoretically required to produce one unit of output.**

Work content means the amount of work "contained in" a given product or process measured in man-hour or machine-hour. The basic work content is the time the product would take to manufacture or the operation to perform, if the design or specification was perfect, if the process or method of manufacture or operation was perfectly carried out, and if there was no loss of working time from any cause whatsoever during the period of the operation (other than legitimate rest pauses permitted to the operative).

Except in some cases like in processing industries, actual operation times are far in excess of this theoretical minimum. Thus, we have:

> **Total Work Content = Basic Work Content + Excess time**

As explained above, this excess time is contributed by the improper design of the product or the method:

> **Excess time =**
>
> **Time due to defects in design or specification of the product**
>
> **+**
>
> **time due to inefficient methods of manufacture or operation**

The above assumes uninterrupted working. In practice, however, interruptions occur, causing the worker or machine or both to remain idle. Leaving aside the interruptions due to sudden powercuts etc., which may not be under the control of the organisation, the other interruptions which are caused due to shortcomings on the part of either the management or the workers, need to be examined for raising productivity. The time lost due to such interruptions are termed as ineffective time.

> **Ineffective time is the time for which the worker or machine or both are idle due to the shortcomings of the management or the worker.**

From the above, it is apparent that the total time of operation under existing conditions is basically the summation of the total work content and the ineffective time.

Figure 2 provides a diagrammatic summary of the above concepts. Figures 3 and 4 present the details on how the excess work content and ineffective time arise in practice. In the next two sections, we present a brief overview of different methods that have been developed in the past, to reduce the work content and the ineffective time.

## 3.3    Improving productivity by reducing work content

We have observed that improper product and process design lead to excess work content. We now present in brief, the methods for removing such defects.

If the design of the product is such that it is not possible to use the most economical processes and methods of manufacture, this is usually because designers are not familiar enough with these processes. The weakness can be overcome if the design and production staffs work closely together from the beginning. If the product is to be produced in large quantities or is one of a range of similar products produced by the firm, improvements to make it easier to produce can be introduced at the product development stage, when production staff can examine the components and assemblies and call for changes before money has been spent on production tools and equipment. At this time also, alterations in design can be made to avoid making it necessary to remove too much material, and tests can be made in running the product to ensure that it meets the technical specifications demanded. The equivalent to the product development stage in the chemical and allied industries is the pilot plant. In transport (a non-manufacturing industry) the equivalent is the experimental service or the proving flights which are carried out on airliners.

Specialisation and standardisation are the techniques by which the variety of products or components can be reduced and batch sizes increased so that use can be made of high-production processes.

The other factor affecting the design is the product quality. Quality standards must be geared to requirements. They should be set neither too high nor too low, and they should be consistent. The management must be sure of the requirements of the market and of the customer, and of the technical requirements of the product itself. The first two may be established by market research and consumer research. Where the quality level is set by technical considerations, product research may be necessary to establish what it should be. Ensuring that quality requirements are met in the production shops is the concern of the quality control or inspection function. The men who perform this function must be properly informed of the quality level required and should be able to advise the designers which quality standards can safely be altered to achieve higher productivity.

Figure 5 shows the effect of applying these techniques to reduce the work content of the product. Yet another technique, which is used also to reduce the work content due to the process or method, is value analysis, the systematised investigation of the product and its manufacture to reduce cost and improve value.

If proper steps are taken to remove features that cause unnecessary work in the product before production actually starts, effort can be concentrated on reducing the work content of the process. In industries which have developed their practice from engineering, it is usual nowadays for the process planning function to be responsible for specifying the machines on which the product and its components shall be made, the

## Figure 2. How manufacturing time is made up

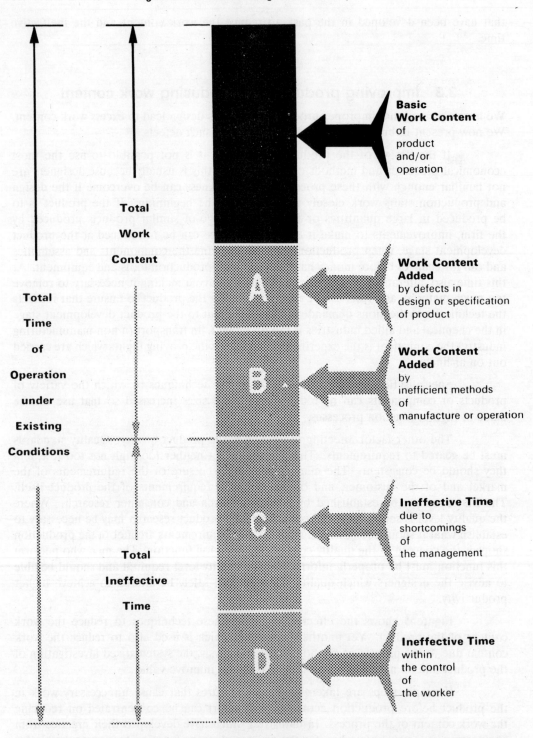

**Basic Work Content** of product and/or operation

**Total Work Content**

**Total Time of Operation under Existing Conditions**

A

B

C

D

**Work Content Added** by defects in design or specification of product

**Work Content Added** by inefficient methods of manufacture or operation

**Ineffective Time** due to shortcomings of the management

**Total Ineffective Time**

**Ineffective Time** within the control of the worker

*Note:* In the B.S. *Glossary* the terms "work content" and "ineffective time" are accorded precise technical meanings which differ slightly from those used here. The *Glossary* definitions are intended for use in applying work measurement techniques, and are not strictly relevant to the present discussion. In this chapter and the next, "work content" and "ineffective time" are used with their ordinary common meanings, as defined in the text.

*Figure 3. Work content due to the product and processes*

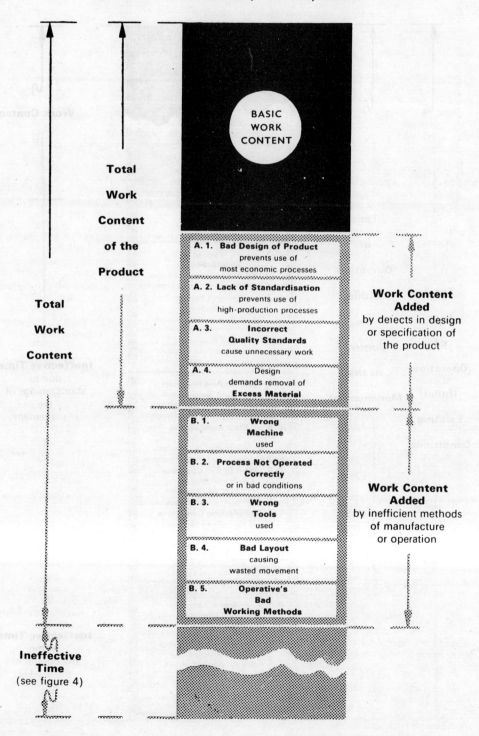

Total Work Content of the Product

Total Work Content

**BASIC WORK CONTENT**

**A. 1. Bad Design of Product** prevents use of most economic processes

**A. 2. Lack of Standardisation** prevents use of high-production processes

**A. 3. Incorrect Quality Standards** cause unnecessary work

**A. 4.** Design demands removal of **Excess Material**

**Work Content Added** by defects in design or specification of the product

**B. 1. Wrong Machine** used

**B. 2. Process Not Operated Correctly** or in bad conditions

**B. 3. Wrong Tools** used

**B. 4. Bad Layout** causing wasted movement

**B. 5. Operative's Bad Working Methods**

**Work Content Added** by inefficient methods of manufacture or operation

**Ineffective Time** (see figure 4)

*Figure 4. Ineffective time due to shortcomings on the part of management and workers*

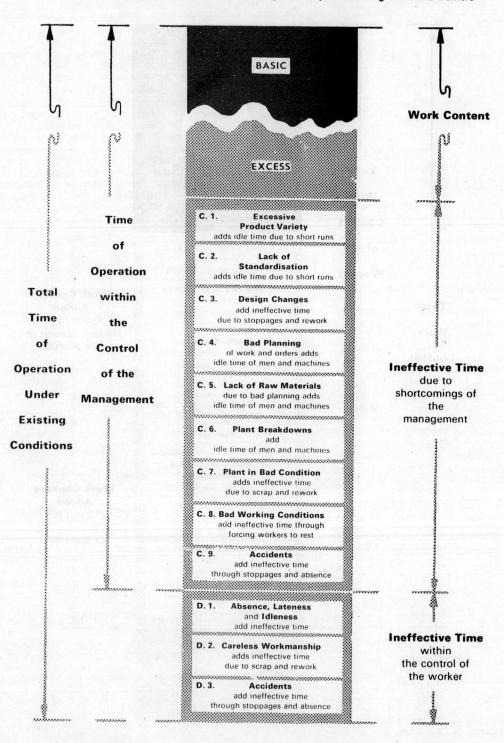

Note:   "Idle time" is used here in the ordinary sense of the term, not that defined in the B.S. *Glossary*.

*Figure 5. How management techniques can reduce excess work content*

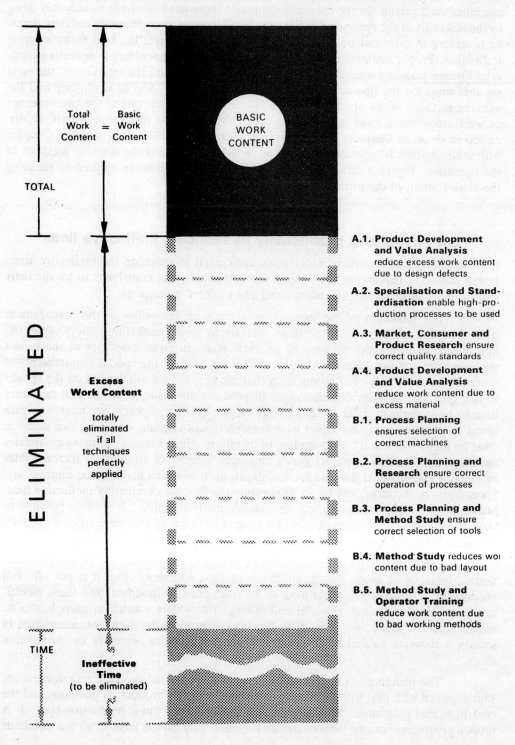

Total Work Content = Basic Work Content

TOTAL

BASIC WORK CONTENT

ELIMINATED

TOTAL

Excess Work Content

totally eliminated if all techniques perfectly applied

TIME

Ineffective Time (to be eliminated)

**A.1. Product Development and Value Analysis** reduce excess work content due to design defects

**A.2. Specialisation and Standardisation** enable high-production processes to be used

**A.3. Market, Consumer and Product Research** ensure correct quality standards

**A.4. Product Development and Value Analysis** reduce work content due to excess material

**B.1. Process Planning** ensures selection of correct machines

**B.2. Process Planning and Research** ensure correct operation of processes

**B.3. Process Planning and Method Study** ensure correct selection of tools

**B.4. Method Study** reduces work content due to bad layout

**B.5. Method Study and Operator Training** reduce work content due to bad working methods types of tools necessary and the speeds and feeds and other conditions under which the machines shall be run. In the chemical industries these conditions are usually laid down by the scientists in the research department. In all types of manufacturing industry it may be necessary to carry out process research in order to discover the best manufacturing techniques. Proper maintenance will ensure that plant and machinery is operating properly. Process planning combined with method study will ensure the selection of the most suitable tools for the operative. The layout of the factory, shop or workplace and the working methods of the operative are the task of method study, one of the two branches or work study which form the main subject of this book. As method study will be discussed in detail in Chapters 7 to 12 nothing more will be said about it here. Coupled with method study is 'operator training' as an aid to improving the working methods of the operative. Figure 5 shows the effect of these techniques when applied to reducing the work content of the process.

## 3.4    Improving productivity by reducing ineffective time

Both managerial and workers' efficiencies are crucial in reducing the ineffective time. In the following paragraphs, we discuss the techniques that contribute to productivity improvement by increasing the managerial and workers' efficiencies.

The reduction of ineffective time starts with the policy of the management concerning the markets which the firms shall try to serve (marketing policy). Shall the firm specialise in a small number of products made in large quantities at the lowest possible price and sell them cheaply, or shall it try to meet the special requirements of every customer? The level of productivity that can be achieved will depend on the answer to this question. If many different types of products are made, it means that machines have to be stopped in order to change one type to another; workers are unable to gain speed on work because they never have enough practice on any one job. This decision must be taken with a full understanding of its effects. Unfortunately, in many companies the range and variety of product grows unnoticed because of attempts to increase sales by meeting every special demand for variations, most of which may well be unnecessary. Specialisation, therefore, can be an important step towards eliminating ineffective time. Standardisation of components will also reduce ineffective time. It is often possible to standardise most of the components in a range of models of the same type of product; this gives longer runs and reduces the time spent in changing over machines.

Much ineffective time is caused by failing to ensure that the product is functioning correctly or meets the requirements of the customers before it is put into full production. Consequently, parts have to be redesigned or modified, and these modifications mean wasted time, material and money. Every time a batch of parts has to be remade there is ineffective time. The function of product development, mentioned in section 3 above, is to make these modifications before work begins in the production shops.

The planning of proper programmes of work so that the plan and workers are kept supplied with jobs without having to wait is known as production planning, and the control of that programme to ensure that it is being carried out is production control. A proper programme can be worked out and applied only on the basis of sound standards of performance. These are set by the use of work measurement, the second technique of work study. The importance of knowing accurately how long each job may be expected to take is discussed at length in the chapters on work measurement.

Workers and machines may be made idle because materials or tools are not ready for them when they are needed. Material control ensures that these requirements are foreseen and fulfilled in time, and at the same time that materials are bought as economically as possible and that the stocks maintained are not excessive. In this way the cost of holding stocks of materials is kept down.

Improper maintenance of plant and machinery may also lead to high ineffective time due to breakdowns and rejections, calling for a separate maintenance function. Finally, management should provide for proper precautions for the safety of the workers and good working conditions. In the absence of these, loss of time may occur through accidents and absenteeism and workers having to take more rest to overcome fatigue.

The proper use of the available time depends on the worker as well. Most people who have been doing a job for a long time have a certain pace at which they work best and at which they will normally work. Any attempt to speed up the rate of working, except by proper training, will tend to increase the number of errors made. The worker can save time mainly by reducing the amount of time when he is not working, that is, when he is talking to his fellow workers, smoking, waiting to clock off, or is late or absent. In order to reduce this ineffective time the following factors are often helpful:

a) bad working conditions make it difficult to work for long stretches at a time without frequent periods of rest, and produce a negative attitude in the worker;

b) if the worker feels that he is simply looked upon by the management as a tool of production, without any regard being paid to his feelings as a human being, he will not want to make a greater effort than he has to in order to keep his job;

c) if the worker does not know what he is doing or why he is doing it, if he knows nothing of the work of the firm as a whole, he can hardly be expected to give of his best;

d) if the worker feels that he does not receive justice from the management, the feeling of grievance will hinder him from doing his best.

The willingness of the worker to get on with the job and reduce this ineffective time depends very much on the personnel policy of the management and its attitude towards him. Personnel policy involves the whole relationship between the management and employees; if this relationship is not a good one, it is very difficult to make any techniques work satisfactorily. A sound personnel policy includes the training of managers and supervisors of all ranks in proper attitudes to and relations with the workers.

A motivating climate, a job that allows for variety and a soundly based wage structure, including where appropriate, incentive schemes, can motivate the worker to reduce ineffective time and hence will make for high productivity. Careless workmanship and the carelessness which leads to accidents can be overcome by a suitable personnel policy and proper training. Thus, management has a great responsibility for reducing the ineffective time due to the action or inaction of workers. This reduction is shown diagrammatically in Figure 6.

*Figure 6. How management techniques can reduce ineffective time*

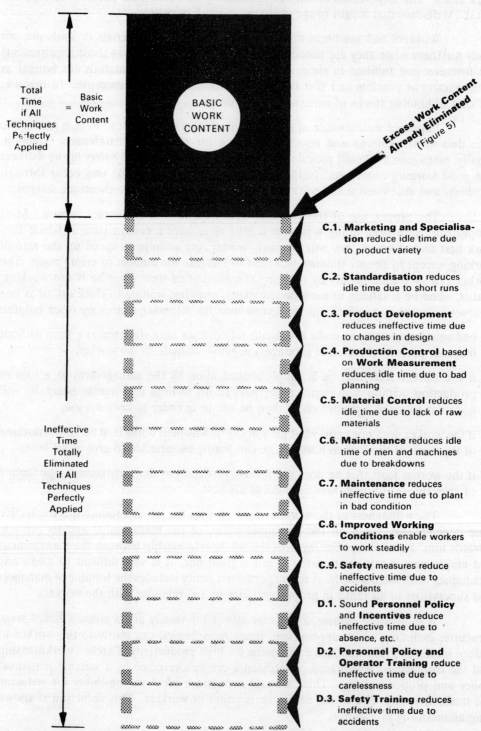

Total Time if All Techniques Perfectly Applied = Basic Work Content

BASIC WORK CONTENT

Excess Work Content Already Eliminated (Figure 5)

Ineffective Time Totally Eliminated if All Techniques Perfectly Applied

**C.1. Marketing and Specialisation** reduce idle time due to product variety

**C.2. Standardisation** reduces idle time due to short runs

**C.3. Product Development** reduces ineffective time due to changes in design

**C.4. Production Control** based on **Work Measurement** reduces idle time due to bad planning

**C.5. Material Control** reduces idle time due to lack of raw materials

**C.6. Maintenance** reduces idle time of men and machines due to breakdowns

**C.7. Maintenance** reduces ineffective time due to plant in bad condition

**C.8. Improved Working Conditions** enable workers to work steadily

**C.9. Safety** measures reduce ineffective time due to accidents

**D.1.** Sound **Personnel Policy** and **Incentives** reduce ineffective time due to absence, etc.

**D.2. Personnel Policy and Operator Training** reduce ineffective time due to carelessness

**D.3. Safety Training** reduces ineffective time due to accidents

## 3.5   Management of productivity

In the foregoing sections, we have noted the different productivity improvement techniques that may be useful at different stages of the manufacturing process. In this context, we have established Work Study as a method of raising productivity. Successful application of any such technique requires not only knowledge specific to a skill, but also commitments from management and employees at all levels. Management of productivity, rather than improvement *per se*, thus becomes relevant. Techniques form only a part of the total process of managing, which include planning, measurement, improvement and control of productivity.

In this book, we will be concerned with managing productivity through Work Study. The subject of Work Study is presented in the next chapter. In the subsequent chapters of this part, we discuss the importance of human factors and working conditions in successful application of Work Study.

# Chapter 4
# Work study

## 4.1  Introduction

Work Study was widely known for years as 'time and motion study'. With the development of the subject and its subsequent application to a very wide range of activities, it was felt by many that the older title was both too narrow and insufficiently descriptive. Work Study, as it stands today, provides us with a scientific approach to investigate into all forms of work, with a view to increase productivity. While many techniques for raising productivity are available today that qualify as a scientific approach, not all of them fall under domain of Work Study. Among the techniques that we talked of in the last chapter, Method Study and Work Measurement are the two that are covered in Work Study. The objective of this chapter is to introduce the reader to the subject of Work Study. The formal definition and the procedure are presented in the next two sections, and the conditions necessary for successful application of the subject are discussed in the concluding section.

## 4.2  Definition

The definition of Work Study as given in the British Standard Glossary is as follows:

> *Work Study* is a generic term for those techniques, particularly *method study* and *work measurement*, which are used in the examination of human work in all its contexts, and which lead systematically to the investigation of all the factors which affect the efficiency and economy of the situation being reviewed, in order to effect improvement.

While the linkage between Work Study and Productivity is apparent from the above, it is not exactly clear as to what makes it different from other productivity raising techniques. It is thus necessary to examine the definitions of Method Study and Work Measurement, the two major techniques of Work Study.

The definitions from the B.S. glossary are given below:

---

*Method study* **is the systematic recording and critical examination of existing and proposed ways of doing work, as a means of developing and applying easier and more effective methods and reducing costs.**

---

*Work Measurement* **is the application of techniques designed to establish the time for a qualified worker to carry out a specified job at a defined level of performance.**

---

Thus, Method Study is concerned with the reduction of the work content of a job or operation, while Work Measurement is mostly concerned with the investigation and reduction of any ineffective time associated with it; and with the subsequent establishment of time standards for the operation when carried out in the improved fashion, as determined by Method Study. Obviously in such cases, Work Measurement is applied only after Method Study has been conducted. This relationship is shown in Figure 7. However, Work Measurement may precede Method Study, wherever the basis of conducting Method Study is not apparent a priori. In such cases, Work Measurement results may call for a Method Study by focussing on the ineffective time.

Finally, it should be noted that the techniques mentioned here, usually contribute towards increasing productivity with little or no capital expenditure. A rough guide on the role of Work Study in raising productivity is given in Table 1. The Table illustrates as to how much we can expect to gain by using Work Study to improve the use of existing resources as against investing in new plant and equipment.

We will be dealing with the details on Method Study and Work Measurement in Parts 2 and 3 of this book. For the present, we move over to the basic procedure of Work Study that applies to every study, whatever the operation or process being examined, in whatever industry. This procedure is fundamental to the whole of Work Study.

## 4.3  Basic procedure

There are eight steps in performing a complete Work Study. They are:

Step 1. Select the job or process to be studied.

Step 2. Record from direct observation everything that happens, using the most suitable of the recording techniques (to be explained later), so that the data will be in the most convenient form to be analysed.

Step 3. Examine the recorded facts critically and challenge everything that is done, considering in turn: the purpose of the activity; the place where it is performed; the

Figure 7. Work study

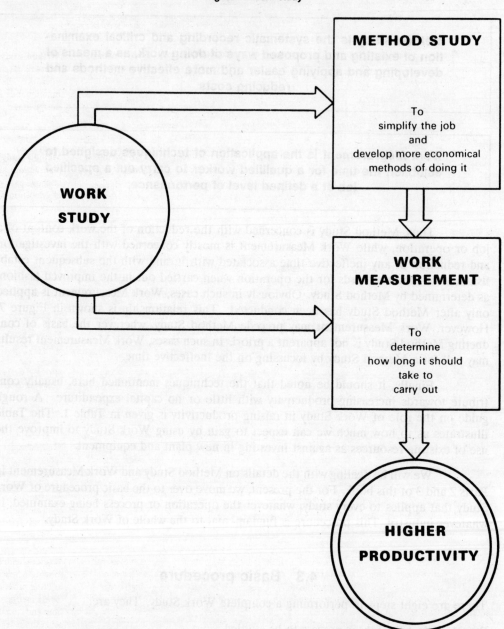

sequence in which it is done; the person who is doing it; the means by which it is done.

**Step 4.** Develop the most economic method, taking into account all the circumstances.

**Step 5.** Measure the quantity of work involved in the method selected and calculate a standard time for doing it.

*Table 1. Direct means of raising productivity*

| Approach | Type of improvement | Means | Cost | How quickly can results be achieved? | Extent of improvement in productivity | The role of work study |
|---|---|---|---|---|---|---|
| Capital investment | 1. **Development of new basic process** or fundamental improvement of existing ones | Basic research Applied research Pilot plant | High | Generally years | No obvious limit | **Method study** to improve ease of operation and maintenance at design stage |
| | 2. **Install more modern or higher-capacity plant or equipment** or modernise existing plant | Purchase Process research | High | Immediately after installation | No obvious limit | **Method study** in plant layout and to improve ease of of operation when modernising |
| Better management | 3. **Reduce the work content of the product** | Product research Product development Quality management **Method study** Value analysis | Not high compared with 1 and 2 | Generally months | Limited—of the same order as that to be expected from 4 and 5. Should *precede* action under those heads | **Method study** (and its extension, value analysis) to improve design for ease of production |
| | 4. **Reduce the work content of the process** | Process research Pilot plant Process planning **Method study** Operator training Value analysis | Low | Immediate | Limited, but often of a high order | **Method study** to reduce wasted effort and time in operating the process by eliminating unnecessary movement |
| | 5. **Reduce ineffective time** (whether due to management or to workers) | **Work measurement** Marketing policy Standardisation Product development Production planning and control Material control Planned maintenance Personnel policy Improved working conditions Operator training Incentive schemes | Low | May start slowly but effect grows quickly | Limited, but often of a high order | **Work measurement** to investigate existing practice, locate ineffective time and set standards of performance as a basis for— A. Planning and control B. Utilisation of plant C. Labour cost control D. Incentive schemes |

Step 6. Define the new method and the related time so that it can always be identified.

Step 7. Install the new method as agreed standard practice with the time allowed.

Step 8. Maintain the new standard practice by proper control procedures.

Steps 1, 2 and 3 occur in every study, whether the technique being used is Method Study or Work Measurement. Step 4 is part of Method Study practice, while step 5 calls for the use of Work Measurement.

These eight steps will all be discussed in detail in the chapters devoted to Method Study and Work Measurement. Before doing so, however, we shall discuss the background and conditions necessary for Work Study to operate effectively.

## 4.4  Prerequisites of conducting a work study

Work Study is not a substitute for good management and never can be. It is one of the tools in the manager's tool kit. By itself it will not make bad industrial relations good, although, wisely applied, it may often improve them. If Work Study is to contribute seriously to the improvement of productivity, relations between the management and the workers must be reasonably good, and the workers must have confidence in the sincerity of the management towards them. The type of relationship is often determined by the working conditions and environment provided by the management over time. Good human relations and working conditions are thus the two important prerequisites for successful application of Work Study. These are discussed separately in the next two chapters.

# Chapter 5

# The human factor in the application of work study

## 5.1 Introduction

Successful application of any technique, in general, is dependent on the people who apply them as also on those for whom they are applied. This is particularly true for Work Study where the major focus is the investigation of human work, with an aim to improve the efficiency of the same. Investigation by nature involves different questioning related to the effectiveness and efficiency of the work under consideration, and as such may not be always pleasant for the worker or the group of workers working on the job. It is thus necessary that right at the beginning, the objectives of conducting such a study be made very clear by the management and the work study man to all concerned workers and supervisors. At the study stage, the work study man has to be objective in recording all pertinent data related to the work. Successful implementation, amongst other factors, depends on the prevalent relationship between the workers and the management. It is much easier to implement any change in an organisation where mutual trust between the worker and the management exists.

Thus, the human factor plays a very important role in the successful conducting and implementation of Work Study in organisations. The management who is responsible for spelling out the objectives and planning of work, the supervisor who translates these plans into day-to-day operations and continuously monitor them, the worker who carries out the operations, and the work study man who conducts the study, all of them have to contribute positively if the study has to succeed. The objective of this chapter is to look at the ways in which these different interacting groups can contribute. Accordingly, the roles of the management, the supervisors and the workers are discussed in the next section. In the final section, we discuss the qualities necessary in a work study man.

## 5.2 Management and supervisor: Their roles in work study

Before we embark on a discussion of the roles, it is first necessary to have an understanding of how the management and the supervisor form two distinct groups in an organisation. While the management's tasks are to put forward the objectives of the organisation and planning of the different activities, the supervisor's jobs are to translate these tasks into day-to-day operations and monitor the progress by ensuring that the workers perform the work as desired, and by making relevant tools available to them. The supervisor thus acts as the liaison between the management and the worker. The nature of tasks, in turn, defines the respective roles they play in a work study situation. We now discuss below the roles under the different heads.

### Role of management

The role of management for successful application of Work Study can be summarised as follows:

a) The management should define the organisational objectives clearly. This is important, as without clarity, the workers or supervisors may set their own objectives, which may be in conflict with those of the management. Work Study aims at identifying a better method of doing work, and this better method is defined with respect to the organisational objectives.

b) The management must try to plan in a way so as to minimise certain ineffective time that may crop up otherwise. In case there is some ineffective time due to management (as described in an earlier chapter), and it is identified as a part of a Work Study, management should be open to suggestion and criticism.

c) Finally, the management should consciously attempt to maintain good relationship with the workers and provide a good work environment for them. This helps building up a mutual trust which is useful for any study to succeed.

### Role of the supervisor

The supervisory role in Work Study applications are summarised here:

a) As the liaison between the management and the workers, the supervisor should clearly communicate to the workers, the organisational objectives as laid down by the management. On the other hand, he should be able to give a clear picture to the management about the practical problems of the shop floor, to enable the management to set realistic goals.

b) As a person who is much closer to the actual jobs than the management, he should be fully aware of the different aspects of the work including its limitations. This should help him in identifying potential areas of improvement, and he could be of help in selecting the proper work for study.

c) As a person who is responsible for operationalising the plans, the supervisor has to be associated with the study right from selection of the job to be studied, to its implementation. This necessitates that he is open to ideas. *Status quo* is normally a favourite choice in our shop floor, more so to the supervisor, as he himself is in the job of operationalisation of the work. The supervisor should be conscious of this and contribute to the study by sharing his expertise on the work with the work study man.

### Role of the worker

As the person who is actually doing the job, the worker plays a crucial role in successful application of any study. His values and attitude, his behaviour as an individual and also in group should be taken into consideration for the purpose of the study. His role in the study is summarised below:

a) The worker should not neglect his work or waste time unnecessarily. As we have seen in an earlier chapter, ineffective time due to a worker may crop up because of negligence on his part. He needs to be aware of the fact that the resulting lowering of productivity will affect him in the long run (if not immediately).

b) The worker should take interest in the work and take the initiative with work related factors. Quite often, it is possible to select the job to be studied through the initiative taken by the worker. Individual, formal or informal groups or sometimes unions can be used as a platform to initiate a study.

## 5.3   The works study man

In the last section we have examined how the management, the supervisors and the workers can contribute in the context of Work Study. The objective of this concluding section is to examine the qualities necessary in the work study man. While the other three groups are insiders, so far as the organisation is concerned, the work study man is the only outsider. His job thus becomes the most difficult one in terms of coordinating among all three groups and effecting change in the system.

a) The problem of raising productivity should be approached in a balanced way without too great an emphasis being placed on productivity of labour. In most enterprises in developing countries, and even in industrialised countries, great increases in productivity can generally be effected through the application of Work Study to improve plan utilisation and operation, to make more effective use of space and to secure greater economy of materials before the question of increasing the productivity of the labour force need be raised. The importance of studying the productivity of all the resources of the enterprise and of not confining the application of Work Study to productivity of labour alone cannot be overemphasised. It is only natural that workers should resent efforts being made to improve their efficiency while they can see glaring inefficiency on the part of the management. What is the use of halving the time a worker takes to do a certain job or of imposing a production output on him by well applied Work Study if he is held back by a lack of materials or by frequent machine breakdowns resulting from bad planning by his superiors?

b) It is important that the work study man be open and frank as to the purpose of his study. Nothing breeds suspicion like attempts to hide what is being done; nothing dispels it like frankness, whether in answering questions or in showing information obtained from studies. Work Study, honestly applied, has nothing to hide.

c) Workers' representatives should be kept fully informed of what is being studied, and why. They should receive induction training in Work Study so that they can understand properly what is being attempted. Similarly, involving the workers in the development of an improved method of operation can win them over to the new method and can sometimes produce unexpected results. Thus, by asking workers the right questions and by inviting them to come forward with explanations or proposals, several work study specialists have been rewarded by dues or ideas that had never occurred to them. After all, a worker has an intimate knowledge of his own job and of details that can escape a work study man. One tried and tested practice is to invite the workers in a section to be studied to nominate one of their number to join the work study specialist and together with the foreman, to form a team that can review the work to be done, discuss the results achieved and agree on steps for implementation. While interacting with the worker, the work study man should ensure that he in no way undermines the authority of the supervisor in the eyes of the worker.

d) Although asking for a worker's suggestions and ideas implicitly serves to satisfy his need for recognition, this can be achieved in a more direct way by giving proper credit where it is due. In many instances a foreman, a worker or a staff specialist contributes useful ideas that assist the work study man to develop an improved method of work. This should be acknowledged readily, and the work study man should resist the temptation of accumulating all the glory for himself.

e) It is important that the work study man should remember that his objective is not merely to increase productivity but also to improve job satisfaction, and that he should devote enough attention to this latter issue by looking for ways to minimise fatigue and to make the job more interesting and more satisfying. In recent years several enterprises have developed new concepts and ideas to organise work to this end and to attempt to meet the workers' need for fulfilment.

The above guideline on the job requirements of a work study man can be translated into certain qualities and qualifications that one should look for in him. These can be summarised as follows:

a) **Education:** The very minimum standard of education for anyone who is to take charge of Work Study application in an enterprise is a good secondary education with matriculation or the equivalent school-leaving examination. It is unlikely that anyone who has not had such an education will be able to benefit fully from a full Work Study training course, although there may be a few exceptions. However, if a work study man is also to be involved in studying other production management problems, a university degree in engineering or management or the equivalent becomes an important asset.

b) **Experience:** It is desirable that candidates for posts as work study specialists should have had practical experience in the industries in which they will be working. This experience should include a period of actual work at one or more of the processes of the industry. This will enable them to understand what it means to do a day's work under the conditions in which the ordinary workers with whom they will be dealing have to work. Practical experience will also command respect from foreman and workers, and an engineering background enables a man to adapt himself to most other industries.

c) **Personal qualities:** Anyone who is going to undertake improvements in methods should have an inventive turn of mind, be capable of devising simple mechanisms and devices which can often save a great deal of time and effort, and be able to gain the co-operation of the engineers and technicians in developing them. The type of man who is good at this is not always so good at human relations, and in some large companies the methods department is separated from the work measurement department, although both are under the same chief. The following are essential qualities:

i) **Sincerity and honesty**
The work study man must be sincere and he must be honest; only if he is will he gain the confidence and respect of those with whom he has to deal.

ii) **Enthusiasm**
He must be really keen on his job, believe in the importance of what he is doing and be able to transmit his enthusiasm to the people around him.

iii) **Interest in and sympathy with people**
He must be able to get along with people at all levels. To get along with people it is necessary to be interested in them, to be able to see their points of view and to understand the motives behind their behaviour.

iv) **Tact**
Tact in dealing with people comes by understanding people and not wishing to hurt their feelings by unkind or thoughtless words, even when these may be justified. Without tact no work study man is going to get very far.

v) **Good appearance**
He must be neat and tidy and look efficient. This will inspire confidence in him among the people with whom he has to work.

vi) **Self-confidence**
This can only come with good training and experience of applying Work Study successfully. The work study man must be able to stand up to top management, foreman, trade union officials or workers in defence of his opinions and findings, and do so in such a way that he will win respect and not give offence.

The personal qualities, particularly the ability to deal with people, can all be further developed with the right training. Far too often this aspect of the training of work study man is neglected, the assumption being that, if the right man is selected in the first place, that is all that needs to be done. In most work study courses more time should be given to the human side of applying Work Study.

It will be seen from these requirements that the results of Work Study, however scientifically arrived at must be applied with art, just like any other management technique. In fact, the qualities which go to make a good work study man are the same qualities as go to make a good manager. Work Study is an excellent training for young men destined for higher management. People with these qualities are not easy to find, but the careful selection of men for training as work study specialists will repay itself in the results obtained, in terms both of increased productivity and of improved human relations in the factory.

Having described the background against which Work Study is to be applied, we can now turn to the question of applying it, starting with method study. Before we do so, however, some attention must be given to some general factors which have considerable bearing on its effect, namely the conditions under which the work is done in the area, factory or workshop concerned.

# Chapter 6

# The influence of working conditions on work study

## 6.1 Introduction

The interaction between working conditions and Work Study has been one of the factors the criticality of which has been accepted rather late. The effects were however evident long before the integrated set of cause i.e. the vector of working condition was recognised and identified. The influence can be broadly categorised into two groups; the first being of immediate nature leading to reduction of production raise or increase in number of rejections or increase in costs while the other being of long drawn nature like causing disability varying from purely temporary to near permanent or permanent nature.

Though the first group of effects is of immediate concern to the practice of Work Study yet the relevance of the second group cannot also be neglected particularly in the backdrop of overall economic productivity of the organisation.

The unfavourable working conditions may lead to creation, sustenance and regeneration of dissatisfaction which ultimately get manifested in the form of reduced labour productivity, increased rejections, reduced overall economy through increased turnover and absenteeism. The combined effect of all such manifestations may further lead to socio-political tension.

## 6.2 Factors affecting working conditions

The set of elements constituting the working conditions vector primarily consists of the following:

    a) Occupational Safety and Health elements
    b) Fire Prevention and Protection elements
    c) Layout and Housekeeping elements
    d) Lighting and Ventilation elements
    e) Noise and Vibration elements
    f) Ergonomics elements and
    g) Arrangement of Working time

The factors listed are those which are directly connected with and have direct impact on the task that any worker has to perform. Some of these factors when present in the adverse form also lead to the occurrence of temporary/near permanent/permanent disableness in the workers.

The other elements of the working condition set which have not been listed are, those which have only indirect influence on the performance of the worker, like amenities offered to the worker and to his/her families etc. which are appropriately formed as service conditions.

## 6.3   Occupational safety and health

Occupational safety aspects have become an integrated part of management function today. The level of occupational safety is primarily designated by the occurrence of accidents in an organisation. The safety of the worker performing to complete any given task is determined by a rather complex interaction of in-built product safeties of the different aids being used by the worker, layout and housekeeping of the workplace, work posture and other elements of working conditions vector. The level of in-built product safety of the different aids being used are primarily dictated by the price of the products but can be increased substantially by marginal modification to adopt to the situation in which it will ultimately be used, such as, providing guards, by modifying lever handles, illumination provided on the machine and providing improvised tool/work trays.

However, for aids which are to be used by immobile mode, that is, at different places and may be by different persons the scope of improving the safety features gets limited considerably.

Statistics, however, indicate that the most frequent type of accidents are not connected with the so-called 'highly dangerous machines' like machine saws and power presses nor with the 'highly dangerous working materials' like, explosives or volatile inflammable liquids. These statistics indicate that very ordinary mundane actions like stumbling, falling, being struck by a falling object, by a crane or any projected part of a moving object do cause a very large portion of accidents. These statistics further indicate that young workers considered to be possessing a very high level of physical and psycho sensorial fitness are a most accident prone segment of the workforce.

To prevent an occurrence of accidents and thereby to increase the occupational safety level is to exercise the highest possible control on the potential causes both of human and technical nature aiming at their total elimination. Since any accident requires the presence of both man and the potential cause(s) inherited by a device which is essential for the task to be performed by the man, four basic strategies for controlling occupational hazards i.e. improving occupational safety has been suggested by Gniza. The strategies and their relative effectiveness in controlling occupational hazards or in improving occupational safety are given in Figure 8.

The most potential source of accidents is the task involving manual material handling which is found to be involved in as many as 30% of all accidents. Adopting

*Figure 8. Four basic strategies to control occupational hazards*

| Strategy | (Increasing) Decreasing order of effectiveness for controlling hazards (improving occupational safety). | | | |
|---|---|---|---|---|
| a) Elimination of Hazards | Hazards | → | Individual | 1 |
| b) Removed of the individual from the exposure | Hazard | → | Individual | 2 |
| c) Isolation of Hazard | [Hazard] | → | Individual | 3 |
| d) Isolation of Individual | Hazard | → | [Individual] | 4 |

(Hazards ↑ / Safety ↓)

safe ergonomic means for such tasks besides reducing the number of such tasks and the distance of transportation can contribute significantly to the improvement of occupational safety conditions leading the reduction in the number of accidents.

Rapid technological progress seen during the last few decades has often created new and totally unrecognised occupational health hazards. Any strategy to tackle occupational health problems must involve medical as well as technical knowledge and such an imperative has caused a markedly slow pace of progress in this field as in other fields calling for a truly interdisciplinary thrust. However, as short-term bypass alternative most suitable solutions of specific nature with very limited scope for universalisation are recommended.

As in other 'health' problems 'hygiene' dictum; to be more specific 'industrial hygiene' dictum pertaining to the scenario, also provides significant preventive measures. Industrial hygiene principles aim at providing effective means of prevention sight at the design stage. Potentially dangerous operations or working with potentially hazardous materials are always recommended to be replaced whenever feasible. Industrial hygiene principles also call for reduction of exposure duration of individuals or group of workers to risky environment or else to provide personal protective equipment in cases where reduction of exposure duration is infeasible.

The impact of lack of occupational health strategies is seen to be more severe on persons already suffering from ailments of different natures, however, the number of such persons in a country like ours is not very large.

The latest types of occupational health/industrial hygiene problems make the distinction between occupational and non-occupational categorisation rather illusory.

The efforts of providing a high level of occupational safety and health standard must contain sufficient emphasis on training, educating and retraining of the workers with the statistics, analysis of past accidents, health hazards and methods already adopted as well as required to be adopted. As an effective means of supplementing such efforts, tangible rewards/incentive systems also should be planned and implemented.

## 6.4    Fire prevention and protection

Occurrence of accidents involving fire and/or explosions always result in direct and indirect costs. The extent of damage depends on the intensity and spread of fire as well as on the nature of work material involved, and on the climatic condition existing at the time of occurrence of such activities.

The first principle of prevention of fire is to adopt design and construction strategies of buildings which offer in-built fire resistance capabilities. The second principle is to give adequate training to workers and impose bans on smoking and prohibiting usage of potential ignition source and naked flames, particularly in areas which can be considered to prove fatal to the occurrence of fire accidents. The third principle is to make available fire-fighting devices like extinguishers, not constituting supplementary hazards, in easily and commonly accessible locations. The fourth principle is to adopt a suitable reliable alarm and system or warning system capable of emitting audible warning signals throughout the building. The fifth principle is to have the management (both supervisory and operational) to be well acquainted and vested with their role in the event of a fire taking place and the kind of leadership they should provide to the work-force.

To have the principles enunciated and translated into reality and to yield effective results in case of fire, the following supplementary activities play a very important role:

a) Regular mock fire-fighting exercises at different levels involving the fire-fighting team already trained.

b) A system of periodic inspection by full-time fire inspectors of the potentially hazardous (fire prone) work areas and the fire-fighting equipment and systems as well as the fire alarm/warning systems installed.

c) A suitable liaison with the local fire brigade office and with central fire brigade authorities.

## 6.5    Layout and housekeeping

To control the extent of damage caused and the spread of fire in case of occurrence of a fire accident, isolation of fire-prone work areas through optimum layout of the workplace usually provides an effective strategy.

The general principles to be adhered to for arriving at a reasonably good workplace layout design are as follows:

a) Sufficient window area, close to 20% of the floor area should be provided.

b) Minimum ceiling height should be 3 metres and each worker should be provided with at least 10 cubic metres of air and a minimum free floor area of 2 square metres.

c) Traffic aisle should be wide enough to allow free movement of vehicles and of workers at peak hours as well as it should be enough for the fastest possible evacuation in emergency situation.

d) Walls and ceilings should have a finish which prevent accumulation of dirt, dust and absorption of moisture. Floor coverings should be of non-slip, non-dust forming and easy to clean naturally.

Besides the factors already listed, an overall study of the neighbourhood and environmental protection are very important and as a matter of fact are imperative musts in a majority cases. Such an overall study also makes the integration of the plant with its surrounding most efficient and cost effective.

Housekeeping, the general state of affairs existing in the workplace comprising its cleanliness, tidiness, systematic location and in-process storing of materials, production aids as well as the state of repairs of the walls, ceilings, floors, not only contribute to accident prevention, effective fire prevention and/or fighting but also is an influencing factor of productivity. A good housekeeping also ensures faster movement of workers, material transferring vehicles reducing the time involved and thereby shortening the cycle time. Such a practice also helps in very fast location of production aids since they are systematically stored.

Cleanliness supplements good housekeeping by providing a high standard of occupational health environment. Clean workplace has the inherent capability of providing protection to workers against infection, infestation and accidents.

Disposal of process wastes and residues, particularly in situations where they act as the sources of dangerous emissions of vapours, gases or dust should be carefully monitored and implemented so that the workers as well as the environment into which such things are ultimately discharged do not get exposed to the harmful effects.

Provisions for sufficient amount of water, cooled appropriately for drinking purposes by the workers as and when necessary and for washing purposes must be made. Availability of inadequate showers and washing facility may prevent some workers from getting cleaned as soon as they come out of their workplaces. Such a condition affects the industrial hygiene standard adversely particularly in cases where such workers are exposed to dangerous or toxic work materials.

Clean working dresses significantly reduce the occurrences of skin diseases, skin sensitisation and chronic or acute irritation.

## 6.6   Lighting and climate conditioning

The mostly used communication medium for the data/information relevant for performing a task is of visual nature. So to make the information transfer effective a high level of visibility of the equipment, the product and the information are absolutely essential. Presence of good visibility ensures an accelerated rate of production and hence increased productivity. Its absence may cause occupational health hazards like persistent headache, frequent visual fatigue and also lack of occupational safety causing avoidable accidents.

Visibility depends on the quality of illumination, which in turn, depends on a number of factors, like the workpiece size and colour, its distance from the eyes, contrasts of colour and lighting levels with the background and the flare.

Table 2. *Recommended levels of illumination for different categories of work with modifications suggested for particular circumstances.*

| Task Group | Standard Service Illumination (LUX) | Reflectance Contrast usually low (LUX) | Errors will have serious consequences (LUX) | Task of short duration (LUX) |
|---|---|---|---|---|
| 1. Storage area and plant room with no continuous work | 150 | – | – | – |
| 2. Casual work | 200 | – | – | – |
| 3. Rough work, rough machining and assembly | 300 | 500 | 500 | 300 |
| 4. Routine work, offices, control room medium machining and assembly | 500 | 750 | 750 | 300 |
| 5. Demanding work, inspection of medium machining | 750 | 1000 | 1000 | 500 |
| 6. Fine work: Colour discrimination, textile processing, fine machining and assembly | 1000 | 1500 | 1500 | 750 |
| 7. Very fine work: hand engravings, inspection of fine machining and assembly | 1500 | 3000 | 3000 | 1000 |
| 8. Minute work, inspection of very fine assembly | 3000 | 3000 | 3000 | 1500 |

However, the level of illumination should be fixed not only on the basis of Task group categorisation but also in relation to the age of the worker as shown below in Table 3.

Table 3. *A qualitative variation of suggested illumination level with the age of the worker.*

| Age (years) | Qualitative level of illumination |
|---|---|
| 10 | 1/3 |
| 20 | 1/2 |
| 30 | 2/3 |
| 40 | 1 |
| 50 | 2 |
| 60 | 5 |

The installed level of illumination rapidly falls by 10 to 25% and then more slowly until it is only 50% or less of the original level mainly because of accumulation of dust and wear of the lighting elements. As a general rule the light should be uniformly diffused (Figures 9, 10, 11 and 12) yielding shadows to the extent that help to distinguish objects. The recommended maximum lighting intensity ratio for providing the optimum level of contrast is given in the following Table 4.

*Figure 9. Mounting of general lighting units*

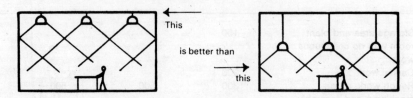

General lighting units should preferably be mounted as high as possible

*Source:* ILO, CIS: *Artificial lighting . . .,* op. cit.

*Figure 10. Need for general lighting*

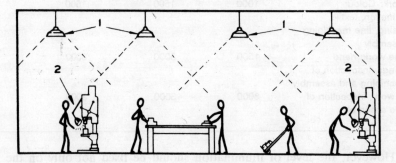

Some general lighting is always needed even when tasks are locally lit. (1) Uniform general lighting (2) Local supplementary lighting.

*Source:* ILO, CIS: *Artificial lighting . . .,* op. cit.

*Table 4. Recommended maximum lighting intensity ratios.*

| Ratios between | Ratio value |
| --- | --- |
| Work and immediate environment | 5 to 1 |
| Work and distant surfaces | 20 to 1 |
| Light source or the sky and adjacent surfaces | 40 to 1 |
| All points in the worker's immediate vicinity | 80 to 1 |

Natural lighting should be used wherever possible, but should always be supplemented with artificial light to make up variations in the intensity of natural light. Provisions should always be made to avoid glare associated with either natural light or the supplementary artificial light. Flourescent light backed up with antiglare provisions

**Figure 11. *Maximum recommended spacing for industrial type units***

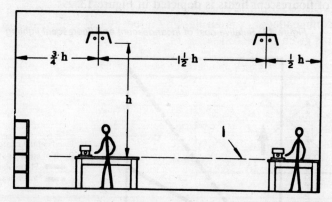

Measurements are to the centre point of the unit in all cases, and are expressed as a multiple of the mounting height *h* above the work plane (I). The ¾ *h* figure applies when there is a gangway next to the wall, whilst the ½ *h* figure is used when people work close to the wall. For louvred units, maximum spacing between fittings should be reduced to 1¼ *h*.

*Source:* ILO, CIS: *Artificial lighting . . .,* op. cit.

**Figure 12. *Factors influencing the degree of glare produced by a given diffusing fitting (or a bare fluorescent lamp unit)***

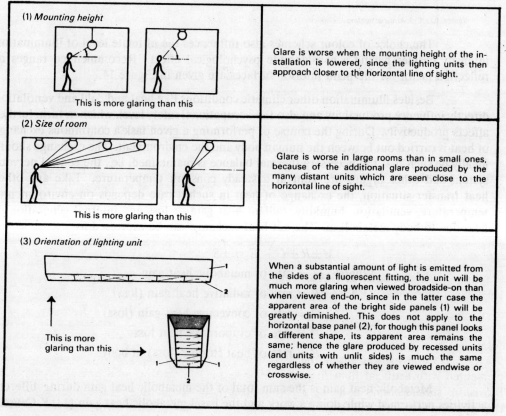

| | |
|---|---|
| (1) *Mounting height*<br><br>This is more glaring than this | Glare is worse when the mounting height of the installation is lowered, since the lighting units then approach closer to the horizontal line of sight. |
| (2) *Size of room*<br><br>This is more glaring than this | Glare is worse in large rooms than in small ones, because of the additional glare produced by the many distant units which are seen close to the horizontal line of sight. |
| (3) *Orientation of lighting unit*<br><br>This is more glaring than this | When a substantial amount of light is emitted from the sides of a fluorescent fitting, the unit will be much more glaring when viewed broadside-on than when viewed end-on, since in the latter case the apparent area of the bright side panels (1) will be greatly diminished. This does not apply to the horizontal base panel (2), for though this panel looks a different shape, its apparent area remains the same; hence the glare produced by recessed units (and units with unlit sides) is much the same regardless of whether they are viewed endwise or crosswise. |

*Source:* ILO, CIS: *Artificial lighting . . .,* op. cit.

offer an economically effective potential for extensive use in such situations. The cost effectiveness of flourescent lights is depicted in Figure 13.

*Figure 13. Relative cost of incandescent and fluorescent lighting*

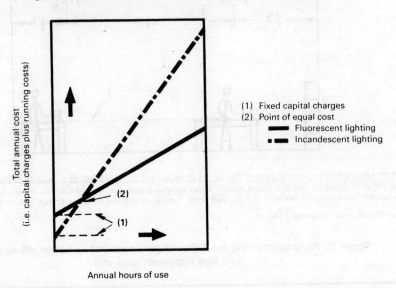

(1)  Fixed capital charges
(2)  Point of equal cost
━━━  Fluorescent lighting
▪━▪━  Incandescent lighting

Annual hours of use

*Source:* ILO, CIS: *Artificial lighting . . .,* op. cit.

The choice of colour schemes also influences the ultimate level of illumination intensity besides having a considerable psychological effect. Recommended ranges of reflection factor for the main interior surfaces are given in Figure 14.

Besides illumination other climatic conditions like heat and cold and ventilation directly influence productivity and also the occupational safety level, which in turn, further affects productivity. During the course of performing a given task a continuous exchange of heat is carried out between the human body and the environment. The exchange should ideally be such that the necessary thermal balance is maintained, i.e., the central nervous system and other organs are kept at a steady constant temperature. Like any other heat transfer situation, the exchange of heat in such a case depends on environmental temperature, ventilation, humidity, radiant heat gain or loss and internal generation of heat due to body metabolism. The net heat exchange can be expressed as

$$M \pm R \pm C - E = \pm S$$

Where  $M$ = rate of metabolic heat gain
$R$ = rate of radiative heat gain (loss)
$C$ = rate of convective heat gain (loss)
$E$ = rate of evaporative heat loss
$S$ = rate of heat storage (loss) in body

Metabolic heat gain is the sum total of the metabolic heat gain during different activities performed while doing a work and the basal metabolic heat gain (1.0 K Cal/min

*Figure 14. Recommended ranges of reflection factor for main interior surfaces*

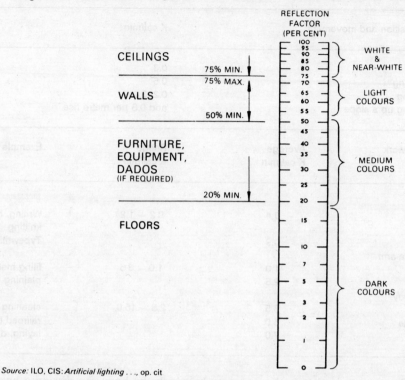

Source: ILO, CIS: *Artificial lighting . . .,* op. cit

= 4.186 KJ) required to keep a man just alive at rest at fasting. The assessment of work load/energy expenditure during different activities are given in Table 5:

Radiative heat gain or loss depends on temperature difference between bodies, emissivity of radiating surfaces and sources, surface area of the body receiving radiative heat and the clothings used.

Convective heat gain or loss depends on the dry-bulb temperature of the work environment and the wind velocity therein which is significantly influenced by the ventilation, mean body surface temperature, surface area of the body and the clothings used.

Evaporative heat loss depends on the wet-bulb and dry-bulb temperature and wind velocity in the work environment and on the ability to produce sweat, body surface area and the clothings used.

The metabolic heat generation rate can be controlled by adopting mechanisation wherever possible and sharing work-load with others. The radiative heat gain can be controlled by interposing line of sight barrier, insulating radiative source, wearing heat reflecting clothes. The convective heat gain for cases where air temperature is more than 35°C can be controlled by reducing air temperature and air speed.

For situations wherein the air temperature is less than 35°C the convective heat gain can be reduced by reducing air temperature and increasing air speed. The evaporative heat loss can be increased by increasing air speed and decreasing humidity.

*Table 5. Average values of metabolic rate during different activities.*

| A. Body position and movement | K cal/min |
| --- | --- |
| Sitting | 0.3 |
| Standing | 0.6 |
| Walking | 0.2–3.0 |
| Walking up a slope | add 0.8 per metre rise |

| B. Type of work | Average K cal/min | Range K cal/min | Example |
| --- | --- | --- | --- |
| *Hand work* | | | |
| Light | 0.4 | 0.2 – 1.2 | Writing, hand knitting |
| Heavy | 0.9 | | Typewriting |
| *Work with one arm* | | | |
| Light | 1.0 | 1.0 – 3.5 | filing metal, |
| Heavy | 2.5 | | plaining wood |
| *Work with body* | | | |
| Light | 3.5 | 2.5 – 15.0 | cleaning a floor |
| Moderate | 5.0 | | railroad track |
| Heavy | 7.0 | | laying, digging |

Sweat vaporisation is the only way by which the body can dissipate heat in a hot working environment. This method becomes more and more effective and refreshes the worker increasingly when the ventilation is adequate as well as relative humidity is less. Naturally situations wherein inadequate ventilation and high relative humidity is in existence like deep mines, textile mills and sugar refineries, the working conditions are most difficult to bear. Wet Bulb Globe Temperature (WBGT) index is the most comprehensive index used commonly for evaluating such conditions.

The severity of such working conditions can suitably be controlled by providing adequate ventilation, controlling humidity and providing optimum work-rest proportion.

Similarly, workers working in low temperature environmental condition must be provided with suitable clothing and footwear and frequent periodic exposure to normal temperature as well as hot drinks at regular intervals to prevent dehydration. Modern devices like infrared heaters can also be suitably used to allow more prolonged exposure to low temperature, thereby sacrificing productivity to the minimum extent without jeopardising the occupational health standards.

Where the work-load is at significantly high levels of relative humidity such situations are also not liked by the workers. Typical examples are textile industries, laundries and various chemical plants. Increased ventilation and removal of steam by exhaust ventilation improves the situation.

Excessive dry air also has its ill effects like causing respiratory tract diseases and thereby endangering the occupational health standards.

Though individual preferences to particular temperature and climatic conditions vary rather widely, the following air temperatures have been recommended for various types of work:

|  | °C |
|---|---|
| Sedentary work | 20–22 |
| Light Physical Work in a seated condition | 19–20 |
| Light work in a standing condition | 17–18 |
| Moderate work in a standing position | 16–17 |
| Heavy work in a standing position | 14–16 |

Since the airspace i.e., the volume of workplaces can never be made large enough to such an extent that ventilation is not required, replacement of contaminated air by fresh air or in other words ventilation is almost always a must. Ventilation essentially has a wider scope than air circulation which primarily makes the air to circulate.

The primary functions of ventilation include:

• dispersal of heat generated by machines and men at work to maintain the thermal balance

• dilution of atmospheric contamination

• maintaining levels of oxygen and carbon dioxide and carbon monoxide at pre-specified level to allow feeling of air freshness.

The recommended minimum air flow is 50 cubic metres per hour and air is usually required to be changed at least 3 times every hour. However, the number of changes recommended vary depending on the number of people working, humidity and atmospheric pollution.

Natural ventilation obtained through window opening or wall or roof air vents is found to be effective in relatively mild climatic and work conditions. The drawbacks of natural ventilation are wide in variation with difficult regulation procedures.

Artificial ventilation procedure serves as a complementary means to natural ventilation in situations where the latter is inadequate. Blown air ventilation systems are capable of heating/cooling the workplace besides providing the required ventilation effect. But such systems also result in dust deposition at a workplace, free surfaces and light sources. Exhaust ventilation or combined blown-and-exhaust air systems are very suitable in workplaces where large volumes of fumes, gases, vapours and mists or dusts are emitted. But in exhaust ventilation system the choice between downwards and upwards systems must be made after giving due consideration to the specific gravity of the pollutant relative to air. The material for the exhaust air systems should be carefully chosen so that they are corrosion-resistant and non-combustible which characteristics are present.

## 6.7  Noise and vibration

Noise is defined to be unwanted sound to which an individual is exposed. In industry virtually all machine generated sound can be classified as noise and as such undesirable and needs to be reduced to the lowest level.

The situation has become more precarious with larger utilisation of high levels of mechanisation, increased speeds and larger number of machines in a workplace.

The general effects of noise can be listed as follows:-

a) disturbs concentration, causes annoyance;

b) interferes with speech communication;

c) interferes with audible warning system;

d) causes noise-induced hearing losses e.g.

    i) temporary threshold shift,

    ii) permanent threshold shift,

    iii) acoustic trauma or physical damage to the components of ear, and

    iv) fatigue, reduction in gastric activity rise in breathing, heart rate and increased blood pressure.

All the above effects significantly influence productivity and occupational health and safety levels.

The effect of impeding speech communication is depicted in Figures 15 and 16.

Exposure to continuous noise levels of 90 dB(A) or above is dangerous to hearing but the level of 85 dB(A) is an accepted warning threshold value. Special care should be taken in case of impulse noises i.e. noises of very short duration at a level of at least 3 dB above the background noise and separated by intervals of at least one second which the more rudimentary type of instruments may not be able to detect. Frequency-wise the most dangerous frequencies are those around 4,000 Hz.

The sound pressure doubles for each increase in sound level by 6 dB which also quadruples the acoustic energy. So for every increase of 6 dB in sound level the exposure should be halved to keep the biological effect unchanged. The resulting fatigue due to a noise level depends on the nature of the work and the effect is found to be more intense if the work has intellectual content or calls for concentration.

In a systematic noise control approach the first step is the assessment of the existing noise level. The actual strategies listed below or any combination thereof are decided upon only after such assessment.

1) Control at source which can be implemented by replacing noisy machines or equipment or process by less noisiness, using dynamically balanced rotating parts or using resilient mounting pads or using rubber and plastic components in place of metallic ones etc.

2) Control of transmission by increasing the distance of the source from the background noise, segregation of the noise source by brick walls or persons walls with plastering on both sides, double wall with 10 cm intervening space, damping with structures. The effect of increasing distance of noise source from the background noise is given in Table 6.

3) Protection at the receiver ensured through use of ear plugs made from glass fibre or foam plastics, use of ear muffs also provide sufficient protection.

The recommended exposure duration which should not be exceeded to ensure the prevention of occupational deafness is given in Table 7.

Figure 15. *Distance at which the normal voice can be heard against background noise*

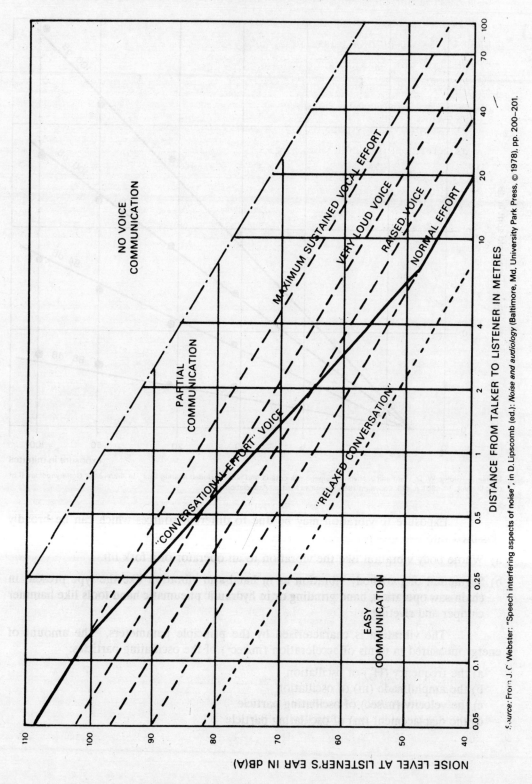

NO VOICE
COMMUNICATION

PARTIAL
COMMUNICATION

MAXIMUM SUSTAINED VOCAL EFFORT

VERY LOUD VOICE

RAISED VOICE

NORMAL EFFORT

"CONVERSATIONAL-EFFORT" VOICE

"RELAXED CONVERSATION"

EASY
COMMUNICATION

DISTANCE FROM TALKER TO LISTENER IN METRES

NOISE LEVEL AT LISTENER'S EAR IN dB(A)

*Source:* From J. C. Webster: "Speech interfering aspects of noise", in D. Lipscomb (ed.): *Noise and audiology* (Baltimore, Md, University Park Press, © 1978), pp. 200–201.

*Figure 16. Temporary hearing threshold shift in dB as a function of duration of exposure to wide-band noise*

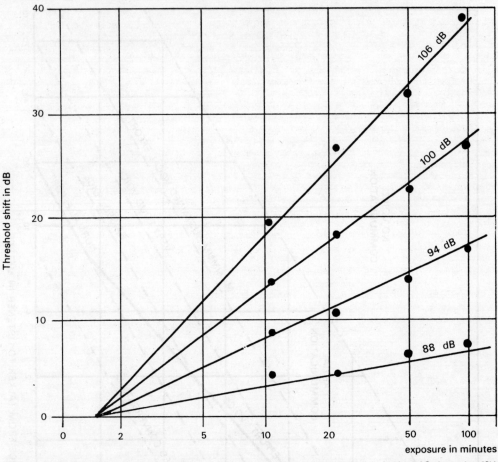

*Source:* A. Glorig, W. D. Ward and J. Nixon: "Damage risk criteria and noise-included hearing loss", in *Archives of Otolaryngology* (Chicago III.), Vol. 74, 1961, p. 413. Copyright 1961, American Medical Association.

Exposure to vibration may be due to different sources which can be broadly classified into two groups:

a) Whole body vibration like the vibration to an operator of a fork lift.

b) Segmental or part body vibration as in hand-arm vibration like the type present in chain-saw operation, hand grinding or in hydraulic pneumatic hand tools like hammer chipper and riveting.

The vibration is characterised by the principle parameters, *the amount of energy measured in terms of acceleration ($m/sec^2$) of the oscillating particles:

a) the frequency ($H_3$) of oscillation
b) the amphilmade (m) of oscillation
c) the velocity (m/sec) of oscillating particle
d) the displacement (m) of oscillating particle

*Table 6. Calculation of noise level obtained by removing a source of noise from the background noise.*

| Difference in dB between the two noise levels | Reduction in dB of the higher noise level |
|---|---|
| 1 | 7.0 |
| 2 | 4.4 |
| 3 | 3.0 |
| 4 | 2.2 |
| 5 | 1.8 |
| 6 | 1.3 |
| 7 | 1.0 |
| 8 | 0.8 |
| 9 | 0.6 |
| 10 | 0.5 |

*Table 7. Duration of continuous noise exposure which should not be exceeded to ensure the prevention of occupational deafness amongst the majority of workers.*

| Daily duration of noise in hours | Noise level in dB(A) (measured "slow") |
|---|---|
| 16 | 80 |
| 8 | 85 |
| 4 | 90 |
| 2 | 95 |
| 1 | 100 |
| 1/2 | 105 |
| 1/4 | 110 |
| 1/8 | 115 |

Source: American Conference of Governmental Industrial Hygienists (ACGIH):
*Threshold limit values for chemical substances and physical agents in the workroom environment adopted by the ACGIH for 1977* (Cincinnati, Ohio).

Whole body vibration, transmitted to the entire body through the supporting surfaces like the feet of a standing person, buttocks of a seated person or supporting areas of a reclining person, causes fatigue and thereby adversely affects productivity.

Part body vibration limited to the portions of a body in immediate contact with the source may have the effects like loss of sense of touch and temperature in fingers, muscular weakness, loss of mental dexterity leading to possible adverse effect on performance, pain and stiffness in joints of finger and hands, degenerative changes to bones and joints and lastly Raynaud's syndrome in which blood circulation becomes impaired in hands used to guide the vibrating tools.

Exposure to vibration can be controlled by relying on the techniques on substitution and isolation. The strategies are similar to those adopted in case of noise i.e.

1) Control at source by balancing all revolving and rotating parts, reducing speed of moving parts, reducing the number of impacts, damping the joints of moving parts, providing a resilient bumper at the point of impact and regular maintenance of machinery and tools etc.

2) Control of transmission by isolating vibrating machinery from floors and walls through use of damping foundation, using vibration absorbing materials in floor joints, damping handles and guide arms on hand held power tools etc.

3) Protection at the receiver end by minimising exposure duration, providing rest breaks between periods of exposures etc.

## 6.8  Ergonomics

Ergonomics a relatively recent interdisciplinary domain provides tools and techniques which can usefully integrate working condition vector elements with the physical characteristics and physiological and psychological capabilities. Such an integration goes a long way to evaluate the combined effects of all these interacting factors on productivity. Successful application of the tools and techniques of the domain of ergonomics may lead to the development of the most comfortable working conditions in terms of illumination, climatic condition, noise level which ensures minimisation of physical work-load, to improve work postures as well as to reduce the effort of certain movements, to facilitate psycho-sensorial functions in reading instrument displays (Figure 17), to make the handling of machine levers and controls easier (Figure 18), to make better and more effective use of spontaneous and stereotyped reflexes and to avoid unnecessary information recall efforts.

Such ergonomic measures can be incorporated with very little additional marginal cost sight at the design stage since otherwise the associated modifications may call for substantial increase in cost in the event of attempting such measures at any later stage. These measures should be incorporated in items affecting production and also to items expected to be attended for maintenance activities rather frequently.

Manual material handling, the singularly most potential source of occupational accidents and by far the mostly followed material transfer mode in this country can be made more comfortable, less exhausting and more safe by applying simple but effective ergonomic recommendations. Such recommendation can be effectively adapted by the workers through training and instruction programmes in Kinetic techniques and safe lifting and handling postures leading to the prevention of low back pain and injuries to the lumber spine.

Postural techniques for optimal use of effort and simultaneously maintaining high occupational safety levels are depicted in Figure 19.

The recommended composition, for introducing and subsequently maintaining an effective ergonomic programme, in medium-sized and large organisation, is Work

Figure 17. Ergonomic display design

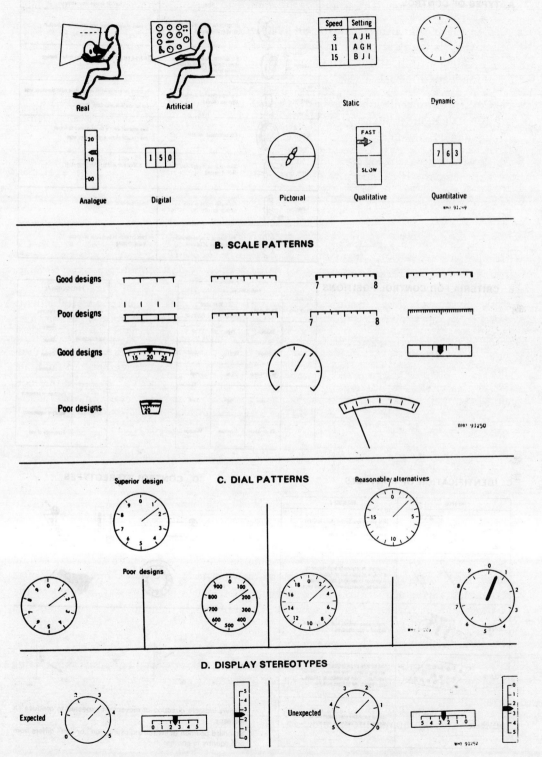

Source: W. T. Singleton: *Introduction to ergonomics* (Geneva, World Health Organization, 1972), pp. 79-80.

## Figure 18. Ergonomic design of controls

### A. TYPES OF CONTROL

| | | Use | Special design requirements |
|---|---|---|---|
| Button | | In arrays for rapid selection between alternatives | To avoid slipping finger and accidental activation |
| Toggle | | For definite, rarely used action involving only choice of two (normally on  off) | To avoid excessive finger pressure or nail damage |
| Selector | | For more than two and less than ten choices | To avoid excessive wrist action make total movement less than 180°. Do not use simple circular shape |
| Knob | | For continuous variables | Size depends mainly on resistance to motion. Use circular shape with serrated edge |
| Crank | | When rotation through more than 360° is needed | Grip handle should turn freely on shaft |
| Lever | | For higher forces or very definite activity | Identification of neutral or zero |
| Wheel | | For precise activity involving large angles or rotation | Identification of particular positions Avoid slipping |

### B. CRITERIA FOR CONTROL POSITIONS

| ANATOMICAL | | | | | PSYCHOLOGICAL | |
|---|---|---|---|---|---|---|
| Which limb? | Which joint? | When used | | | | |
| | | Force | Precision | | | |
| Hand | Shoulder | High | Low | | Identification | Position |
| | | | | | | Size |
| | Elbow | Medium | Medium | | | Shape |
| | | | | | | Colour |
| | Wrist/finger | Low | High | | | Legend |
| Foot | Ankle | High | Low | | Selectivity | Position in sequence |
| | | | | | | Relative importance |
| | Thigh  knee | Maximum | Minimum | | | Frequency of use |

### C. IDENTIFICATION OF CONTROLS

| METHOD | EXAMPLE | WHEN USED |
|---|---|---|
| POSITION | | Best placed near to controlled function (mechanical device or dial) provided access is reasonable and high speed is not essential. |
| SIZE | | Important in relation to high frequency of use (e.g. space bar on typewriter) or in relation to operating force.  Size should be increased with required force. |
| SHAPE | | Useful when controls must be operated without visual attention. |
| COLOUR | | Useful only as secondary cue or as warning unless there are many controls together.  Even so number of colours used should not be more than five. |
| LEGEND | MAIN   STAND-BY | Useful secondary cue. Should not be obscured when control is operated. |

### D. CONTROL STEREOTYPES

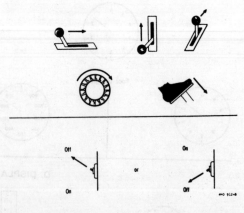

Arrows indicate direction of movement expected to produce an increase.

Standard position of switch indicating "on" or "off" differs from one country to another.

Source: W. T. Singleton, op. cit., pp. 69-70.

**Figure 19. Optimal use of physical effort**

## A. ASPECTS OF WEIGHT DISTRIBUTION

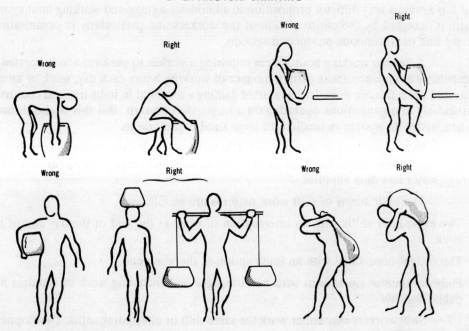

**B. LIFTING AND CARRYING**

Study specialists, a safety specialist, a medical officer and representatives of personnel department and workers.

## 6.9   Arrangement of working time

Working hours are usually dictated by Law. The duration is found to vary between 40—48 hours per week. In organisations where overtime provision is in existence, a maximum limit should be set to ensure production at acceptable quality and quantity in the long run, though the workers involved may be inclined to work overtime beyond such limit. Usually workers below 18 years of age and expectant mothers should not be allowed to work overtime.

With the increase in psycho-physical component of the work-load of an average worker in the recent years due to increasing use of mechanised and/or automated aids and facilities, the importance of introduction of breaks during the working hours has assumed greatest heights. Such breaks are absolutely essential for dissipation of fatigues of different nature and proper restoration of physical and nervous energy. The break duration should be utilised by workers involved with hard physical work in a manner such that he can sit and even lie down while persons doing intellectual work should move around, listen to light music and even perform some light gymnastics or yoga during such breaks. Routine interruptions for meals or tiffins or resulting from accidents should not be considered as breaks.

Systems of staggered hours of work are also adopted in certain organisations. The workers prefer such systems since they can avoid peak hour road and rail traffic and they also permit them to make use of public services as well as fulfil social obligations. But it is always a very difficult proposition to administer a staggered working hour system which is accepted by the entire section of the workers and particularly in organisations having line or continuous production systems.

A flexible working hour system requiring a worker to perform a 'core' period of compulsory attendance and a certain number of working hours each day, week or month in agreed upon flexible modules has started gaining a foothold in industrialised countries, particularly in organisations operating on a single shift pattern. But this is also difficult to administer the system in medium to large sized organisations.

Shift work is common in several industries particularly industries relying on process operations, requiring huge capital investments and certain utility functions like energy, water and milk supplies.

The basic forms of shift work patterns are as follows:

a) Two eight-hour shifts with an interruption of work at the end of the day and of the week.

b) Three eight-hour shifts with an interruption at the week end.

c) Fully continuous operations with no stoppages and including work on sundays and public holidays.

Shift workers may either work the same shift or alternating shifts. Development of nervous, digestive or circulatory problems are rather common with workers working in fully continuous operations, on alternating shift pattern. To compensate such ill factors proper distribution of works in different shifts should be done supplemented by compensatory measures. Better amenities like canteen facilities, transport and housing facilities are also found to be effective mitigating agents for the drawbacks of shift work particularly on the family and social life aspects of the shift workers.

## 6.10  Conclusion

Working condition is a multi-attribute function involving occupational health and safety elements, fire protection and prevention elements, layout and housekeeping elements, illumination and climate conditioning elements, noise and vibration elements. Concept and practice of ergonomics present the possibility of synthesising all such interacting factors with anatomical, physiological and psychological characteristics of workers. Such an integrated analysis will allow a better and closer insight in the analysis and estimation of individual as well as combined effects of these factors on productivity. The reliable estimates thus arrived will subsequently help in identifying the optimum status of each of the elements of the multi-attribute function i.e. the working condition.

Such an approach has become more important specially in view of the marked shift of the major component of average work-load of workers which has become more and more interesting in certain cases while becoming more and more mundane in some other cases, both resulting from the ever increasing adoption of mechanisation and automation practices.

# Part B
## Method study

# Chapter 7

# Introduction to method study and the selection of job

## 7.1 Introduction

Resource required to produce goods and services would be from the following: (a) Man (b) Materials (c) Machines (d) Money (e) Technology (f) Time. They are to be deployed in the most effective and efficient manner. This process of deployment is a continuous one since the best available combination of the resources at some point would not necessarily coincide with the best available combination at some later point of time. This emphasises that there is a continuing need for analysing existing working methods to develop more efficient working methods for the future.

## 7.2 Definition and objectives of method study

Method study has already been defined in Chapter 4, but the definition is repeated at this point to facilitate easy reading of the text.

> **Method study is the systematic recording and critical examination of existing and proposed ways of doing work, as a means of developing and applying easier and more effective methods and reducing costs.**

The objectives of Method study are:

i) Improvement of processes and procedures.

ii) Improvement in the design of plant and equipment.

iii) Improvement of plant layout.

iv) Improvement in the use of men, materials and machines.

   v) Efficient materials handling.

  vi) Improvement in the flow of production and processes.

 vii) Economy in human effort and the reduction of unnecessary fatigue.

viii) Method standardisation.

  ix) Improvement in safety standards.

   x) Development of a better physical working environment.

## 7.3   Procedure

The solution of any problem follows the following sequence of phases in that order:

1) DEFINE the problem.

2) RECORD all the facts relevant to the problem.

3) EXAMINE the facts critically but impartially.

4) CONSIDER the courses of actions (possible solutions) and decide which to follow.

5) IMPLEMENT the solution.

6) FOLLOW UP the development.

    We have already discussed the basic procedure for the whole of work study, which embraces the procedures of both method study and work measurement. Let us now examine the basic procedure for method study.

    They are as follows:-

a) SELECT the work to be studied.

b) RECORD all the relevant facts about the present method by direct observation.

c) EXAMINE those facts critically and in an ordered sequence, using the techniques best suited to the purpose.

d) DEVELOP the most practical, economic and effective method, having due regard to all contingent circumstances.

e) DEFINE the new method so that it can always be identified.

f) INSTALL that methods as standard practice.

g) MAINTAIN that standard practice by regular routine checks.

    These are the seven essential stages in the application of method study; none can be excluded. Strict adherence to their sequence, as well as to their content, is essential for the success of an investigation. They are shown diagrammatically on the chart in Figure 20.

    Do not be deceived by the simplicity of the basic procedure into thinking that method study is easy. On the contrary, method study may on occasion be very complex, but for the purposes of descriptions it has been reduced to these few simple steps.

Figure 20. Method study

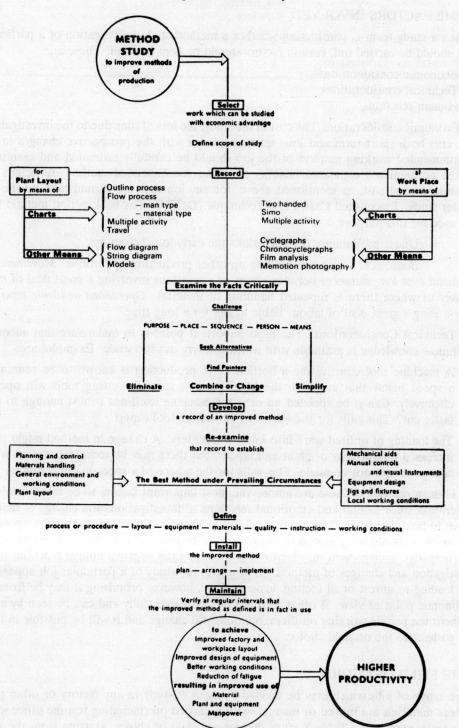

## 7.4  Selection of job

### SOME FACTORS INVOLVED

When a study team is considering whether a method study investigation of a particular job should be carried out, certain factors should be kept in mind. These are:-

1) Economic considerations
2) Technical considerations
3) Human reactions

1) Economic consideration: The cost of the study, the loss of time due to the investigation, the cost both short-term and long-term associated with the prospective changes in the recommended working method of the job should be carefully estimated and examined. If the accumulated estimated benefits from the recommended method outweighs the estimated total cost, as mentioned above, for any job then we should take up the job under study. Discounted Cash Flow Technique (DCF) or Pay-back period method may be used for this purpose.

Under preliminary considerations the early job choices are:

*Bottlenecks* which are holding up other production operations. *Movements of material over long distances* between shops or operations involving a great deal of manpower or where there is repeated handling of material. *Operations involving repetitive work* using a great deal of labour liable to run for a long time.

2) Technical Considerations: The most important point is to make sure that adequate technical knowledge is available with which to carry out the study. Examples are:-

a) A machine tool constituting a bottleneck in production is known to be running at a speed below that at which the high-speed or ceramic cutting tools will operate effectively. Can it be speeded up or is the machine itself not robust enough to take faster cut? This calls for the advice of a machine tool expert.

b) The loading of unfired ware into kilns in a pottery. A change in method might bring increased productivity of plant and labour, but there may be technical reasons why a change should not be made. This calls for the advice of a specialist in ceramics.

3) Human reactions: These are among the most important factors to be taken into consideration since mental and emotional reactions to investigation, and change of method have to be anticipated. Trade union official workers' representatives and the operators themselves should be educated in the general principles and objectives of method study. Participative management may facilitate overcoming the negative human reactions to investigation and changes of method. If, however, the study of a particular job appears to be leading to unrest or ill feeling, leave it alone, however promising it may be from the economic point of view. If other jobs are tackled successfully and can be seen by all to benefit the people working on them, opinions will change and it will be possible in time, to go back to the original choice.

### THE FIELD OF CHOICE

The range of jobs which may be tackled by method study in any factory or other place where materials are moved or manual work is carried on (including routine office work) is usually very wide. Table 8 gives the general field of choice, starting from the most comprehensive investigation covering, possibly, the whole operation of the plant and working down to the study of the movements of the individual worker. Beside each type of job are listed the recording techniques with which it may be tackled. It should be pointed out that, in the course of a single investigation, two or more of these techniques may be used. These techniques will be described in subsequent chapters.

*Table 8. Typical industrial problems and appropriate method study techniques.*

| Type of job | Examples | Recording technique |
| --- | --- | --- |
| Complete sequence of manufacture | Manufacture of an electric motor from raw material to dispatch<br>Transformation of thread into cloth from preparation to inspection<br>Receipt, packing and dispatch of fruit. | Outline process chart<br>Flow process chart<br>Flow diagram |
| Factory layout: movement of materials | Movements of a diesel engine cylinder head through all machining operations.<br>Movements of grain between milling operations | Outline process chart<br>Flow process chart—material type<br>Flow diagram<br>Travel chart<br>Models |
| Factory layout: movement of workers | Labourers servicing spinning machine with bobbins<br>Cooks preparing meals in a restaurant kitchen | Flow process chart—man type<br>String diagram<br>Travel chart |
| Handling of materials | Putting materials into and taking them out of stores<br>Loading lorries with finished products | Flow process chart—material type<br>Flow diagram<br>String diagram |
| Workplace layout | Light assembly work on a bench<br>Typesetting by hand | Flow process chart—man type<br>Two-handed process chart<br>Multiple activity chart<br>Simo chart<br>Cyclegraph<br>Chronocyclegraph |
| Gang work or automatic machine operation | Assembly line<br>Operator looking after semi-automatic lathe | Multiple activity chart<br>Flow process chart—equipment type |
| Movements of operatives at work | Female operatives on short-cycle repetition work<br>Operations demanding great manual dexterity | Films<br>Film analysis<br>Simo chart<br>Memotion photography<br>Micromotion analysis |

When selecting a job for method study it will be found helpful to have a standardised list of points to be covered. This prevents factors from being overlooked and enables the suitability of different jobs to be easily compared. A sample list [1] is given which is fairly full, but lists should be adapted to individual needs.

1) Product and operation.

2) Person who proposes investigation.

3) Reason for proposal.

4) Suggested limits of investigation.

5) Particulars of the job.

   a) How much is [2] (many are) produced or handled per week?

   b) What percentage (roughly) is this of the total produced or handled in the shop or plant?

   c) How long will the job continue?

   d) Will more or less be required in future?

   e) How many operatives are employed on the job—

      i) directly?

      ii) indirectly?

   f) How many operatives are there in each grade and on each rate of pay?

   g) What is the average output per operative (per team) per day?

   h) What is the daily output compared with the output over a shorter period? (e.g. an hour)

   i) How is payment made? (team-work, piece-work, premium bonus time rate, etc.)

   j) What is the daily output—

      i) of the best operative?

      ii) of the worst operative?

   k) When were production standards set?

   l) Has the job any especially unpleasant or injurious features? Is it unpopular (a) with workers? (b) with supervisors?

6) Equipment.

   a) What is the approximate cost of plant and equipment?

   b) What is the present machine utilisation index?[1]

7) Layout.

   a) Is the existing space allowed for the job enough?

   b) Is extra space available?

8) Product.

   a) Are the frequent design changes causing modifications?

   b) Can the product be altered for easier manufacture?

   c) What quality is demanded?

---

[1] Machine utilisation index = the ratio of Machine Running Time to Machine Available Time.

d) When and how is the product inspected?

9. What savings or increase in productivity may be expected from a method improvement?

   a) Through reduction in the work content of the product or process.

   b) Through better machine utilisation.

   c) Through better use of labour.

   (Figures may be given in money, man-hours or machine-hours or as a percentage).

Item 4 deserves some comment. It is important to set clearly defined limits to the scope of the investigation. Method study investigations so often reveal scope for even greater savings that there is a strong temptation to go beyond the immediate objective. This should be resisted, and any jobs shown up as offering scope for big improvements through method study should be noted and tackled separately.

Such a list will prevent the work study man from going first for a small bench job which will entail a detailed study of the worker's movements and yield a saving of a few seconds per operation, unless the job is one that is being done by a large number of operatives, so that the total saving will significantly affect the operating costs of the factor. It is no use playing around with split seconds and inches of movement when a great waste of time and effort is taking place as a result of bad shop layout and the handling of heavy materials.

Finally, remember the adage: "Do not use a spoon when a steam shovel is needed."

Subject to the considerations listed above, tackle first the job most likely to have the greatest overall effect on the productivity of the enterprise as a whole.

**NOTES**

1. This list has been adapted from one given in Anne G. Shaw: The purpose and practice of motion study (Buxton, (United Kingdom), Columbine Press, 2nd ed., 1960).

2. For bulk materials measured in tons, pounds, feet, kilograms, metres, etc.

# Chapter 8
# Record, examine, develop

## 8.1  Introduction

After selecting the work to be studied systematic recording of all the facts of the existing method and critical examination of these are carried out to eliminate every unnecessary element or operation and to develop the quickest and best method by having an improved sequence of doing the work, omitting the redundant elements, selecting more appropriate person and more suitable place for doing the work.

The next step in the basic procedure, after selecting the work to be studied, is to record all the facts relating to the existing method. The success of the whole procedure depends on the accuracy with which the facts are recorded, because they will provide the basis of both the critical examination and the development of the improved method. It is therefore essential that the record be clear and concise.

The usual way of recording facts is to write them down. Unfortunately, this method is not suited to the recording of the complicated processes which are so common in modern industry. This is particularly so when an exact record is required of every minute detail of a process or operation. To describe exactly everything that is done in even a very simple job which takes perhaps only a few minutes to perform would probably result in several pages of closely written script, which would require careful study before anyone reading it could be quite sure that he had grasped all the details.

To overcome this difficulty other techniques or 'tools' of recording have been developed, so that detailed information may be recorded precisely and at the same time in a standard form, in order that it may be readily understood by all method study men, in whatever factory or country they may be working.

The most commonly used of these recording techniques are charts and diagrams. There are several different types of standard charts available, each with its own special purposes. They will be described in turn later in this chapter and in subsequent chapters. For the present it will be sufficient to note that the charts available fall into two groups—

a) those which are used to record a process sequence, i.e. a series of events or happenings in the order in which they occur, but which do not depict the events to scale; and

b) those which record events, also in sequence, but on a time scale, so that the interaction of related events may be more easily studied.

The names of the various charts were listed in Table 8 in the last chapter against the types of job for which they are most suitable. They are shown again in Table 9, which lists them in the two groups given and also lists the types of diagram commonly used.

*Table 9. The most commonly used method study charts and diagrams*

| | |
|---|---|
| **A. CHARTS** | indicating process SEQUENCE |
| | Outline Process Chart |
| | Flow Process Chart—Man Type |
| | Flow Process Chart—Material Type |
| | Flow Process Chart—Equipment Type |
| | Two-Handed Process Chart |
| **B. CHARTS** | using a TIME SCALE |
| | Multiple Activity Chart |
| | Simo Chart |
| **C. DIAGRAMS** | indicating movement |
| | Flow Diagram |
| | String Diagram |
| | Cyclegraph |
| | Chronocyclegraph |
| | Travel Chart |

Charts indicating process sequence provide a systematic description of a process or work cycle with details for the analyst to develop method improvements.

Diagrams are used to indicate movement more clearly than charts can do. They usually do not show all the information recorded on charts, which they supplement rather than replace. Among the diagrams is one which has come to be known as the Travel Chart, but despite its name it is classed as a diagram.

## PROCESS CHART SYMBOLS

The recording of the facts about a job or operation on a process chart is made much easier by the use of a set office standard [1] symbols, which together serve to represent all the different types of activity or event likely to be encountered in any factory or office. They thus serve as a very convenient, widely understood type of shorthand, saving a lot of writing and helping to show clearly just what is happening in the sequence being recorded.

The two principal activities in a process are operation and inspection. These are represented by the following symbols:

## OPERATION

> **Indicates the main steps in a process, method or procedure. Usually the part, material or product concerned is modified or changed during the operation.**

It will be seen that the symbol for an operation is also used when charting a procedure, as for instance a clerical routine. An operation is said to take place when information is given or received, or when planning or calculating takes place.

## INSPECTION

> **Indicates an inspection for quality and/or check for quantity.**

The distinction between these two activities is quite clear—

An operation always takes the material, component or service a stage further towards completion, whether by changing its shape (as in the case of a machined part) or its chemical composition (during a chemical process) or by adding or subtracting material (as in the case of an assembly). An operation may equally well be a preparation for any activity which brings the completion of the product nearer.

An inspection does not take the material any nearer to becoming a completed product. It merely verifies that an operation has been carried out correctly as to quality and/or quantity. Were it not for human shortcomings, most inspections could be done away with.

Often a more detailed picture will be required than can be obtained by the use of these two symbols alone. In order to achieve this, three more symbols are used—

## TRANSPORT

> **Indicates the movement of workers, materials or equipment from place to place.**

A transport thus occurs when an object is moved form one place to another, except when such movements are part of an operation or are caused by the operative at the work station during an operation or an inspection. This symbol is used throughout this book whenever material is handled on or off trucks, benches, storage bins, etc.

## TEMPORARY STORAGE OR DELAY

> **Indicates a delay in the sequence of events: for example, work waiting between consecutive operations, or any object laid aside temporarily without record until required.**

Examples are work stacked on the floor of a shop between operations, cases awaiting unpacking, parts waiting to be put into storage bins or a letter waiting to be signed.

## PERMANENT STORAGE

> **Indicates a controlled storage in which material is received into or issued from a store under some form of authorisation; or an item is retained for reference purposes.**

A permanent storage thus occurs when an object is kept and protected against unauthorised removal.

The difference between a "permanent storage" and a "temporary storage or delay" is that a requisition, chit or other form of formal authorisation is generally required to get an article out of permanent storage but not out of temporary storage.

In this book, for the sake of simplicity, temporary storage or delay will be referred to in brief as 'delay', and permanent storage as just 'storage'.

## COMBINED ACTIVITIES

When it is desired to show activities performed at the same time or by the same operative at the same work station, the symbols for those activities are combined, e.g. the circle within the square represents a combined operation and inspection.

## THE OUTLINE PROCESS CHART

It is often valuable to obtain a 'bird's-eye' view of a whole process or activity before embarking on a detailed study. This can be obtained by using an outline process chart.

> **An outline process chart is a process chart giving an overall picture by recording in sequence only the main operations and inspections.**

In an outline process chart, only the principal operations are carried out and the inspections made to ensure their effectiveness are recorded, irrespective of who does them and where they are performed. In preparing such a chart, only the symbols for 'operation' and 'inspection' are necessary.

In addition to the information given by the symbols and their sequence, a brief note of the nature of each operation or inspection is made beside the symbol, and the time allowed for it (where known) is also added.

An example of an outline process chart is given in Figure 22. In order that the reader may obtain a firm grasp of the principles involved, the assembly represented on this chart is shown in a sketch (Figure 21) and the operations charted are given in some detail on the next page.

### EXAMPLE OF AN OUTLINE PROCESS CHART: ASSEMBLING A SWITCH ROTOR[2]

The assembly drawing (Figure 21) shows the rotor for a slow make-and-break switch.

*Figure 21. Switch rotor assembly*

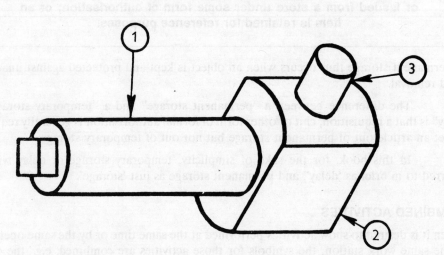

In making an outline process chart it is usually convenient to start with a vertical line down the right-hand side of the page to show the operations and inspections undergone by the principal unit or component of the assembly (or compound in chemical processes)—in this case the spindle. The time allowed per piece in hours is shown to the left of each operation. No specific time is allowed for inspections as the inspectors are on time work.

The brief descriptions of the operations and inspections which would normally be shown along side the symbols have been omitted so as not to clutter the Figure.

The operations and inspections carried out on the spindle which is made from 10 mm diameter steel rod are as follows :

Operation 1  Face, turn, undercut and part off on a capstan lathe (0.025 hours).

Operation 2  Face opposite end on the same machine (0.010 hours).

After this operation the work is sent to the inspection department for—

Inspection 1 Inspect for dimensions and finish (no time fixed). From the inspection department the work is sent to the milling section.

Operation 3 Straddle-mill four flats on end on a horizontal miller (0.070 hours). The work is now sent to the burring bench.

Operation 4 Remove burrs at the burring bench (0.020 hours). The work is returned to the inspection department for—

Inspection 2 Final inspection of machining (no time). From the inspection department the work goes to the plating shop for—

Operation 5 Degreasing (0.0015 hours).

Operation 6 Cadmium plating (0.008 hours). From the plating shop the work goes again to the inspection department for—

Inspection 3 Final check (no time). The plastic moulding is supplied with a hole bored concentric with the longitudinal axis.

Operation 7 Face on both sides, bore the cored hole and ream to size on a capstan lathe (0.080 hours).

Operation 8 Drill cross-hole (for the stop pin) and burr on two-spindle drill press (0.022 hours). From the drilling operation the work goes to the inspection department for—

Inspection 4 Final check dimensions and finish (no time). It is then passed to the finished-part stores to await with-drawal for assembly.

It will be seen from the chart that the operations and inspections on the moulding are on a vertical line next to that of the spindle. This is because the moulding is the first component to be assembled to the spindle. The stop-pin line is set further to the left, and if there were other components they would be set out from right to left in the order in which they were to be assembled to the main item.

## NOTE ESPECIALLY THE METHOD OF NUMBERING THE OPERATIONS AND INSPECTIONS

It will be seen that both operations and inspections start from 1. The numbering is continuous from one component to another, starting from the right, to the point where the second component joins the first. The sequence of numbers is then transferred to the next component on the left and continues through its assembly to the first component until the next assembly point, when it is transferred to the component about to be assembled. Figure 23 makes this clear. The assembly of any component to the main component or assembly is shown by a horizontal line from the vertical operation line of the minor component to the proper place in the sequence of operations on the main line. (Sub-assemblies can, of course, be made up of any number of components before being assembled to the principal one; in that case the horizontal line joins the appropriate vertical line which appears to the right of it). The assembly of the moulding to the spindle, followed by the operation symbol and number, is clearly shown in the Figure.

Operation 9    Assemble the moulding to the small end of the spindle and drill the stop-pin hole right through (0.020 hours).

Once this has been done the assembly is ready for the insertion of the stop pin (made from 5 mm diameter steel rod), which has been made as follows:

Operation 10   Turn 2 mm diameter shank, chamfer end and part off, on a capstan lathe (0.025 hours).

Operation 11   Remove the 'pip' on a finisher (0.005 hours).
                The work is then taken to the inspection department.

Inspection 5   Inspect for dimensions and finish (no time).
                After inspection the work goes to the plating shod for—

Operation 12   Degreasing (0.0015 hours).

Operation 13   Cadmium plating (0.006 hours).
                The work now goes back to the inspection department for—

Inspection 6   Final check (no time).
                It then passes to the finished-part stores and is withdrawn for—

Operation 14   Stop pin is fitted to assembly and lightly riveted to retain it in position (0.045 hours).

Inspection 7   The completed assembly is finally inspected (no time).
                It is then returned to the finished-parts store.

In practice, the outline process chart would bear against each symbol, beside and to the right of it, an abbreviated description of what is done during the operation or inspection. These entries have been left out of Figure 23 so that the main sequence of charting may be seen more clearly.

Figure 23 shows some of the conventions used when drawing outline process charts. In this instance the subsidiary component joins the main part after inspection 3, and is assembled to it during operation 7. The assembly undergoes two more operations, numbers 8 and 9, each of which is performed four times in all, as is shown by the 'repeat' entry. Note that the next operation after these repeats bears the number 16, not 10.

Outline process chart which gives an overall view of the entire process is designed to give a quick understanding of the work which must be done to produce a given product. It makes possible a study of the operations and inspections so that the best sequence may be developed. The analyst questioning on the outline process chart may discover significant cost reductions by combining or eliminating certain operations and inspections.

Figure 22. Outline process chart: switch rotor assembly

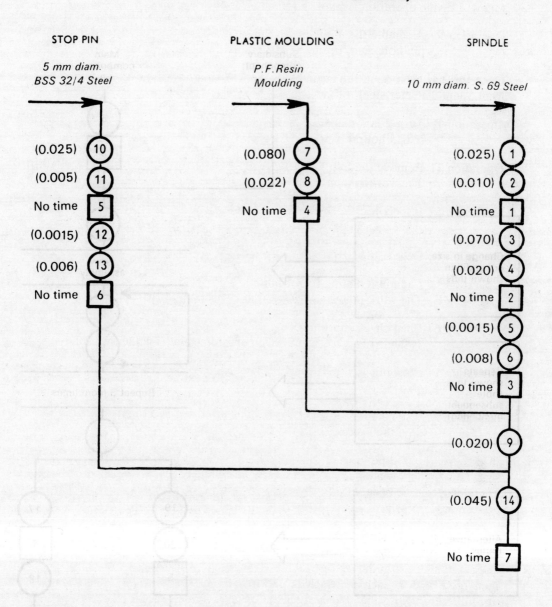

STOP PIN

5 mm diam.
BSS 32/4 Steel

PLASTIC MOULDING

P. F. Resin
Moulding

SPINDLE

10 mm diam. S. 69 Steel

*Figure 23. Some charting conventions*

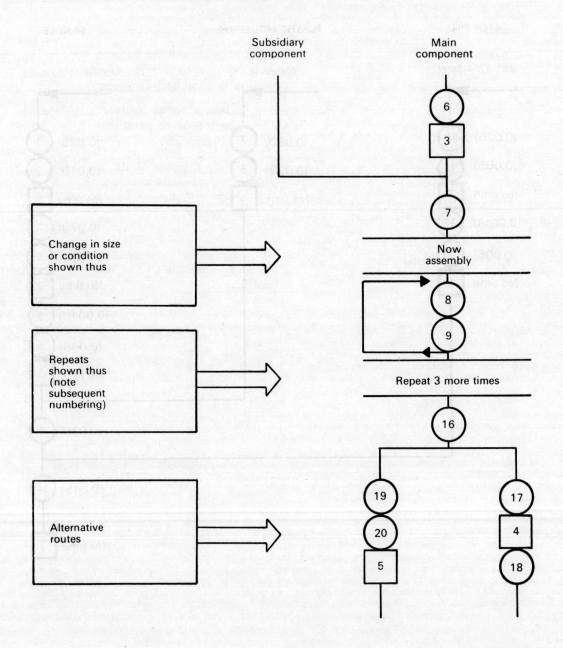

## FLOW PROCESS CHARTS

Once the general picture of a process has been established, it is possible to go into greater detail. The first stage is to construct a flow process chart.

---

**A flow process chart is a process chart setting out the sequence of the flow of a product or a procedure by recording all events under review using the appropriate process chart symbols.**

**Flow process chart—man type:**
A flow process chart which records what the worker does

**Flow process chart—material type:**
A flow process chart which records how material is handled or treated.

**Flow process chart—equipment type:**
A flow process chart which records how the equipment is used.

---

A flow process chart is prepared in a manner similar to that in which the outline process chart is made, but using, in addition to the symbols for 'operation' and 'inspection', those for 'transport', 'delay' and 'storage'. Whichever type of flow process chart is being constructed, the same symbols are always used and the charting procedure is very similar. (It is customary to use the active voice of verbs for entries on man type charts, and the passive voice on material type and equipment type charts. This convention is more fully explained in Chapter 10, section 3.) In fact, it is usual to have only one printed form of chart for all three types, the heading bearing the words "Man/Material/Equipment Type", the two words not required being deleted.

Flow process charts contain more information than outline process charts because they indicate additionally, storage, delay and transportation also which represent a major portion of the product cost. Because of this, it may be desirable to analyse a work in detail and to study with flow process chart the events which occur between operations, as well as the operations themselves. And in this connection it is worthwhile to mention that like outline process charts, flow process charts also aid in finding out means of combining or eliminating operations and inspections.

Because of its greater detail, the flow process chart does not usually cover as many operations per sheet as may appear on a single outline process chart. It is usual to make a separate chart for each major component of an assembly, so that the amount of handling, delays and storages of each may be independently studied. This means that the flow process chart is usually a single line.

An example of a material type flow process chart constructed to study what happened when a bus engine was stripped, degreased and cleaned for inspection is given in Figure 24. This is an actual case recorded at the workshop of a transport authority in a developing country. After discussing the principles of flow process charting and the means of using them in the next few pages, we shall go on to consider this example in detail. Man type charts are discussed in Chapter 10.

When flow process charts are being made regularly, it is convenient to use printed or stencilled sheets similar to that shown in Figure 25 (In charts of this kind the five symbols are usually repeated down the whole length of the appropriate columns. This has not been done in the charts presented in this book, which have been simplified to improve clarity). This also ensures that the study man does not omit any essential information. In Figure 26 the operation just described on the chart in Figure 25 is set down again.

Before we go on to discuss the uses of the flow process chart as a means of examining critically the job concerned with a view to developing an improved method, there are some points which must always be remembered in the preparation of process charts. These are important because process charts are the most useful tools in the field of method improvement; whatever techniques may be used later, the making of a process chart is always the first step.

1) Charting is used for recording because it gives a complete picture of what is being done and helps the mind to understand the facts and their relationship to one another.

2) The details which appear on a chart must be obtained from direct observation. Once they have been recorded on the chart, the mind is freed from the task of carrying them, but they remain available for reference and for explaining the situation to others. Charts must not be based on memory but must be prepared as the work is observed (except when a chart is prepared to illustrate a proposed new method).

3) A high standard of neatness and accuracy should be maintained in preparing fair copies of charts constructed from direct observation. The charts will be used in explaining proposals for standardising work or improving methods. An untidy chart will always make a bad impression and may lead to errors.

4) To maintain their value for future reference and to provide as complete information as possible, all charts should carry a heading giving the following information (see Figure 25):

   a) The name of the product, material or equipment charted, withdrawing numbers or code numbers.

   b) The job or process being carried out, clearly stating the starting point and the end point, and whether the method is the present or the proposed one.

   c) The location in which the operation is taking place (department, factory, site, etc.).

   d) The chart reference number, sheet number and the total number of sheets.

   e) The observer's name and, if desired, that of the person approving the chart.

   f) The date of the study.

   g) A key to the symbols used. This is necessary for the benefit of anyone who may study the chart later and who may have been accustomed to using different symbols. It is convenient to show these as part of a Table summarising the activities in the present and proposed methods (see Figure 26).

*Figure 24. Flow process chart: engine stripping, cleaning and degreasing*

CHART No. *1*      SHEET No. *1*    OF *1*     METHOD:    *Original*

PRODUCT:   *Bus Engines*                      OPERATIVE (S):

LOCATION:    *Degreasing shop*

PROCESS:   *Stripping, degreasing and*        CHARTED BY:

*cleaning used engines*                       APPROVED BY:                    DATE:

| DISTANCE (m) | SYMBOL | ACTIVITY | TYPE OF ACTIVITY |
|---|---|---|---|
| | ▽ | In old-engine stores . . . . . . . . . . . . | |
| | 1 ⇨ | Picked up engine by crane (electric) . . . . . . . . | *Non-productive* |
| *24* | 2 ⇨ | Transported to next crane . . . . . . . . . . . | *"* |
| | 3 ⇨ | Unloaded to floor . . . . . . . . . . . . | *"* |
| | 4 ⇨ | Picked up by second crane (electric) . . . . . . | *"* |
| *30* | 5 ⇨ | Transported to stripping bay . . . . . . . . . . | *"* |
| | 6 ⇨ | Unloaded to floor . . . . . . . . . . . . | *"* |
| | ① | Engine stripped . . . . . . . . . . . . . | *Productive* |
| | ② | Main components cleaned and laid out . . . . . . . | *"* |
| | ☐1 | Components inspected for wear; inspection report written . | *Non-productive* |
| *3* | 7 ⇨ | Parts carried to degreasing basket . . . . . . . . | *"* |
| | 8 ⇨ | Loaded for degreasing by hand-operated crane . . . . . | *"* |
| *1.5* | 9 ⇨ | Transported to degreaser . . . . . . . . . . | *"* |
| | 10 ⇨ | Unloaded into degreaser . . . . . . . . . . | *"* |
| | ③ | Degreased . . . . . . . . . . . . . . | *Productive* |
| | 11 ⇨ | Lifted out of degreaser by crane . . . . . . . . | *Non-productive* |
| *6* | 12 ⇨ | Transported away from degreaser . . . . . . . . | *"* |
| | 13 ⇨ | Unloaded to ground . . . . . . . . . . . | *"* |
| | ☐1 | To cool . . . . . . . . . . . . | *"* |
| *12* | 14 ⇨ | Transported to cleaning benches . . . . . . . . . | *"* |
| | ④ | All parts completely cleaned . . . . . . . . . | *Productive* |
| *9* | 15 ⇨ | All cleaned parts placed in one box . . . . . . . . | *Non-productive* |
| | ☐2 | Awaiting transport . . . . . . . . . . . . | *"* |
| | 16 ⇨ | All parts except cylinder block and heads loaded on trolley | *"* |
| *76* | 17 ⇨ | Transported to engine inspection section . . . . . . | *"* |
| | 18 ⇨ | Parts unloaded and arranged on inspection table . . . . | *"* |
| | 19 ⇨ | Cylinder block and head loaded on trolley . . . . . . | *"* |
| *76* | 20 ⇨ | Transported to engine inspection section . . . . . . | *"* |
| | 21 ⇨ | Unloaded on ground . . . . . . . . . . . | *"* |
| *237.5* | ☐3 | Stored temporarily awaiting inspection . . . . . . . | *"* |

(Adapted from an original)

*Figure 25. Flow process chart–material type: engine stripping, cleaning and degreasing (original method)*

| FLOW PROCESS CHART | | | | MAN/MATERIAL/EQUIPMENT TYPE | | | | | | |
|---|---|---|---|---|---|---|---|---|---|---|

| CHART No. *1*    SHEET No. *1*    OF *1* | | S  U  M  M  A  R  Y | | |
|---|---|---|---|---|
| | | ACTIVITY | PRESENT | PROPOSED | SAVING |
| Subject charted: *Used bus engines* | | OPERATION ◯ | 4 | | |
| | | TRANSPORT ⇨ | 21 | | |
| ACTIVITY: | | DELAY ◱ | 3 | | |
| *Stripping, cleaning and degreasing prior to inspection* | | INSPECTION ▢ | 1 | | |
| | | STORAGE ▽ | 1 | | |
| METHOD: PRESENT/PROPOSED | | DISTANCE (m) | 237.5 | | |
| LOCATION: *Degreasing Shop* | | TIME *(man-min)* | — | — | — |
| OPERATIVE(S):       CLOCK Nos. *1234* | | COST | — | | |
|                         *571* | | LABOUR | — | | |
| CHARTED BY: | | MATERIAL | — | | |
| APPROVED BY:       DATE: | | TOTAL | — | — | — |

| DESCRIPTION | QTY. | DIST-ANCE (m) | TIME (min) | ◯ | ⇨ | ◱ | ▢ | ▽ | REMARKS |
|---|---|---|---|---|---|---|---|---|---|
| Stored in old-engine store | | | | | | | | | |
| Engine picked up | | | | | | | | | |
| Transported to next crane | | 24 | | | | | | | Electric crane |
| Unloaded to floor | | | | | | | | | " " |
| Picked up | | | | | | | | | |
| Transported to stripping bay | | 30 | | | | | | | " " |
| Unloaded to floor | | | | | | | | | " " |
| Engine stripped | | | | | | | | | |
| Main components cleaned and laid out | | | | | | | | | |
| Components inspected for wear; inspection report written | | | | | | | | | |
| Parts carried to degreasing basket | | 3 | | | | | | | |
| Loaded for degreasing | | | | | | | | | |
| Transported to degreaser | | 1.5 | | | | | | | Hand crane |
| Unloaded into degreaser | | | | | | | | | |
| Degreased | | | | | | | | | |
| Lifted out of degreaser | | | | | | | | | " " |
| Transported away from degreaser | | 6 | | | | | | | " " |
| Unloaded to ground | | | | | | | | | |
| To cool | | | | | | | | | |
| Transported to cleaning benches | | 12 | | | | | | | By hand |
| All parts cleaned completely | | | | | | | | | |
| All cleaned parts placed in one box | | 9 | | | | | | | By hand |
| Awaiting transport | | | | | | | | | |
| All parts except cylinder block and heads loaded on trolley | | | | | | | | | |
| Transported to engine inspection section | | 76 | | | | | | | Trolley |
| Parts unloaded and arranged on inspection table | | | | | | | | | |
| Cylinder block and head loaded on trolley | | | | | | | | | |
| Transported to engine inspection section | | 76 | | | | | | | Trolley |
| Unloaded to ground | | | | | | | | | |
| Stored temporarily awaiting inspection | | | | | | | | | |
| | | | | | | | | | |
| | | | | | | | | | |
| | | | | | | | | | |
| | | | | | | | | | |
| | | | | | | | | | |
| | | | | | | | | | |
| | | | | | | | | | |
| TOTAL | | 237.5 | | 4 | 21 | 3 | 1 | 1 | |

(Adapted from the original)

h) A summary of distance, time and, if desired, cost of labour and material, for comparison of old and new methods.

5) Before leaving the chart, check the following points:

a) Have the facts been correctly recorded?

b) Have any over-simplifying assumptions been made (e.g. is the investigation so incomplete as to be inaccurate)?

c) Have all the factor contributing to the process been recorded?

So far we have been concerned only with the record stage. We must now consider the steps necessary to examine critically the data recorded.

## 8.2   Examine critically: The questioning technique

> **The questioning technique is the means by which the critical examination is conducted, each activity being subjected in turn to a systematic and progressive series of questions.**

The five sets of activities recorded on the flow process chart fall naturally into two main categories, namely—

□ those in which something is actually happening to the material or workpiece under consideration, i.e. it is being worked upon, moved or examined; and

□ those in which it is not being touched, being either in storage or at a standstill owing to the delay.

Activities in the first category may be subdivided into three groups:

□ MAKE READY activities required to prepare the material or workpiece and set it in position ready to be worked on. In the example in Figure 25 these are represented by the loading and transporting of the engine to the degreasing shop, transporting it to the cleaning benches, etc.

□ DO operations in which a change is made in the shape, chemical composition or physical condition of the product. In the case of the example these are the dismantling, cleaning and degreasing operations.

□ PUT AWAY activities during which the work is moved aside from the machine or workplace. The put away activities of one operation may be the make ready activities of the next—as, for example, transport between operations from the degreaser to the cleaning benches. Putting parts into storage, putting letters into an Out tray and inspecting finished parts are other examples.

It will be seen that, while make ready and put away activities may be represented by transport and inspection symbols, do operations can only be represented by operation symbols.

The aim is obviously to have as high a proportion of do operations as possible, since these are the only ones which carry the product forward in its progress from raw material to completion. (Do operations in non-manufacturing industries are those operations which actually carry out the activity for which the organisation exists, for example the act of selling in a shop or the act of typing in an office). These are productive activities; all others, however necessary, may be considered as non-productive (see Figure 25). The first activities to be challenged must therefore be those which are obviously non-productive, including storages and delays which represent tied-up capital that could be used to further the business.

## THE PRIMARY QUESTIONS

The questioning sequence used follows a well-established pattern which examines—

the PURPOSE for which ⎫
the PLACE at which ⎪
the SEQUENCE in which ⎬ the activities are undertaken.
the PERSON by which ⎪
the MEANS by which ⎭

with a view to ⎧ ELIMINATING
⎪ COMBINING
⎨ REARRANGING ⎬ those activities.
⎪ or
⎩ SIMPLIFYING

In the first stage of the questioning technique, the Purpose, Place, Sequence, Person, Means of every activity recorded is systematically queried, and a reason for each reply is sought.

The primary questions therefore are —

| | | | |
|---|---|---|---|
| PURPOSE: | What | is actually done?[3] | ELIMINATE |
| | Why | is the activity necessary at all? | unnecessary parts of the job. |
| PLACE: | Where | is it being done? Why is it done at the particular place? | COMBINE wherever possible or |
| SEQUENCE: | When | is it done? Why is it done at that particular time? | REARRANGE the sequence of operations for more effective results. |
| PERSON: | Who | is doing it? Why is it done by that particular person? | |
| MEANS: | How | is it being done? Why is it being done in that particular way? | SIMPLIFY the operation. |

## THE SECONDARY QUESTIONS

> The secondary questions cover the second stage of the questioning technique, during which the answers to the primary questions are subjected to further query to determine whether possible alternatives to Place, Sequence, Persons and/or Means are practicable and preferable as a means of improvement upon the existing method.

Thus, during this second stage of questioning (having asked already, about every activity recorded, what is done and why is it done), the method study man goes on to inquire: What else might be done? And, hence: What should be done? In the same way, the answers already obtained on Place, Sequence, Persons and Means are subjected to further inquiry.

Combining the two primary questions with the two secondary questions under each of the headings purpose, place, etc., yields the following list, which sets out the questioning technique in full:

PURPOSE:    What is done?
            Why is it done?
            What else might be done?
            What should be done?

PLACE:      Where is it done?
            Why is it done there?
            Where else might it be done?
            Where should it be done?

SEQUENCE:   When is it done?
            Why is it done then?
            When might it be done?
            When should it be done?

PERSON:     Who does it?
            Why does that person do it?
            Who else might do it?
            Who should do it?

MEANS:      How is it done?
            Why is it done that way?
            How else might it be done
            How should it be done?

These questions, in the above sequence, must be asked systematically every time a method study is undertaken. They are the basis of successful method study.

## EXAMPLE: ENGINE STRIPPING, CLEANING AND DEGREASING

Let us now consider how the method study men who prepared the flow process chart in Figure 25 set about examining the record of facts which they had obtained in order to develop an improved method. Before doing so, we shall transfer the same record to a standard flow process chart from (Figure 26) with the necessary information on the operation, location, etc., duly filled in.

This form, like all the forms in this book, is designed so that it can be prepared on a standard typewriter. The arrangement of the symbols in the columns is to enable those used most to be closest together.

To help the reader to visualise the operations, a flow diagram showing the layout of the degreasing shop and the path taken by the engine in its journey from the old-engine stores to the engine-inspection section is given in Figure 26. It is evident from this that the engine and its parts follow an unnecessarily complicated path.

Examination of the flow process chart shows a very high proportion of non-productive activities. There are in fact only four operations and one inspection, while there are 21 transports and three delays. Out of 29 activities, excluding the original storage only five can be considered as productive.

Detailed examination of the chart leads to a number of questions. For example, it will be seen that an engine being transported from the old-engine stores has to change cranes in the middle of its journey. Let us apply the questioning technique to these first transports:

Q. What is done?

A. The engine is carried part of the way through the stores by one electric crane, is placed on the ground and is then picked up by another which transports it to the stripping bay.

Q. Why is this done?

A. Because the engines are stores in such a way that they cannot be directly picked up by the monorail crane which runs through the stores and degreasing shop.

Q. What else might be done?

A. The engines could be stored so that they are immediately accessible to the monorail crane, which could then pick them up and run directly to the stripping bay.

Q. What should be done?

A. The above suggestion should be adopted.

When this suggestion was adopted, three transports were eliminated (see Figure 27).

Let us continue the questioning technique.

Q. Why are the engine components cleaned before going to be degreased since they are again cleaned after the grease is removed?

A. The original reason for this practice has been forgotten.

Q. Why are they inspected at this stage, when it must be difficult to make a proper inspection of greasy parts and when they will be inspected again in the engine-inspection section?

A. The original reason for this practice has been forgotten.

Figure 26. Flow diagram: engine stripping, cleaning and degreasing

**ORIGINAL METHOD**      **PROPOSED METHOD**

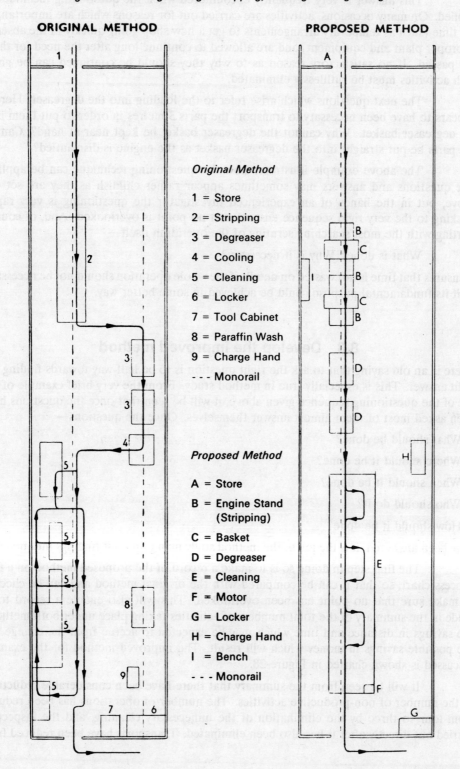

*Original Method*

1 = Store

2 = Stripping

3 = Degreaser

4 = Cooling

5 = Cleaning

6 = Locker

7 = Tool Cabinet

8 = Paraffin Wash

9 = Charge Hand

- - - Monorail

*Proposed Method*

A = Store

B = Engine Stand
(Stripping)

C = Basket

D = Degreaser

E = Cleaning

F = Motor

G = Locker

H = Charge Hand

I = Bench

- - - Monorail

This answer is very frequently encountered when the questioning technique is applied. On many occasions, activities are carried out for reasons which are important at the time (such as temporary arrangements to get a new shop going quickly in the absence of proper plant and equipment) and are allowed to continue long after the need for them has passed. If no satisfactory reason as to why they should be continued can be given such activities must be ruthlessly eliminated.

The next questions which arise refer to the loading into the degreaser. Here it appears to have been necessary to transport the parts 3 metres in order to put them into the degreaser basket. Why cannot the degreaser basket be kept near at hand? Cannot the parts be put straight into the degreaser basket as the engine is dismantled?

The above example illustrates how the questioning technique can be applied. The questions and answers may sometimes appear rather childish as they are set out above, but in the hands of an experienced investigator the questioning is very rapid. Sticking to the very rigid sequence ensures that no point is overlooked. And, of course, starting with the most searching scrutiny of the operation itself—

What is done? Why is it necessary?

It ensures that time is not wasted on details if the whole operation should not be necessary, or if its fundamental purpose could be achieved in some better way.

## 8.3   Develop the improved method

There is an old saying that to ask the right question is to be half-way towards finding the right answer. This is especially true in method study. From the very brief example of the use of the questioning sequence given above, it will be seen that once the questions have been asked most of them almost answer themselves. Once the questions—

☐ What should be done?

☐ Where should it be done?

☐ When should it be done?

☐ Who should do it?

☐ How should it be done?

have been answered, it is the job of the method study man to put his findings into practice.

The first step in doing so is to make a record of the proposed method on a flow process chart, so that it can be compared with the original method and can be checked to make sure that no point has been overlooked. This will also enable a record to be made in the summary of the total numbers of activities taking place under both methods, the savings in distance and time which may be expected to accrue from the change and the possible savings in money which will result. The improved method for the example discussed is shown charted in Figure 28.

It will be seen from the summary that there have been considerable reductions in the number of non-productive activities. The number of operations has been reduced from four to three by the elimination of the unnecessary cleaning, and the inspection carried out directly after it has also been eliminated. Transports have been reduced from

**Figure 27. Flow process chart–material type: engine stripping, cleaning and degreasing (improved method)**

| FLOW PROCESS CHART | | | | MAN/MATERIAL/EQUIPMENT TYPE | | | | |
|---|---|---|---|---|---|---|---|---|
| CHART No. 2    SHEET No 1    OF 1 | | | | S U M M A R Y | | | | |
| Subject charted: | | | | ACTIVITY | | PRESENT | PROPOSED | SAVING |
| *Used bus engines* | | | | OPERATION ◯ | | 4 | 3 | 1 |
| | | | | TRANSPORT ⇨ | | 21 | 15 | 6 |
| ACTIVITY: | | | | DELAY ◻ | | 3 | 2 | 1 |
| *Stripping, degreasing and cleaning* | | | | INSPECTION ◻ | | 1 | – | 1 |
| *prior to inspection* | | | | STORAGE ▽ | | 1 | 1 | 1 |
| METHOD: PRESENT/PROPOSED | | | | DISTANCE (m) | | 237.5 | 150.0 | 87.5 |
| LOCATION: *Degreasing shop* | | | | TIME *(man-min)* | | — | | |
| OPERATIVE(S):    CLOCK Nos. *1234* | | | | COST | | | | |
| *571* | | | |   LABOUR | | | | |
| CHARTED BY: | | | |   MATERIAL | | | | |
| APPROVED BY:    DATE: | | | | TOTAL | | — | | |

| DESCRIPTION | QTY. | DIST-ANCE (m) | TIME (min) | SYMBOL ◯ ⇨ ◻ ◻ ▽ | REMARKS |
|---|---|---|---|---|---|
| Stored in old-engine store | | — | — | | |
| Engine picked up | | | | | |
| Transported to stripping bay | | 55 | | | Electric |
| Unloaded on to engine stand | | | | | hoist on mono-rail |
| Engine stripped | | | | | |
| Transported to degreaser basket | | 1 | | | |
| Loaded into basket | | | | | By hand |
| Transported to degreaser | | 1.5 | | | Hoist |
| Unloaded into degreaser | | | | | ,, |
| Degreased | | | | | ,, |
| Unloaded from degreaser | | | | | |
| Transported from degreaser | | 4.5 | | | ,, |
| Unloaded to ground | | | | | ,, |
| Allowed to cool | | | | | |
| Transported to cleaning benches | | 6 | | | |
| All parts cleaned | | | | | ,, |
| All parts collected in special trays | | 6 | | | |
| Awaiting transport | | | | | |
| Trays and cylinder block loaded on trolley | | | | | |
| Transported to engine inspection section | | 76 | | | Trolley |
| Trays slid on to inspection benches and blocks on to platform | | | | | |
| **TOTAL** | | 150 | | 3  15  2  —  1 | |

21 and 15 and the distances involved have been cut from 237.5 to 150 metres—a saving of over 37 per cent in the travel of each engine. In order not to complicate this example, times of the various activities have not been given; but a study of the two flow process charts will make it evident that a very great saving in the time of operation per engine has been achieved.

No further example of a flow process chart is given in this chapter because flow process charts will be used later in the book in association with other techniques.

## NOTES

1. The symbols used throughout this book are those recommended by the American Society of Mechanical Engineers and adopted in the B.S. Glossary, op. cit. There is another set of symbols still in fairly common use, an abbreviated form of the set originated by F.B. and L.M. Gilbreth. It is recommended that the ASME symbols should be adopted in preference to those of Gilbreth.

2. This example is adopted from W. Rodgers: Methods engineering chart and glossary (Nottingham (United Kingdom), School of Management Studies Ltd.).

3. Many investigators use the question: What is actually achieved?

# Chapter 9

# Flow and handling of materials

## 9.1 Introduction

Materials handling accounts for a significant portion of the total production cost. In the workers and materials have to cover long distances in the course of the manufacturing process; this leads to a loss of time and energy without anything being added to the value of the product. Through effective plant layout analysis and design, many materials handling operations can be reduced or eliminated. The choice of materials handling methods and equipment is an integral part of the plant layout design.

In developing the plant layout, the possibility of future expansion and changes must be considered. The layout should be flexible enough to accommodate changes in product design, process design and schedule design.

## 9.2 Plant layout

Invariably, in conducting a method study, it becomes desirable at some stage to look critically at the movement of men and materials through the plant or work area and to examine the plant layout. This is so because in many factories either the initial layout was not well thought out or, as the enterprise expanded or changed some of its products or processes, extra machines, equipment or offices were added wherever space could be found. In other cases temporary arrangements may have been made to cope with an emergency situation, such as the sudden increase in demand for a certain product; but then these arrangements remain on a permanent basis even if the situation that provoked them subsequently changes. The net result is that materials and workers often have to make long, round about journeys in the course of the manufacturing process; this leads to a loss of time and energy without anything being added to the value of the product. Improving plant layout is, therefore, part of the job of the work study specialist. There are four major types of layout, although in practice a combination of two or more may be found in the same plant. These are shown in Figure 28 and are as follows:

### Figure 28. Types of layout

#### (a) Layout by fixed position

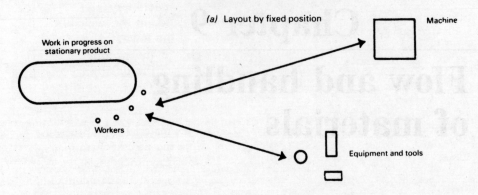

#### (b) Layout by process or function

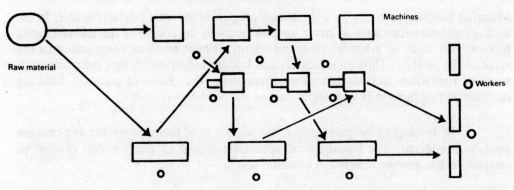

#### (c) Layout by product (line layout)

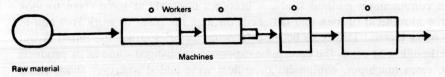

#### (d) Group layout

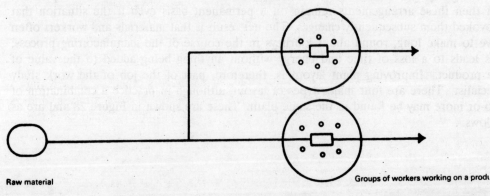

> **Plant layout is the arrangement of the desired machinery and equipment of a plant, established or contemplated, in the way which will permit the easiest flow of materials, at the lowest cost and with the minimum of handling, in processing the product from the receipt of raw materials to the dispatch of the finished product. [1]**

a) Layout by fixed position. This arrangement is used when the material to be processed does not travel round the plant but stays in one place: all the necessary equipment and machinery is brought to it instead. This is the case when the product is bulky and heavy and when only a few units are made at a time. Typical examples are shipbuilding and the manufacture of diesel engines or large motors or aircraft construction.

b) Layout by process or function. Here all operations of the same nature are grouped together: for example, in the garment industries all the cutting of material is carried out in one area, all the sewing or stitching in another area, all the finishing in a third area, and so on. This layout is usually chosen where a great many products which share the same machinery are being made and where any one product has only a relatively low volume of output. Examples are textile spinning and weaving, maintenance workshops and the garment industries.

c) Layout by product, or line layout, sometimes popularly referred to as mass production. In this layout all the necessary machinery and equipment needed to make a given product is set out in the same area and in the sequence of the manufacturing process. This layout is mainly used where there is a high demand for one or several products that are more or less standardised. Typical examples are soft drinks bottling, car assembly and some canning operations.

d) Layout making possible group production methods, or group layout. Recently, in an effort to increase job satisfaction, several enterprises have arranged their operations in a new way, with a group of working together on a given product or on a part of a product and having at hand all the machinery and equipment needed to complete their work. In such cases the workers distribute the work among themselves and usually interchange jobs. Further details of this method of production are given in Chapter 24.

With these various kinds of layout in mind, we may now analyse the flow of materials in the plant. In some situations, rapid changes in output may be realised by switching from one type of layout to another. This is particularly true when a shift is made from a layout by function to a line layout for one or more products of which the output has been increased significantly.

In most cases, however, a careful analysis of the flow is called for before any decision is taken to change a given layout, since this is usually a costly process, and the management has to be convinced that real savings will result before sanctioning the change.

## 9.3  Developing the new layout

The following steps are taken when a layout for a plant or a work area is designed:

a) The equipment and machinery needed for processing is determined by the type of product or products.

b) The number of units of each machine and items of equipment needed to manufacture each product are determined by the volume of expected sales (based on sales forecasts).

c) The space requirements for machinery are determined by calculating the dimensions of each machine and multiplying them by the number of machines needed.

d) Provision is made for the space needed for materials (both for raw materials and for the storage of finished products), for goods-in-process and for material-handling equipment.

e) Provision is also made for additional space for auxiliary services (washrooms, offices, cafeteria, etc.).

f) The total space requirement for the plant is determined by adding the space needed for machinery to the space needed for storage and for auxiliary services.

g) The different departments with their respective areas are so arranged that the most economical flow of work is achieved.

h) The plan of the building is largely determined by the positioning of working areas, storage areas and auxiliary services.

i) The size and design of the site is determined by allocating additional space for parking, receiving and shipping, and landscaping.

However, a work study man is rarely called upon to make a complete design of a plant, starting from the very basic steps described above. This is more the task of the industrial engineer or the production management specialist. It is more common for the work study man to be faced with a problem of modifying an already existing layout. In this case, the major issue becomes that of determining the best possible flow of work, and several diagrams can be helpful here (see Figure 29)[2]. The use of any of these diagrams depend on whether the flow is being studied for one product or process or for a number of products and processes performed simultaneously.

## METHODS OF DETERMINING SPACE REQUIREMENTS

Common used methods of determining space requirements are as follows:

a) Production Centre Method—A production centre consists of a single machine with all the associated equipment and space required for its operation. Work space, maintenance space, material set-down space and access space for aisles are added to the space requirements for the machine. This is then multiplied by the number required of each piece of machinery; and added in space allowances for aisles and in general or for support areas.

b) Conversion Method—The amount of space required currently for each machine, machine group or activity areas is determined. Adjustment is made of this to what should be used to perform the job efficiently. Then the conversion of this is done by some factor or multiplier to determine what would be needed for the new requirements.

Figure 29. Example of various types of flow between work stations, including flow in a multi-storey building

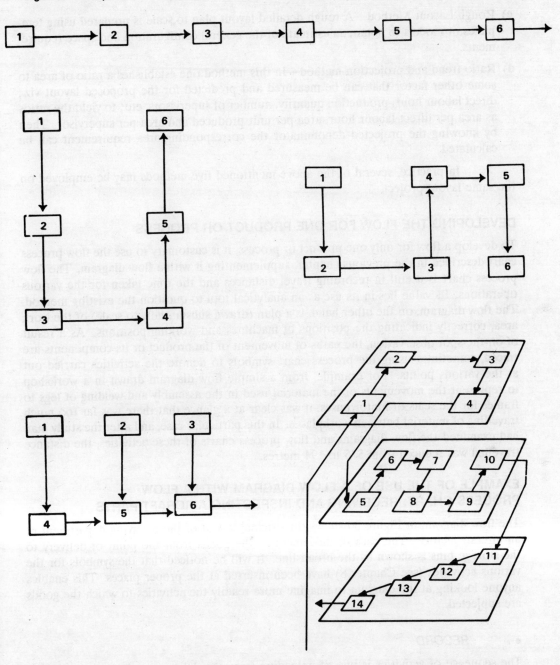

The flow diagram can also be used for the study of movement on several floors of a multi-storey building. Ordinary flow diagrams of each floor can, of course, be made as well.

This method is mostly used to determine space requirements for supporting service and storage areas.

c) Rough Layout Method—A rough detailed layout plan to scale is prepared using templates or models to obtain an estimate of the general configuration and space requirements.

d) Ratio trend and projection method—In this method one establishes a ratio of area to some other factor that can be measured and predicted for the proposed layout viz., direct labour hour, production quantity, number of supervisors, etc., to yield the ratios as area per direct labour hour, area per unit produced and area per supervisor. Then by knowing the projected denominator the corresponding area requirement can be calculated.

In practice, several of the above-mentioned five methods may be employed on the same layout project.

## DEVELOPING THE FLOW FOR ONE PRODUCT OR PROCESS

To develop a flow for only one product or process, it is customary to use the flow process chart described in the previous chapter, supplementing it with a flow diagram. The flow process chart is useful in recording travel distances and the time taken for the various operations. Its value lies in its use as an analytical tool to question the existing method. The flow diagram, on the other hand, is a plan (drawn substantially to scale) of the work area, correctly indicating the positions of machines and working positions. As a result of on-the-spot observation, the paths of movement of the product or its components are traced, sometimes using the process chart symbols to denote the activities carried out at the various points. For example, from a simple flow diagram drawn in a workshop to represent the movements of the material used in the assembly and welding of legs to frames for the seats of motor buses, it was clear at a glance that there was far too much travelling of material between workplaces. In this particular case, and after the study man had examined the flow diagrams and flow process charts of these activities, the distance travelled was reduced from 575 to 194 metres.

## EXAMPLE OF THE USE OF A FLOW DIAGRAM WITH A FLOW PROCESS CHART: RECEIVING AND INSPECTING AIRCRAFT PARTS

The flow diagram in Figure 30 shows the original layout of the receiving department of an aircraft factory. The path of movement of the goods from the point of delivery to the storage bins is shown by the broad line. It will be noticed that the symbols for the various activities (see Chapter 8) have been inserted at the proper places. This enables anyone looking at the diagram to imagine more readily the activities to which the goods are subjected.

•     *RECORD*

The sequence of activities is one of unloading from the delivery truck cases containing aircraft parts (which are themselves packed individually in cartons), checking, inspecting and marking them before storing them. These cases are slid down an inclined plane from the tail of the truck, slid across the floor to the "unpacking space" and there stacked one

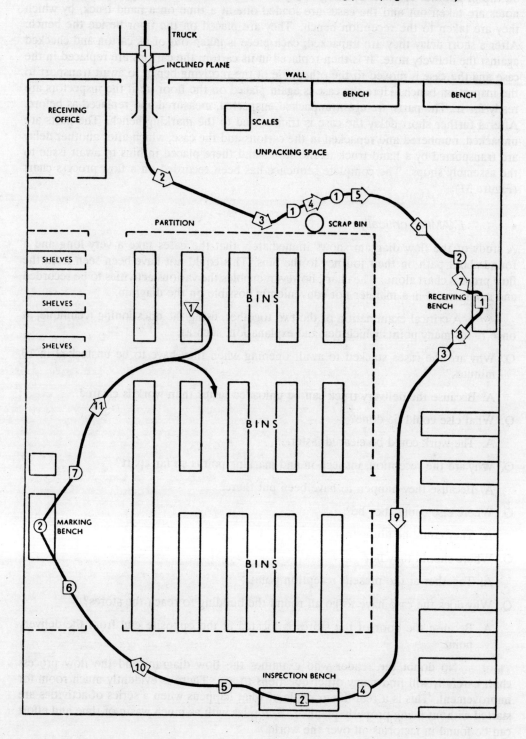

Figure 30. Flow diagram: inspecting and marking incoming parts (original method)

on top of another to await opening. They are then unstacked and opened. The delivery notes are taken out and the cases are loaded one at a time on a hand truck, by which they are taken to the reception bench. They are placed on the floor beside the bench. After a short delay they are unpacked; each piece is taken out of its carton and checked against the delivery note. It is then replaced in its carton; the cartons are replaced in the case and the case is moved to the other side of the receiving bench to await transport to the inspection bench. Here the case is again placed on the floor until the inspectors are ready for it. The parts are again unpacked, inspected, measured and replaced as before. After a further short delay the case is transported to the marking bench. The parts are unpacked, numbered and repacked in the cartons and the case, which after another delay are transported by a hand truck to the stores and there placed in bins to await issue to the assembly shops. The complete sequence has been recorded on a flow process chart (Figure 31).

- *EXAMINE critically*

A study of the flow diagram shows immediately that the cases take a very long and a round about path on their journey to the bins. This could not have been seen from the flow process chart alone. The chart, however, enables the various activities to be recorded and summarised in a manner not conveniently possible on the diagram.

A critical examination of the two together, using the questioning technique, at once raises many points which demand explanation, such as:

Q. Why are the cases stacked to await opening when they have to be unstacked in 10 minutes?

A. Because the delivery truck can be unloaded faster than work is cleared.

Q. What else could be done?

A. The work could be cleared faster.

Q. Why are the reception, inspection and marking points so far apart?

A. Because they happen to have been put there.

Q. Where else could they be?

A. They could be altogether.

Q. Where should they be?

A. Together at the present reception point.

Q. Why does the case have to go all round the building to reach the stores?

A. Because the door of the stores is located at the opposite end from the delivery point.

No doubt the reader who examines the flow diagram and the flow process chart carefully will find many other questions to ask. There is evidently much room for improvement. This is a real-life example of what happens when a series of activities are started without being properly planned. Examples with as much waste of time and effort can be found in factories all over the world.

*Figure 31. Flow process chart: inspecting and marking incoming parts (original method)*

| FLOW PROCESS CHART | | | | MAN/MATERIAL/EQUIPMENT TYPE | | | | |
|---|---|---|---|---|---|---|---|---|
| **CHART No. 3**  SHEET No. 1  OF 1 | | | | S U M M A R Y | | | | |
| Subject charted: | | | | ACTIVITY | | PRESENT | PROPOSED | SAVING |
| *Case of BX 487 Tee-pieces (10 per case in cartons)* | | | | OPERATION ◯ | | 2 | | |
| | | | | TRANSPORT ⇨ | | 11 | | |
| ACTIVITY: *Receive, check, inspect and number* | | | | DELAY D | | 7 | | |
| *tee-pieces and store in case* | | | | INSPECTION ☐ | | 2 | | |
| | | | | STORAGE ▽ | | 1 | | |
| METHOD: PRESENT/PROPOSED | | | | DISTANCE (m) | | 56.2 | | |
| LOCATION: *Receiving Dept.* | | | | TIME (man-h.) | | 1.96 | | |
| OPERATIVE(S):  CLOCK No. | | | | COST | | | | |
| *See Remarks column* | | | | LABOUR | | $10.19 | | |
| CHARTED BY:  DATE: | | | | MATERIAL | | — | | |
| APPROVED BY:  DATE: | | | | TOTAL | | $10.19 | | |

| DESCRIPTION | QTY. | DIST-ANCE | TIME | SYMBOL | | | | | REMARKS |
|---|---|---|---|---|---|---|---|---|---|
| | 1 case | (m) | (min) | ◯ | ⇨ | D | ☐ | ▽ | |
| Lifted from truck: placed on inclined plane | | 1.2 | | | | | | | |
| Slid on inclined plane | | 6 | 10 | | | | | | 2 labourers |
| Slid to storage and stacked | | 6 | | | | | | | 2  „ |
| Await unpacking | | — | | | | | | | 2  „ |
| Case unstacked | | — | 30 | | | | | | |
| Lid removed: delivery note taken out | | — | 5 | | | | | | |
| Placed on hand truck | | 1 | | | | | | | 2  „ |
| Trucked to reception bench | | 9 | 5 | | | | | | |
| Await discharge from truck | | — | 10 | | | | | | 2  „ |
| Case placed on bench | | 1 | 2 | | | | | | |
| Cartons taken from case: opened: checked | | — | 15 | | | | | | 2  „ |
| replaced contents | | | | | | | | | |
| Case loaded on hand truck | | 1 | 2 | | | | | | Storekeeper |
| Delay awaiting transport | | — | 5 | | | | | | 2 labourers |
| Trucked to inspection bench | | 16.5 | 10 | | | | | | 1 labourer |
| Await inspection | | — | 10 | | | | | | Case on truck |
| Tee-pieces removed from case and cartons: | | 1 | 20 | | | | | | Inspector |
| inspected to drawing: replaced | | | | | | | | | |
| Await transport labourer | | — | 5 | | | | | | Case on truck |
| Trucked to numbering bench | | 9 | 5 | | | | | | 1 labourer |
| Await numbering | | — | 15 | | | | | | Case on truck |
| Tee-pieces withdrawn from case and cartons: | | — | 15 | | | | | | Stores labourer |
| numbered on bench and replaced | | | | | | | | | |
| Await transport labourer | | — | 5 | | | | | | Case on truck |
| Transported to distribution point | | 4.5 | 5 | | | | | | 1 labourer |
| Stored | | | | | | | | | |
| | | | | | | | | | |
| **TOTAL** | | 56.2 | 174 | 2 | 11 | 7 | 2 | 1 | |

- *DEVELOP the improved method*

The solution arrived at by the work study men in this factory can be seen in Figures 32 and 33. It is clear that among the questions they asked were those suggested above, because it will be seen that the case is now slid down the inclined plane from the delivery truck and put straight on a hand truck. It is transported straight to the "Unpacking space", where it is opened while still on the truck and the delivery note is taken out. It is then transported to the reception bench, where, after a short delay, it is unpacked and the parts are put on the bench. The parts are counted and checked against the delivery note. The inspection and numbering benches have now been placed besides the reception bench so that the parts can be passed from hand to hand for inspection, measuring and then numbering. They are finally replaced in their cartons and repacked in the case, which is still on the truck.

It is evident that the investigators were led to ask the same question as we asked, namely: "Why does the case have to go all round the building to reach the stores?" Having received no satisfactory answer, they decided to make a new doorway into the stores opposite the benches, so that the cases could be taken in by the shortest route.

It will be seen from the summary on the flow process chart (Figure 33) that the 'inspections' have been reduced from two to one, the 'transports' from eleven to six and the 'delays' (or temporary storages) from seven to two. The distance travelled has been reduced from 56.2 to 32.2 metres.

The number of man-hours involved has been calculated by multiplying the time taken for each item of activity by the number of workers involved, e.g. "truck to reception bench" = 5 minutes × 2 labourers = 10 man-minutes. Delays are not included as they are caused by operatives being otherwise occupied. In the improved method the inspector and stores labourer are considered to be working simultaneously on inspecting and numbering respectively, and the 20 minutes therefore becomes 40 man-minutes. Labour cost is reckoned at an average of US $ 5.20 per hour for all labour. The cost of making a new doorway is not included, since it will be spread over many other products as well.

## DEVELOPING THE FLOW FOR A NUMBER OF PRODUCTS OR PROCESSES

If several products are being made or several processes are being carried out at the same time, another type of chart is used to determine the ideal placing of the machinery or operations. This is the cross chart.

As can be seen from Figure 35, the cross chart is drawn up by listing the various operations (or machinery) through which the different products pass at the various stages of production, on both the horizontal and vertical dimensions of the chart. The example in Figure 34 illustrates the use of the cross chart for a company making decorated metal products. In this case, the company is producing 70 products, each of which passes through some of the operations indicated.

To complete this chart, take one product at a time and enter its sequence of manufacturing in the appropriate square on the chart. If a product moves from 'Form' to 'Normalise', a stroke is made in the square 'Form/Normalise'. If it subsequently moves from 'Normalise' to 'Plate', a stroke is entered in the corresponding square, and so on

*Figure 32. Flow diagram: inspecting and marking incoming parts (improved method)*

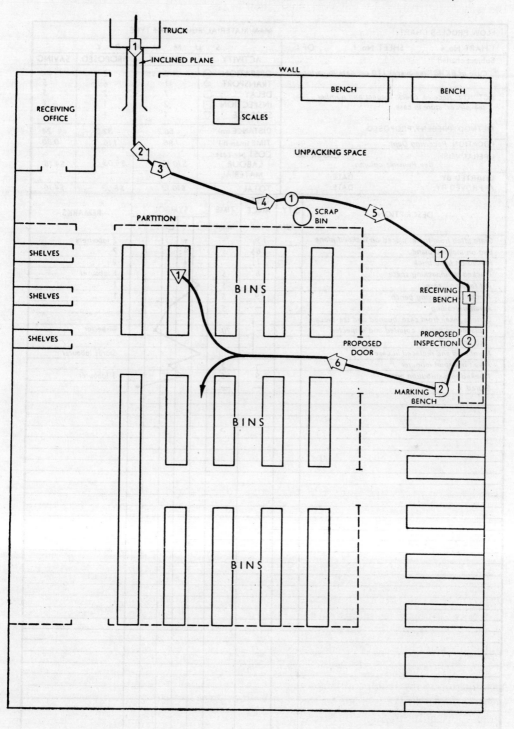

*Figure 33. Flow process chart: inspecting and marking incoming parts (improved method)*

| FLOW PROCESS CHART | | | ~~MAN~~/MATERIAL/~~EQUIPMENT~~ TYPE | | | |
|---|---|---|---|---|---|---|
| CHART No. 4    SHEET No. 1    OF 1 | | | S U M M A R Y | | | |
| Subject charted: | | | ACTIVITY | PRESENT | PROPOSED | SAVING |
| Case of BX 487 tee-pieces (10 per case in cartons) | | | OPERATION ◯ | 2 | 2 | — |
| | | | TRANSPORT ⇨ | 11 | 6 | 5 |
| ACTIVITY: Receive, check, inspect and number | | | DELAY D | 7 | 2 | 5 |
| tee-pieces: store in case | | | INSPECTION ☐ | 2 | 1 | 1 |
| | | | STORAGE ▽ | 1 | 1 | — |
| METHOD: ~~PRESENT~~/PROPOSED | | | DISTANCE (m) | 56.2 | 32.2 | 24 |
| LOCATION: Receiving Dept. | | | TIME (man-h.) | 1.96 | 1.16 | 0.80 |
| OPERATIVE(S)              CLOCK No.    See Remarks column | | | COST per case    LABOUR | $10.19 | $6.03 | $4.16 |
| | | | MATERIAL | — | — | — |
| CHARTED BY:             DATE: | | | TOTAL | $10.19 | $6.03 | $4.16 |
| APPROVED BY:            DATE: | | | | | | |

| DESCRIPTION | QTY. 1 case | DIST-ANCE (m) | TIME (min) | SYMBOL ◯ | ⇨ | D | ☐ | ▽ | REMARKS |
|---|---|---|---|---|---|---|---|---|---|
| Crate lifted from truck: placed on inclined plane | | 1.2 | | | | | | | 2 labourers |
| Slid on inclined plane | | 6 | 5 | | | | | | 2   „ |
| Placed on hand truck | | 1 | | | | | | | 2   „ |
| Trucked to unpacking space | | 6 | 5 | | | | | | 1 labourer |
| Lid taken off case | | — | 5 | | | | | | 1   „ |
| Trucked to receiving bench | | 9 | 5 | | | | | | 1   „ |
| Await unloading | | — | 5 | | | | | | |
| Cartons taken from case: opened and tee-pieces placed on bench: counted and inspected to drawing | | — | 20 | | | | | | Inspector |
| Numbered and replaced in case | | | | | | | | | Stores labourer |
| Await transport labourer | | — | 5 | | | | | | |
| Trucked to distribution point | | 9 | 5 | | | | | | 1 labourer |
| Stored | | — | — | | | | | | |
| TOTAL | | 32,2 | 55 | 2 | 6 | 2 | 1 | 1 | |

*Figure 34. Developing the flow for a number of products, using the cross chart*

| From \ To | Form | Normalise | Machine | Burr/trim | Paint | Plate | Coat | Polish | Wrap | Pack and ship | Total |
|---|---|---|---|---|---|---|---|---|---|---|---|
| Form |  | 14 | 8 | 6 | 14 |  |  |  | 1 | 27 | 70 |
| Normalise |  |  |  | 4 | 17 |  |  |  |  | 1 | 18 |
| Machine |  |  |  | 3 | 2 | 1 |  |  |  | 2 | 8 |
| Burr/trim |  |  |  |  | 11 | 2 |  |  |  | 1 | 10 |
| Paint |  |  |  |  |  | 19 | 11 | 13 | 3 |  | 46 |
| Plate |  |  |  |  | 2 |  | 22 |  | 2 |  | 22 |
| Coat |  |  |  |  |  |  |  | 22 |  |  | 22 |
| Polish |  |  |  |  |  |  |  |  | 33 | 3 | 36 |
| Wrap |  |  |  |  |  |  |  |  |  | 39 | 39 |
| Pack and ship |  |  |  |  |  |  |  |  |  |  | 0 |
| Total | 0 | 18 | 8 | 10 | 46 | 22 | 22 | 36 | 39 | 70 |  |

*Source:* Taken from Richard Muther: "Plant layout", in H. B. Maynard: *Industrial engineering handbook* (New York and London. McGraw-Hill, 3rd ed. 1971), and used by kind permission of the McGraw-Hill Book Company.

until the whole sequence of operations for that particular product is entered. The same process is then repeated for each of the other 70 products. The completed cross chart will appear as in Figure 34.

The next step is to decide which operations should be placed adjacent to each other. From the chart it is clear that 27 products out of 70 (i.e. 39% of the products) pass directly from 'Form' to "Pack and Ship". These two operations should therefore be adjacent. Similarly, all 22 products that were subjected to plating passed from 'Plate' to 'Coat' and from 'Coat' to 'Polish'. Hence, these three operations should follow each other in sequence. By following the same line of reasoning it is possible to reach the preferred sequence of operations.

A variation of this technique is to complete the cross chart by taking a sample of the most frequently produced items. If the plant is producing over 100 different items, it may become cumbersome to follow the method indicated above. However, investigation may reveal that, say, 15 or 20 items account for possibly 80% of the production volume. The sequence of operations of these items would then be entered on the cross chart, and the flow determined in the same way as that described above.

## VISUALISING THE LAYOUT

Once the dimensions and the relative position of machinery, storage facilities and auxiliary services have been determined, it is advisable to make a visual presentation of the proposed layout before proceeding with the actual rearrangement of the workplace, which may be a costly operation. This can be done by the use of 'templates', or pieces of cardboard cut out to scale. Different coloured cards may be used to indicate different items of equipment, such as machines, storage racks, benches or material-handling equipment. When positioning these templates, make sure that gangways are wide enough to allow the free movement of material-handling equipment and goods-in-process.

Alternatively, scale models may be used to provide a three-dimensional display of the layout. Various types of model for many well-known items of machinery and equipment are readily available on the market and are particularly useful for training purposes.

# 9.4   The handling of materials

A good deal of time and effort is often expended in moving material from one place to another in the course of processing. This handling is costly and adds nothing to the value of the product. In essence, therefore, there should ideally be no handling at all. Unfortunately this is not possible. A more realistic aim would be to move material by the most appropriate methods and equipment at the lowest possible cost and with regard to safety. This aim may be met by:

 i) eliminating or reducing handling;

 ii) improving the efficiency of handling; and

 iii) making the correct choice of material-handling equipment.

## ELIMINATING OR REDUCING HANDLING

There is often ample scope for eliminating or reducing handling. In practice, it becomes obvious that there is a need to improve an existing situation when certain symptoms are observed, e.g. too much loading and unloading, repeated manual handling of heavy weights, material travelling considerable distances, non-uniform flow of work with congestion in certain areas, frequent damage or breakage resulting from handling, and so on. These are some of the most frequent phenomena that invite the intervention of the work study specialist. The approach to be followed here is similar to the traditional method study approach, using outline and flow process charts and flow diagrams and asking the same questions as to "where, when, who, how" and, above all, 'why' this handling is done.

However, such a study may frequently have to be preceded by or carried out in conjunction with a study of the layout of the working area, in order to reduce movement to a minimum.

## IMPROVING THE EFFICIENCY OF HANDLING

The observance of certain precepts can improve the efficiency of handling. These precepts are—

a) Increase the size or number of units being handled at any one time. If necessary, review product design and packaging to see if you can achieve this result more readily.

b) Increase the speed of handling if this is possible and economical.

c) Let gravity work for you as much as possible.

d) Have enough containers, pallets, platforms, boxes, etc., available in order to make transportation easier.

e) Give preference in most cases to material-handling equipment that lends itself to a variety of uses and applications.

(f) Try to ensure that materials move in straight lines as much as possible, and ensure that gangways are kept clear.

## MAKING THE CORRECT CHOICE OF HANDLING EQUIPMENT

The work study man should be aware of the different kinds and types of material-handling equipment. Although there are literally hundreds of various types, these may be classified in four major categories.

## • CONVEYORS

Conveyors are useful for moving material between two fixed work stations, either continuously or intermittently. They are mainly used for continuous or mass production operations—indeed, they are suitable for most operations where the flow is more or less steady. Conveyors may be for various types, with either rollers, wheels or belts to help to move the material along: these may be power-driven or may roll freely. The decision to provide conveyors must be taken with care, since they are usually costly to install; moreover, they are less flexible and, where two or more converge, it is necessary to co-ordinate the speeds at which the two conveyors move.

●    *INDUSTRIAL TRUCKS*

Industrial trucks are more flexible in use than conveyors since they can move between various points and are not permanently fixed in one place. They are therefore most suitable for intermittent production and for handling various sizes and shapes of material. There are many types of trucks— petrol-driven, electric, hand-powered, and so on. Their greatest advantage lies in the wide range of attachments available; these increase the trucks' ability to handle various types and shapes of material.

●    *CRANES AND HOISTS*

The major advantage of cranes and hoists is that they can move heavy material through overhead space. However, they can usually serve only a limited area. Here again, there are several types of cranes and hoists and within each type there are various loading capacities. Cranes and hoists may be used both for intermittent and for continuous production.

●    *CONTAINERS*

These are either 'dead' containers (e.g. cartons, barrels, skids and pallets) which hold the material to be transported but do not move themselves, or 'live' containers (e.g. wagons and wheelbarrows). Handling equipment of this kind can both contain and move the material, and is usually operated manually.

Figure 35 shows some types of material-handling equipment.

The choice of material-handling equipment is not easy. In several cases the same material may be handled by various items of equipment (see Figure 36). Nor does the great diversity of equipment available make the problem any easier. In several cases, however, the nature of the material to be handled does narrow the choice.

Among the most important factors to be taken into consideration when choosing material-handling equipment are the following:

a) Properties of the material. Whether it is solid, liquid or gas, and in what size, shape and weight it is to be moved, are important considerations and can lead to a preliminary elimination from the range of available equipment under review. Similarly, if a material is fragile, corrosive or toxic this will imply that certain handling methods and containers will be preferable to others.

b) Layout and characteristics of the building. Another restricting factor is the availability of space for handling. Low-level ceilings may preclude the use of hoists or cranes, and the presence of supporting columns in awkward places can limit the size of the material-handling equipment. If the building is multi-storeyed, chutes, or ramps for industrial trucks, may be used. Finally, the layout itself will indicate the type of production operation (continuous, intermittent, fixed position or group) and can already indicate some items of equipment that will be more suitable than others.

c) Production flow. If the flow is fairly constant between two fixed positions that are not likely to change, fixed equipment such as conveyors or chutes can be successfully used. If, on the other hand, the flow is not constant and the direction changes oc-

*Figure 35. Different types of material-handling equipment*

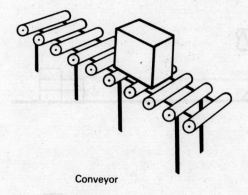

Conveyor

Fork-lift industrial truck

Crane

Hoist

CONTAINERS

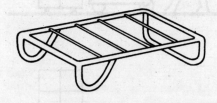

Skid ("dead" container)

Trolley for ceramics or pastries
("live" container)

*Figure 36. Different possibilities of handling the same object*

casionally from one point to another because several products are being produced simultaneously, moving equipment such as trucks would be preferable.

d) Cost considerations. This is one of the most important considerations. The above factors can help to narrow the range of suitable equipment. Costing can help in taking a final decision. Several cost elements need to be taken into consideration when comparisons are made between various items of equipment that are all capable of handling the same load. There is the initial cost of the equipment, from which one can derive the investment cost in terms of interest payment (i.e. if the company has to borrow money to buy the equipment) or opportunity costs (i.e. if the company possesses the funds and does not have to borrow, but the purchase of the equipment would deprive it of an opportunity to invest the funds at a certain rate of return). From the cost of the equipment one can also calculate the depreciation charges per year, to which will be added other charges such as insurance, taxes and additional overheads. Apart from these fixed charges, there are also operating costs, such as the cost of operating personnel, power, maintenance and supervision. By calculating and comparing the total cost for each of the items of equipment under consideration, a more rational decision can be reached on the most appropriate choice.

## NOTES

1. Based on a definition given by R.W. Mallick and A.T. Gaudreau in Plant and Practice (New York, John Wiley, 1966).

2. Readers who wish to go into more detail in the area of plant layout are referred to Richard Muther: Practical plant layout (New York and London, McGraw-Hill 1956) and H.B. Maynard (ed): Industrial engineering handbook (New York and London, McGraw-Hill, 3rd ed., 1971).

# Chapter 10

# Tools for recording the movement of workers

## 10.1 Introduction

Any analysis or inference on the way of carrying out an activity is based on recorded data. Proper recording of all facts relating to an activity thus forms an important step for a successful application of Method Study. Recording of facts for any activity is normally done with the aid of more than one chart and diagram so as to highlight different aspects of the activity. As we have already noted in Chapters 7 and 8, most of these tools serve a specific purpose, their use being determined by the nature of the activity under consideration. Thus, in activities where significant flow of materials is taking place between different points in the workplace, the charts and diagrams described in the last chapter assume importance. Similarly, there are activities where the worker has to move between a number of points with or without material. Both in manufacturing and non-manufacturing organisations, this type of situation may arise as follows:

a) The bulk material is being fed or removed from a continuous process, and is stored around the process.

b) A worker is looking after two or more machines

c) A bearer is serving different tables in a restaurant

d) An attendant is serving different counters in a bank.

In the above and many other similar situations, the workers movements need to be recorded to ascertain whether wasteful movements exist and to find the most effective way of carrying out the activity. The common tools that are used for the purpose are:

a) String diagram

b) Flow process chart—man type

c) Travel chart

d) Multiple activity chart

The objective of this chapter is to discuss the above recording tools. These are now presented in the subsequent sections, in the order listed above.

## 10.2    String diagram

String diagram is a useful tool for recording the distance traversed by a worker in the working area. Construction of the diagram involves the following steps:

Step 1 Preparing a study sheet. The worker under consideration is followed and the different points in the working areas he covers are noted down in a study sheet. The recording of movements is continued till a representative picture of the workers movements is obtained. This implies that the movements are noted down for enough number of cycles so as to capture the actual work cycle in terms of the journey made by the worker with their respective frequencies.

Step 2 Drawing a scale plan of the working area. Once the study sheet is prepared, the next step is to draw a scale plan of the working area. Machines, benches, stores and all other points at which calls are made by the worker are drawn to the scale together with doorways, pillars etc. that affect the paths of movements. The plan is then attached to a softwood or composition board and pins are driven firmly at every stopping point. The heads of the pins are allowed to stay clear of the surface by about 1 cm. Pins are also driven in at all turning points on the route.

Step 3 Combining Steps 1 and 2 to construct the final diagram. In the final Step, a measured length of thread is taken and tied round the pin at the starting point of the movements. It is then led around the pins at the other points of call in the order noted in the study sheet till all the movements as noted in Step 1 have been covered. The resulting diagram is the string diagram.

A sample study sheet and a hypothetical string diagram shown in Figures 37 and 38 respectively.

We can thus define the string diagram as follows:

---

**The string diagram is a scale plan or model on which a thread is used to trace and measure the path of workers, material or equipment during a specified sequence of events.**

---

It should be apparent that the string diagram is similar to the flow diagram discussed in the last chapter. While one is used for recording movements of materials, the other is used for recording movements of workers. Most often these tools are used together with the flow process chart in order to give a better picture. The essential difference between these two diagrams are:

a) It is necessary that the string diagrams be drawn correctly to scale. The ordinary flow diagram can be drawn approximately to scale with pertinent distances marked on it so that scaling off is unnecessary.

b) A flow diagram would look cumbersome if too many to and fro movements between different points are there; however, such movements will not affect a string diagram.

If lesser to and fro movements are involved in material movement *vis-a-vis* workers movement, then obviously a string diagram is a choice for representing the latter.

*Figure 37. A string diagram*

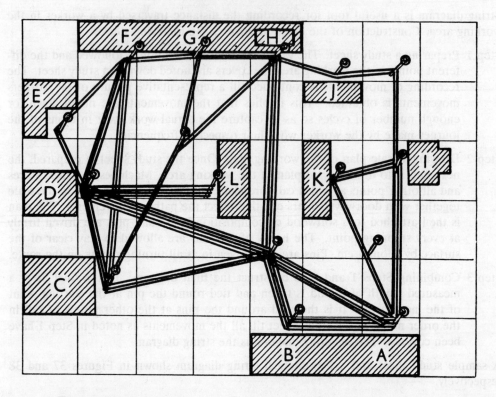

The examination of the diagram and the development of the new layout can now proceed on the same lines as with a flow diagram, with templates being used and the pins and templates being moved around until an arrangement is found by which the same operations can be performed with a minimum movement between them. This can be ascertained by leading the thread around the pins in their new positions, starting from the same point and following the same sequence. When the thread has been led around all the points covered by the study, the length left over can again be measured. The difference in length between this and the thread left over from the original study will represent the reduction in the distance travelled as a result of the improved layout. The process may have to be repeated several times until the best possible layout (i.e. the layout with which the minimum length of thread is used) is achieved. Thus, it can be seen from Figure 38 that certain paths—in particular those between A and D, A and H, and D and L—are traversed more frequently than the others. Since most of these points are at a fair distance from one another, the diagram suggests that critical examination is called for, with a view to moving the work points which they represent closer together. The string diagram is a useful aid in explaining proposed changes to management, supervisors and workers. If two diagrams are made, one showing the original layout and one the improved layout, the contrast is often so vivid—particularly if a brightly coloured thread is used—that the change will not be difficult to sell. Workers especially are interested in seeing the results of such studies and discovering how far they have to walk. The following example shows this technique as applied to the movements of labourers storing tiles after inspection.

*Figure 38. Simple movement study sheet*

# MOVEMENT STUDY SHEET

CHART No. *1*  SHEET No. *1*  OF *2*    OPERATIVE(S):

OPERATION: *Transport biscuit tiles*

*from inspection to storage bins and*    **CHARTED BY:**

*unload into bins*    **DATE:**

LOCATION: *Biscuit warehouse*    CROSS-REFERENCE: *String diagrams*
*1 and 2*

| 1<br><br>TIME<br>DEP. | 2<br><br>TIME<br>ARR. | 3<br><br>TIME<br>ELAPSED | 4<br><br>MOVE TO | 5<br><br>NOTES |
|---|---|---|---|---|
| | | | *Inspection bench (I)* | |
| | | | *to*        *Bin*    *4* | |
| | | | *I*                  *13* | |
| | | | *I*                   *5* | |
| | | | *I*                  *32* | |
| | | | *I*                  *18* | |

Example of a String Diagram: Storing Tiles After Inspection

● *RECORD*

In the operation studied in this example, biscuit tiles (i.e., tiles after first firing and before glazing) are unloaded from kiln trucks on to the bench, where they are inspected. After inspection they are placed on platforms according to size and type. The loaded platforms are taken on hand-lift trucks to the concrete bins where the tiles are stored until required for glazing. The original layout of the store is shown in Figure 39. It was decided to make a study using a string diagram to find out whether the arrangement, which appeared to be a logical one, was in fact the one involving the least transport. Studies were made of a representative number of kiln truck loads. This was because the types of tile on each truck varied somewhat, although 10 cm × 10 cm and 15 cm × 15 cm plain tiles formed by far the largest part of each load.

A study sheet of the type shown in Figure 38 above was used for recording the information. Only a portion is shown, since the nature of the record is obvious. It will be

Figure 39. String diagram storing tiles (original method)

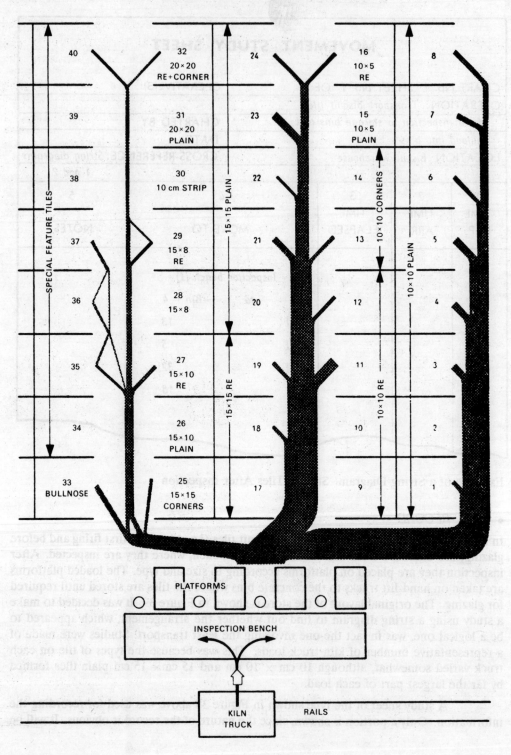

seen that, in this case, times were not recorded. It is more useful to record times when long distances are involved. The string diagram was then drawn up. The width of the shaded bands represents the number of threads between any given points and hence the relative amount of movement between them.

- ### EXAMINE CRITICALLY

A study of the diagram shows at once that the most frequent movement is up the 10 cm × 10 cm and 15 cm × 15 cm rows of bins. The bin into which any particular load of tiles is unloaded depends on which are full or empty (tiles are constantly being withdrawn for glazing). Travel in the case of the 10 cm × 10 cm and 15 cm × 15 cm tiles may therefore by anywhere up or down the rows concerned. It is equally obvious that the special feature tiles (used for decorative purposes in comparatively small numbers) are handled only rarely, and are generally placed by the inspectors on one truck and delivered to several bins at once. Deliveries of tiles other than those mentioned are fairly evenly distributed.

- ### DEVELOP THE NEW LAYOUT

The first step in developing the new layout is to locate the bins containing the most handled tiles as near as possible to the inspection bench and those containing special feature tiles as far away as possible. This certainly spoils the tidy sequence and may, for a time, make tiles a little more difficult to find; however, the bins, which have concrete partitions between them about 1 metre high, can carry cards with the contents marked on them. The cards can be seen from a distance, and the arrangement will soon be memorised by the workers. After a number of arrangements had been tried out, the one shown in Figure 40 proved to be the most economical of transport time. The distances covered were reduced from 520 to 340 metres, a saving of 35 per cent.

## 10.3   Flow process chart: Man type

In Chapter 8 we have defined the flow process chart as a process chart setting out the frequency of the flow of a product or a procedure by recording all events under review using the process chart symbols. We have further defined there the different types of flow process charts. The man type of flow process chart was defined as:

> **A flow process chart which records what the worker does.**

In contrast to the above, a material type chart records how material is handled. The symbols used for both types of charts are the same and it should be apparent that when the worker moves with only one variety of material, both the types of charts will be identical. The difference would be in terms of the use of active voice or passive voice (for a man type chart, we talk about the man handling a thing, whereas for a material type chart, we speak of material being handled). Man type of flow process charts are frequently used in the study of jobs which are not highly repetitive or stan-

*Figure 40. String diagram storing tiles (improved method)*

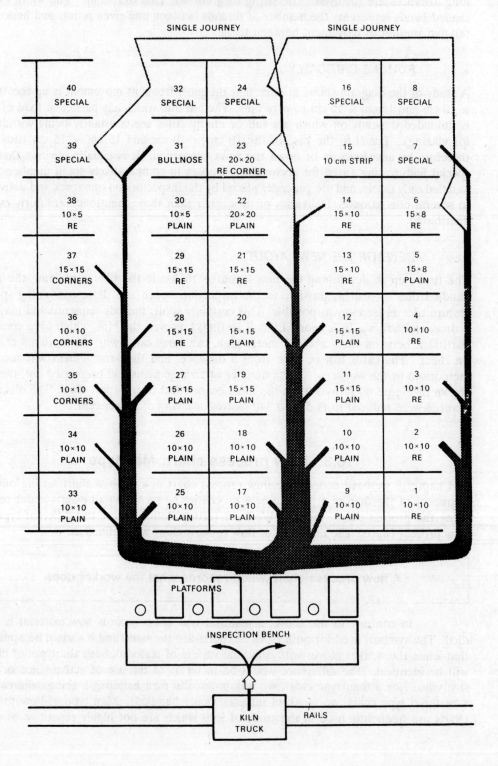

dardised. Service or maintenance work, laboratory procedure and much of the work of supervisors and executives can be recorded on charts of this type. The man concerned may or may not carry material. It is customary to attach to the man type chart a sketch showing the path of movement of the worker while he is carrying out the operation charted. An example of a man type flow process chart applied to hospital activities is given below.

Example: Serving dinners in a hospital ward.

● **RECORD**

Figure 41 shows the layout of a hospital ward containing 17 beds. When dinners were served by the original method, the nurse in charge of the ward fetched a large tray bearing the first course, together with the plates for the patients, from the kitchen. The food was usually contained in three dishes, two of which held vegetables and the third the main dish. The nurse placed the tray on the table marked Serving Table in the diagram. She set the large dishes out on the table, served one plate with meat and vegetables and carried it to bed 1. She returned to the serving table and repeated the operation for the remaining 16 beds. The paths which she followed are shown by the full lines in the diagram. When she had served all the patients with the first course, she returned to the kitchen with the tray and the empty dishes, collected the dishes and plates for the second course and returned to the ward. She then repeated the complete operation, replacing the plates emptied by the patients with plates containing their portions of the second course and returning the used plates to the serving table, where she stacked them. Finally she made a tour of the ward, collecting up the empty plates from the second course, and carried everything on the tray back to the kitchen. (To avoid confusion on the diagram, the final collection of empty plates is not shown. In both the original and the improved method the distance covered and the time taken are the same, since it is possible for her to carry several plate" at a time and move from bed to bed). The operation has been recorded in part on the flow process chart in Figure 42 but only enough has been shown to demonstrate to the reader the method of recording, which as will be seen is very similar to that used for material type flow process charts, bearing in mind that it is a person and not a product that is being followed. As an exercise the reader may wish to work out the serving cycles for himself on the basis provided by the diagram. The dimensions of the ward are given. It is, of course, possible to complete the man type flow process chart in much greater detail if desired.

● **EXAMINE CRITICALLY**

A critical examination of the flow process chart in conjunction with the diagram suggests that there is considerable room for improvement. The first 'Why?' which may come to mind is: Why does the nurse serve and carry only one plate at a time? How many could she carry? The answer is almost certainly: At least two. If she carried two plates at a time, the distance she would have to walk would be almost halved. One of the first questions asked would almost certainly be: Why is the serving table there, in the middle of the ward? Followed, after one or two other questions, by the key questions: Why should it stand still? Why can it not move round? Why not a trolley? This leads straight to the solution which was adopted.

*Figure 41. Flow diagram serving dinners in a hospital ward*

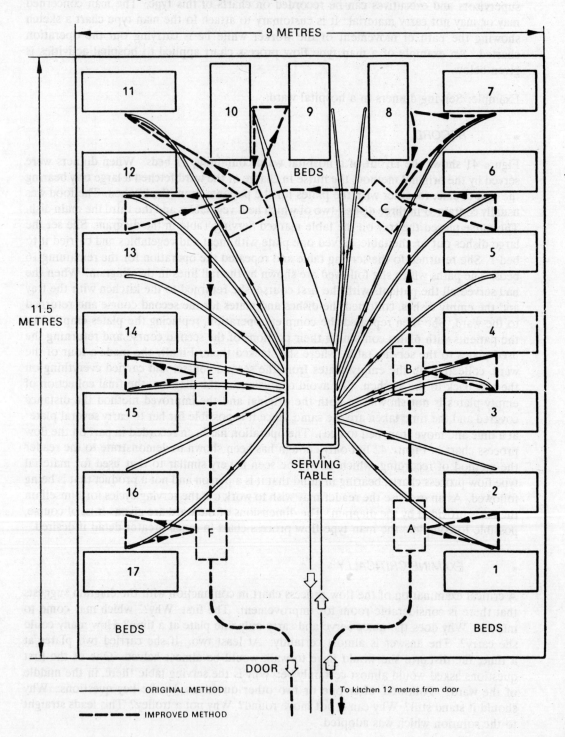

9 METRES

11.5 METRES

11

10

9

8

7

BEDS

12

6

D

C

13

5

14

B

4

E

3

15

16

A

2

17

1

BEDS

SERVING TABLE

BEDS

F

DOOR

To kitchen 12 metres from door

——————— ORIGINAL METHOD

- - - - - - - IMPROVED METHOD

*Figure 42. Flow process chart—man type dinners in a hospital ward*

| FLOW PROCESS CHART | | | | MAN/~~MATERIAL/EQUIPMENT~~ TYPE | | | |
|---|---|---|---|---|---|---|---|
| CHART No. 7   SHEET No. 1   OF 1 | | | | S U M M A R Y | | | |
| Subject charted:<br>  *Hospital nurse* | | | | ACTIVITY | PRESENT | PROPOSED | SAVING |
| | | | | OPERATION ◯ | 34 | 18 | 16 |
| | | | | TRANSPORT ⇨ | 60 | 72 | (—12) |
| ACTIVITY:<br>  *Serve dinners to 17 patients* | | | | DELAY ◗ | — | — | — |
| | | | | INSPECTION ☐ | — | — | — |
| METHOD: PRESENT/PROPOSED | | | | STORAGE ▽ | — | — | — |
| LOCATION: *Ward L* | | | | DISTANCE (m) | 436 | 197 | 239 |
| | | | | TIME (man-h) | 39 | 28 | 11 |
| OPERATIVE(S):     CLOCK No. | | | | COST: | | | |
| | | | | LABOUR | | | |
| CHARTED BY:     DATE: | | | | MATERIAL *(Trolley)* | — | $24 | |
| APPROVED BY:     DATE: | | | | TOTAL *(Capital)* | | $24 | |

| DESCRIPTION<br>ORIGINAL METHOD | QTY.<br>*(plates)* | DIST-<br>ANCE<br>(m) | TIME<br>(min) | ◯ | ⇨ | ◗ | ☐ | ▽ | REMARKS |
|---|---|---|---|---|---|---|---|---|---|
| Transports first course and plates – | | | | | | | | | *Awkward load* |
|   kitchen to serving table on tray | 17 | 16 | .50 | | | | | | |
| Places dishes and plates on table | 17 | — | .30 | | | | | | |
| Serves from three dishes to plate | — | — | .25 | | | | | | |
| Carries plate to bed 1 and return | 1 | 7.3 | .25 | | | | | | |
| Serves | — | — | .25 | | | | | | |
| Carries plate to bed 2 and return | 1 | 6 | .23 | | | | | | |
| Serves | — | — | .25 | | | | | | |
| (Continues until all 17 beds are served. See figure 42 for distances) | | | | | | | | | |
| Service completed, places dishes on tray | | | | | | | | | |
|   and returns to kitchen | — | 16 | .50 | | | | | | |
| Total distance and time, first cycle | | 192 | 10.71 | 17 | 20 | — | — | — | |
| REPEATS CYCLE FOR SECOND COURSE | | 192 | 10.71 | 17 | 20 | — | — | — | |
| Collects empty second course plates | | 52 | 2.0 | — | 20 | — | — | — | |
| TOTAL | | 436 | 23.42 | 34 | 60 | | | | |
| **IMPROVED METHOD** | | | | | | | | | |
| Transports first course and plates – | | | | | | | | | *Serving* |
|   kitchen to position A – trolley | 17 | 16 | .50 | | | | | | *trolley* |
| Serves two plates | — | — | .40 | | | | | | |
| Carries two plates to bed 1; leaves one; | | { 1.5 | | | | | | | |
|   carries one plate from bed 1 to bed 2; | 2 | { 0.6 | .25 | | | | | | |
|   returns to position A | | { 1.5 | | | | | | | |
| Pushes trolley to position B | — | 3.0 | .12 | | | | | | |
| Serves two plates | — | — | .40 | | | | | | |
| Carries two plates to bed 3; leaves one; | | { 1.5 | | | | | | | |
|   carries one plate from bed 3 to bed 4; | 2 | { 0.6 | .25 | | | | | | |
|   returns to position B | | { 1.5 | | | | | | | |
| (Continues until all 17 beds are served. See figure 32 and note variation at bed 11) | | | | | | | | | |
| Returns to kitchen with trolley | — | 16 | .50 | | | | | | |
| Total distance and time, first cycle | — | 72.5 | 7.49 | 9 | 26 | | | | |
| REPEATS CYCLE FOR SECOND COURSE | — | 72.5 | 7.49 | 9 | 26 | | | | |
| Collects empty second course plates | — | 52 | 2.00 | — | 20 | | | | |
| TOTAL | — | 197 | 16.98 | 18 | 72 | | | | |

- *DEVELOP THE NEW METHOD*

It will be seen from the broken line in the diagram (representing the revised path of movement of the nurse when provided with a trolley) and from the flow process chart that the final solution involves the nurse in serving and carrying two plates at a time (which also saves a small amount of serving time). The result, as will be seen from the process chart, is a reduction of over 54% in the total distance walked in serving and clearing away the dinners (the saving is 65% if the distance walked in removing the second-course plates, which is the same in both the old and the new methods, is excluded.) What is important here is not so much the reduction in cost, which is very small, as the fact that the nurse's fatigue, resulting from the considerable distance which she had to walk within the ward and while carrying the loaded tray to and from the kitchen, is lessened.

## 10.4   Travel chart

The string diagram is a very neat and effective way of recording for critical examination the movement of workers or materials about the shop, especially when readily understood before and after models are needed to help in presenting the merits of a proposed change. String diagrams do take rather a long time to construct, however, and when a great many movements along complex paths are involved the diagram may end up looking like a forbidding maze of criss-crossing lines. When the movement patterns are complex, the travel chart is a quicker and more manageable recording technique.

> **A travel chart is a tabular record for presenting quantitative data about the movements of workers, materials or equipment between any number of places over any given period of time.**

Figure 43 shows a typical travel chart. It records the movements of a messenger delivering paper or information to the various desks and work stations in an office. The layout of the office, showing the relative positions of the work stations, is sketched beneath the travel chart. The chart is always a square, having within it smaller squares. Each small square represents a work station—that is, in the present example, a place visited by the messenger. There are ten stations, and so the travel chart is drawn with ten small squares across, numbered 1 to 10 going down. Thus for ten work stations the travel chart contains a total of $10 \times 10 = 100$ small squares, and has a diagonal line drawn across it from top left to bottom right. The squares from left to right along the top of the chart represent the places from where movement or travel takes place: those down the left-hand edge represent the stations to which the movement is made. For example, consider a movement from station 2 to station 9. To record this, the studyman enters the travel chart at the square numbered 2 along the top of the chart, runs his pencil down vertically through all the squares underneath this one until he reaches the square which is horizontally opposite the station marked 9 on the left-hand edge. This is the terminal square, and he will make a mark in that square to indicate one journey from station 2 to station 9. All journeys are recorded in the same way, always starting at the top in the square of departure, always travelling vertically downwards, and always ending in the square opposite the station of arrival, as read from the left-hand edge. Of course, the study man does not actually trace in the path over which his pencil moves but just places a small tick or some other mark in the terminal square to record the journey.

To make the recording method completely clear, let us suppose that the messenger travelled the following route: 2 to 9 to 5 to 3 and back to 2. The journey from 2 to 9 will be marked by a tick as described above. To enter the journey from 9 to 5, the study man will return to the top of the chart, select square 9, move down the column below this until he reaches the square opposite 5 on the left-hand edge, and record the movement by a tick there. To the top again to select square 5, down from there to that opposite 3; another tick for that journey. Finally, up to the top once more to select square 3, and down to that opposite number 2 for the recording of the final leg of the messenger's walk. We give below two examples illustrating the use of the travel chart.

Example 1: Movement of a messenger in an office

● *RECORD*

The first stage of the recording process, that is when the method study man observes the movement of the messenger actually in the office, can be carried out very simply on a study sheet similar to that shown in Figure 44. Once the stations visited have been numbered and keyed to a sketch of the workplace, the entries recording the journeys made require very little writing. The travel chart is then compiled in the method study office. After all the movements have been entered on the chart with ticks, the ticks in each small square are added up, the total being entered in the square itself. The movements are then summarised, in two ways. Down the right-hand side of the chart, the number or movements into each station is entered against the square representing the station, as read from the left-hand edge. Underneath the chart, the number or movements from each station is recorded, this time under the relevant squares as read off the top of the chart.

In the chart in Figure 43 there were two movements into station 1, as can be seen by running an eye across the line of squares against station 1 on the left-hand edge. Similarly, in the next horizontal line of squares, that opposite station 2, there are altogether 10 movements shown, into station 2. For the movements from stations, the totalling is carried out vertically: it will be seen that there were 10 movements from station 2, as shown in the column of squares under station 2 at the top of the chart. With very little practice, the chart and its summaries can be compiled extremely quickly—much quicker than it takes to describe what is done. In Figure 43 the summary of movements into each station shows the same number of movements as those recorded at the bottom as being made from that station, indicating that the messenger ended his travels at the same station as he started out from when the study commenced. If he had finished somewhere else (or if the study had been broken off when he was somewhere else), there would have been one station where there was one more movement in than the number of movements out, and this would be where the study finished.

**Figure 43. Travel chart: movements of messenger in office**

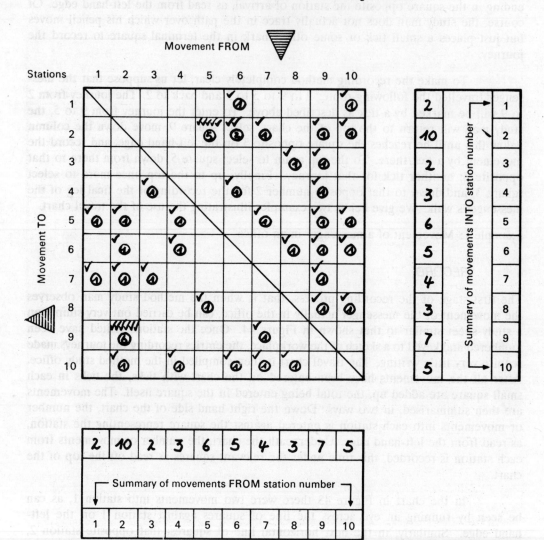

Movement FROM

| Stations | 1 | 2 | 3 | 4 | 5 | 6 | 7 | 8 | 9 | 10 |

Summary of movements INTO station number

| | 1 |
| 2 | 1 |
| 10 | 2 |
| 2 | 3 |
| 3 | 4 |
| 6 | 5 |
| 5 | 6 |
| 4 | 7 |
| 3 | 8 |
| 7 | 9 |
| 5 | 10 |

Summary of movements FROM station number

| 2 | 10 | 2 | 3 | 6 | 5 | 4 | 3 | 7 | 5 |
| 1 | 2 | 3 | 4 | 5 | 6 | 7 | 8 | 9 | 10 |

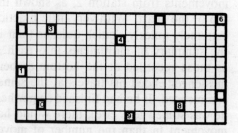

Layout sketch
of office
showing location
of stations

*Figure 44. Simple study sheet*

## STUDY SHEET

| Department: | *Mixing* | | | | | | Section | *1* | | Study No. | *147* | | |
|---|---|---|---|---|---|---|---|---|---|---|---|---|---|
| Equipment: | *Lift Truck: Pallets* | | | | | | | | | Sheet: *1* of *2* | | | |
| Operation: | *Move 25-litre cans of material to mixing machines and then to inspection (Station 6)* | | | | | | | Taken by: *CBA* | | | | | |
| | | | | | | | | | | Date: | | | |
| *From* | 2 | 9 | 7 | 4 | 3 | 9 | 6 | 1 | 9 | 6 | 3 | 2 | 9 |
| *To* | 9 | 7 | 4 | 3 | 9 | 6 | 1 | 9 | 6 | 3 | 2 | 9 | 7 |
| *No. of cans* | 10 | – | 20 | 10 | – | 30 | 10 | – | 30 | 10 | – | 30 | – |
| | | | | | | | | | | | | | |
| *From* | 7 | 1 | 6 | 4 | 9 | 8 | 2 | 5 | 9 | 7 | 2 | 5 | 9 |
| *To* | 1 | 6 | 4 | 9 | 8 | 2 | 5 | 9 | 7 | 2 | 5 | 9 | 6 |
| *No. of cans* | 10 | 20 | – | 30 | 40 | 20 | 30 | 40 | 10 | 20 | 10 | 30 | 40 |
| | | | | | | | | | | | | | |
| *From* | 6 | 1 | 9 | | | | | | | | | | |
| *To* | 1 | 9 | 6 | | | | | | | | | | |
| *No. of cans* | – | 30 | 30 | | | | | | | | | | |

## EXAMINE CRITICALLY

An examination of the chart shows that ten journey have been made into station 2, seven into station 9, and six into station 5. These are the busiest stations. A scrutiny of the body of the chart helps to confirm this: there were six journeys from station 2 to station 9, and five from station 5 to station 2. The busiest route is 5-2-9. This suggests that it would be better to locate these stations next to each other. It might then be possible for the clerk at station 5 to place finished work directly into the in-tray at station 2, and the clerk there to pass his work on to station 9, thus relieving the messenger of a good deal of his travelling.

## Example 2: Material Handling

An example of a travel chart compiled as part of a material-handling study is shown in Figure 45. In the shop in which the study was made, eight mixing machines were used to mix materials in different proportions, the final mixtures being taken to an inspection bench (station 6). The mixes were transported in 25-litre cans, which were placed on pallets and moved by a low lift truck.

*Figure 45. Travel chart material*

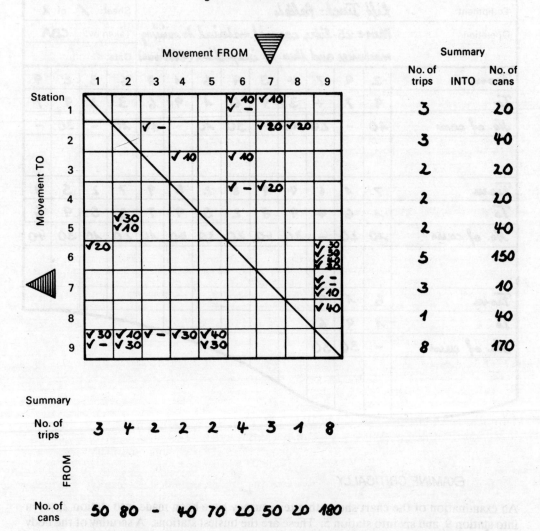

• **RECORD**

Movements were recorded on the shop floor on a study sheet of the type shown in Figure 44. The entries show not only the journeys made but also the number of cans carried on each trip. In the travel chart show in Figure 45, there are nine stations, the eight mixing machines and the inspection bench. The travel chart was made exactly as in the previous example, except that in this instance the number of cans delivered was also entered in the destination squares, beside the ticks for the journeys, and both journeys and cans delivered have been summarised. It will be seen that, for instance, two journeys were made from station 5 to station 9, one with a load of 40 cans and the other with 30.

● *EXAMINE CRITICALLY*

Not much can be learned from the study sheet, except that seven of the 29 trips made were run without any load, and that the size of load varied from 10 to 40 cans. The travel chart, however, shows at once that stations 6 and 9 are busy ones. Five trips were made to station 6, with a total of 150 cans being delivered. (Station 6 was the inspection bench.) Four of these trips were from station 9, bringing in a total of 130 cans. The largest number of trips, and the greatest quantity of cans, was from station 9 to the inspection bench, suggesting that this route might be laid out so that it would be as short as possible. It might be possible to install a roller conveyor between these points, thus relieving the lift truck of a great deal of work.

Eight trips were made into station 9, to deliver 170 cans. The cans came from stations 1, 2, 4 and 5, one trip without load being made from station 3. Stations 1, 2, 4 and 5 appear to feed station 9, which sends its work on to the inspection bench (longer study might be necessary to confirm this). If so, there would be a case for modifying the layout of the shop in order to bring these stations closer together, when roller conveyors might allow gravity to do most of the transporting between them. In this example there is no sketch of the shop layout or table of distances between stations, both of which are essential complements to a travel chart. It is interesting to note that four trips were made from station 2, but only three into the station; and that only four were made from station 6, although five were made into it. This is because the study started at station 2 and finished at the inspection bench.

## 10.5  Multiple activity chart

Multiple activity chart is a useful recording tool for situations where the work involves interactions of different subjects. One or more workers looking after different machines or a group of workers working on loading materials at one point and dumping the same at a different point are some examples where this type of chart can be used effectively. The fundamental difference between this tool and the other charts described in the previous section are as follows:

a) In multiple activity chart a time scale is used. No such time scales are used in the charts described earlier in this chapter.

b) Multiple activity charts can be used equally effectively even if there is no movement of workers involved in the work under consideration. The primary focus of this chart, for situations where the workers are moving as a part of their work, is to identify the idle time on the part of either the workers or the machines. The focus of other charts described so far were primarily to identify excess distances traversed by the worker, which is only indirectly related to the time.

Before we give some examples of a multiple activity chart, we first present a definition of the same:

> **A multiple activity chart is a chart on which the activities of more than one subject (worker, machine or item of equipment) are each recorded on a common time scale to show their interrelationship.**

By using separate vertical columns, or bars, to represent the activities of different operatives or machines against a common time scale, the chart shows very clearly periods of idleness on the part of any of the subjects during the process. A study of the chart often makes it possible to rearrange these activities so that such ineffective time is reduced. The multiple activity chart is extremely useful in work involving repetitive operations. For a situation involving a worker handling different machines, this chart can be used to find the number of machines the worker can look after so as to minimise the cost. This type of situations will be dealt with in Chapter 19. Here, we present two examples to illustrate how idle time of workers can be detected with the help of this chart. In the first example given here, no movement of the workers is involved, the objective being to illustrate the charting procedure. The second example involves work where the workers are moving in the working area as a part of the work.

**Example 1: Inspection of catalyst in a converter.**

- ### RECORD

This is an application in the field of plant maintenance and is useful in showing that method study is not confined to repetition or production operations. During the running-in period of a new catalytic converter in an organic chemical plant, it was necessary to make frequent checks on the condition of the catalyst. In order that the converter should not be out of service for any longer than was strictly necessary during these inspections, the job was studied. In the original method the removal of the top of the vessel was not started until the heaters had been removed, and the replacement of the heaters was not started until the top had been completely fixed. The original operation, with the relationships between the working times of the various workers, is shown in Figure 46.

- ### EXAMINE CRITICALLY

It will be seen from this chart that, before the top of the vessel was removed by the fitter and his mate, the heaters had to be removed by the electrician and his mate. This meant that the fitters had to wait until the electricians had completed their work. Similarly, at the end of the operation the heaters were not replaced until the top had been replaced, and questioning of the existing procedure revealed that in fact it was not necessary to wait for the heaters to be removed before removing the top.

Figure 46. Multiple activity chart: inspection of catalyst in a converter (original method)

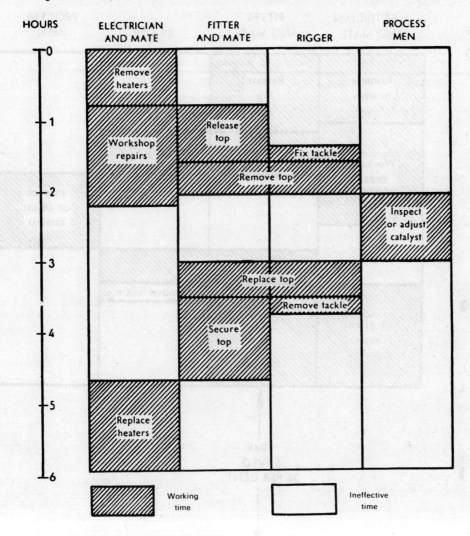

### DEVELOP THE NEW METHOD

Once this had been determined, it was possible to arrange for the top to be unfastened while the heaters were being removed and for the heaters to be replaced while the top was being secured in place. The result appears on the chart in Figure 47. It will be seen that the idle time of the electrician and fitter and their respective mates has been substantially reduced, although that of the rigger remains the same. Obviously the rigger and the process men will be otherwise occupied before and after performing their sections of the job and are not, in fact, idle while heaters and cover are being removed or replaced. The saving effected by this simple change was 32% of the total time of the operation. The simple form of multiple activity chart shown here can be constructed on any piece of paper having lines or squares which can be used to form a time scale. It is more usual, however, to use printed or duplicated forms, similar in general layout to the standard

*Figure 47. Multiple activity chart: inspection of catalyst in a converter (improved method)*

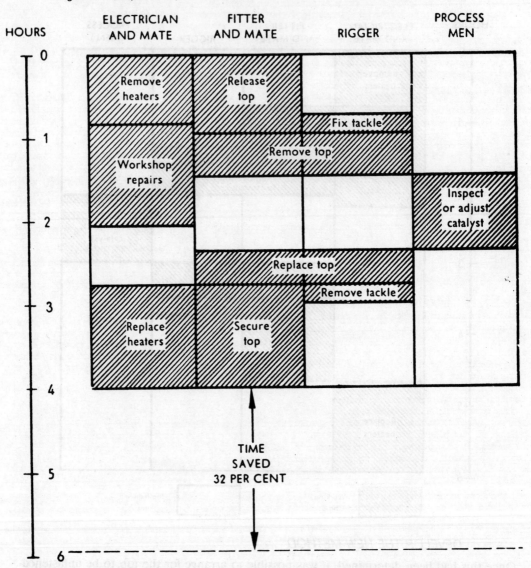

flow process charts, and to draw vertical bars to represent the activities charted.

Example 2: Material handling

The situation presented here is common occurrence in mining industries where workers are involved in extraction of ores. Consider two points A and B in the work area. The extracted ores are first loaded on to trucks at point A. The trucks then travel up to the point B where the material is unloaded after which they come back to point A for further loading. Currently one worker is employed in loading trucks and the company has employed two drivers for driving the two trucks that are operating for this job.

● **RECORDING**

The activities involved together with their time are recorded as follows :

| Activity | Time in minutes |
|---|---|
| Load a truck at point A | 10 |
| Go to point B with material | 15 |
| Unload the material at B | 5 |
| Return to point A | 10 |

● **CRITICAL EXAMINATION**

A close look at the timing reveals that in a cycle, the loader is busy for (10+10)=20 minutes, while each of the trucks will be busy for (10+15+5+10)= 40 minutes. This would imply that in a cycle of 40 minutes, 2 truckloads of material is removed and the loader is idle for 50% of the time. Unloading does not require any extra person, provision for automatic tilting of the truck base (where the material is loaded) being there. Taking a truck and its driver as a single unit and the loader as the other subject, the multiple activity chart can be drawn as shown below in Figure 48.

*Figure 48. Multiple activity chart:*

| MINUTES | LOADER | TRUCK 1 | TRUCK 2 |
|---|---|---|---|
|  | LOAD TRUCK 1 | BEING LOADED | RETURN TO A |
| 10 | LOAD TRUCK 2 | GO TO B | BEING LOADED |
| 20 | IDLE |  |  |
|  |  | BEING UNLOADED | GO TO BE |
| 30 |  | RETURN TO A |  |
|  |  |  | BEING UNLOADED |
| 40 |  |  |  |

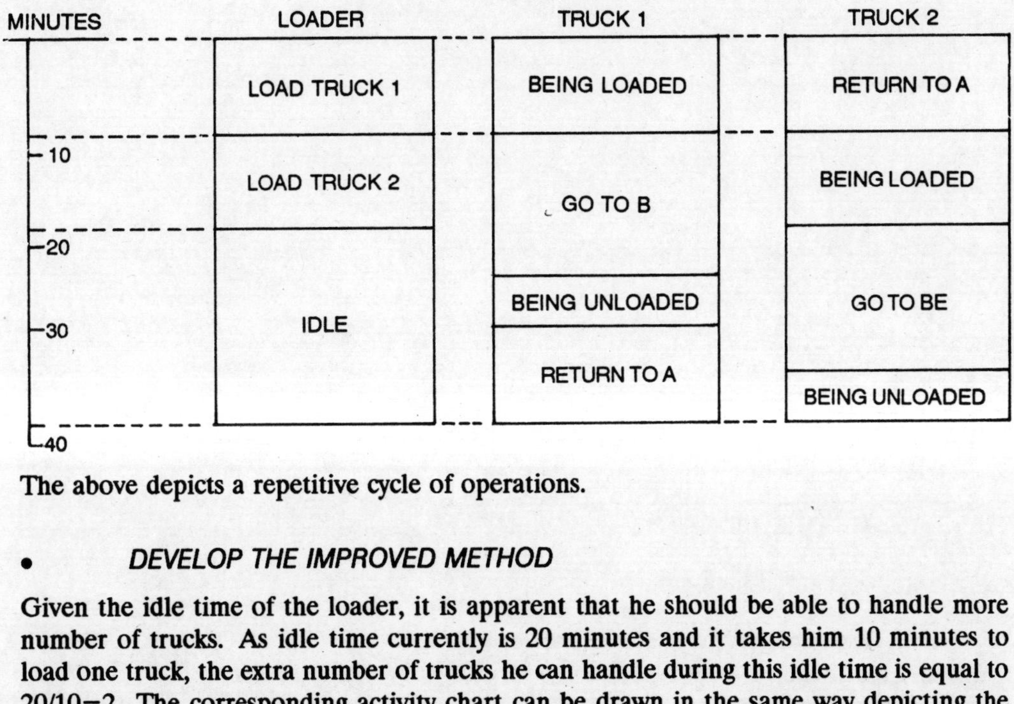

The above depicts a repetitive cycle of operations.

● **DEVELOP THE IMPROVED METHOD**

Given the idle time of the loader, it is apparent that he should be able to handle more number of trucks. As idle time currently is 20 minutes and it takes him 10 minutes to load one truck, the extra number of trucks he can handle during this idle time is equal to 20/10=2. The corresponding activity chart can be drawn in the same way depicting the repetitive cycle of operations. The loader loading 4 trucks now, both the trucks and the loader will be fully occupied. In this way, every 40 minutes 4 truckloads of materials will be removed.

## 10.6  Conclusion

In this chapter we have examined four different tools for recording facts related to the workers movements in the shop area. These are valuable aids in identifying inefficiencies in movements, layout, and in bringing out the idle time of workers or machines. In considering the movements, our objectives have been to attain effective operation of processes and better utilisation of wasteful movements and unnecessary idle times. However, it should be remembered that the fatigue factor plays an important role in any manual work. In example 2, given in the last section, we find that the loader is busy for all the time. Unless relaxation is provided to him, his output is bound to fall. Alternatively, the time to load should account for such relaxation or some such allowances. This will be taken up in a subsequent chapter. Having discussed the broad method of operation in this and the last chapter, we now examine the more detailed method of operation where a worker is working above on a bench or table, in the next chapter.

# Chapter 11

# Methods and movements at the workplace

## 11.1  Introduction

In this book we have gradually moved from the wide field of the productivity of industry as a whole to considering in a general way how the productivity of men and machines can be improved through the use of Work Study. Still moving from the broader to the more detailed approach, we have also examined procedures of a general nature for improving the effectiveness with which complete sequences of operations are performed and with which material flows through the working area. Turning from material to men, we have discussed methods of studying the movements of men around the working area and the relationships between men and machines or of men working together in groups. We have done so foilowing the principle that the broad method of operation must be put right before we attempt improvements in detail.

The time has now come to look at one man working at a workplace, bench or table and to apply to him the principles which have been laid down and the procedures shown in the examples given.

In considering the movements of men and materials on the larger scale, we have been concerned with the better utilisation of existing plant and machinery (and, where possible, materials) through the elimination of unnecessary idle time, the more effective operation of processes and the better utilisation of the services of labour through the elimination of unnecessary and time-consuming movement within the working area of factory, department or yard.

When we come to study the operative at the workplace, the way in which he applies his effort and the amount of fatigue resulting from his manner of working become primary factors affecting his productivity.

Before embarking on a detailed study of an operative doing a job at a single workplace, it is important to make certain that the job is in fact necessary and is being done as it should be done. The questioning technique must be applied as regards.

● *PURPOSE*

to ensure that the job is necessary;

- *PLACE*

to ensure that it is being done where it should be done;

- *SEQUENCE*

to ensure that it is in its right place in the sequence of operations;

- *PERSON*

to ensure that it is being done by the right person.

Once these have been verified and it is certain that the job cannot be eliminated or combined with another operation, it is possible to go on to determine the

- *MEANS*

by which the job is being done
and to simplify them as much as is economically justified.

Later in this chapter we shall consider the recording techniques adopted to set out the detailed movements of an operative at his workplace in ways which facilitate critical examination and the development of improved methods, in particular the two-handed process chart.

## 11.2   The principles of motion economy[1]

There are a number of 'principles' concerning the economy of movements which have been developed as a result of experience and which form a good basis for the development of improved methods at the workplace. They were first used by Frank Gilbreth, the founder of motion study, and have been amplified by other workers, notably Professor Barnes.[2] They may be grouped under three headings—

A. Use of the human body
B. Arrangement of the workplace
C. Design of tools and equipment

They are useful in shop and office alike and, although they cannot always be applied, they do form a very good basis for improving the efficiency and reducing the fatigue of manual work. The ideas expounded by Professor Barnes are described here in a somewhat simplified fashion.

### A. Use of the human body

When possible—

1) The two hands should begin and complete their movements at the same time.

2) The two hands should not be idle at the same time except during periods of rest.

3) Motions of the arms should be symmetrical and in opposite directions and should be made simultaneously.

4) Hand and body motions should be made at the lowest classification at which it is possible to do the work satisfactorily (see section 3).

5) Momentum should be employed to help the worker, but should be reduced to a minimum whenever it has to be overcome by muscular effort.

6) Continuous curved movements are to be preferred to straight-line motions involving sudden and sharp changes in direction.

7) 'Ballistic' (i.e. free-swinging) movements are faster, easier and more accurate than restricted or controlled movements.

8) Rhythm is essential to the smooth and automatic performance of a repetitive operation. The work should be arranged to permit easy and natural rhythm whenever possible.

9) Work should be arranged so that eye movements are confined to a comfortable area, without the need for frequent changes of focus.

## B. Arrangement of the workplace

1) Definite and fixed stations should be provided for all tools and materials to permit habit formation.

2) Tools and materials should be pre-positioned to reduce searching.

3) Gravity feed, bins and containers should be used to deliver the materials as close to the point of use as possible.

4) Tools, materials and controls should be located within the maximum working area (see Figure 49) and as near to the worker as possible.

5) Materials and tools should be arranged to permit the best sequence of motions.

6) "Drop deliveries" or ejectors should be used wherever possible, so that the operative does not have to use his hands to dispose of the finished work.

7) Provision should be made for adequate lighting, and a chair of the type and height to permit good posture should be provided. The height of the workplace and seat should be arranged to allow alternate standing and sitting.

8) The colour of the workplace should contrast with that of the work and thus reduce eye fatigue.

## C. Design of tools and equipment

1) The hands should be relieved of all work of 'holding' the workpiece where this can be done by a jig, fixture or foot-operated device.

2) Two or more tools should be combined wherever possible.

3) Where each finger performs some specific movement, as in typewriting, the load should be distributed in accordance with the inherent capacities of the fingers.

4) Handles such as those on cranks and large screwdrivers should be so designed that as much of the surface of the hand as possible can come into contact with the handle. This is especially necessary when considerable force has to be used on the handle.

5) Levers, crossbars and handwheels should be so placed that the operative can use them with the least change in body position and the greatest "mechanical advantage".

These 'principles', which reflect those discussed in Chapter 6, can be made the basis of a summary 'questionnaire' which will help, when laying out a workplace, to ensure that nothing is overlooked.

*Figure 49. Normal and maximum working areas*

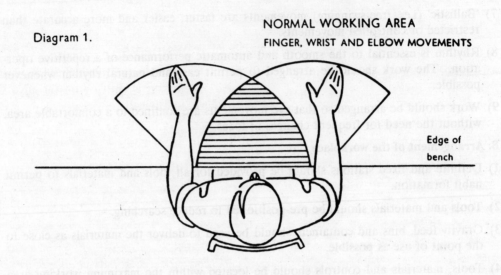

Diagram 1.

NORMAL WORKING AREA
FINGER, WRIST AND ELBOW MOVEMENTS

Edge of
bench

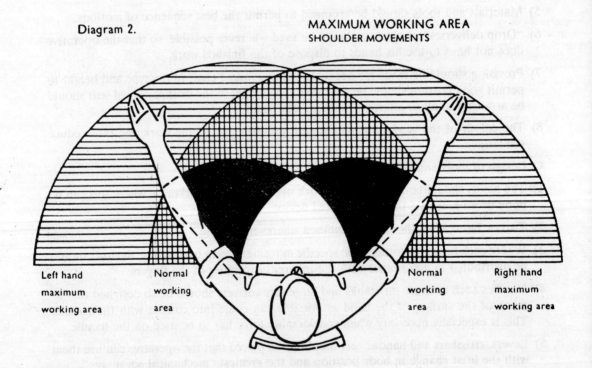

Diagram 2.

MAXIMUM WORKING AREA
SHOULDER MOVEMENTS

Left hand
maximum
working area

Normal
working
area

Normal
working
area

Right hand
maximum
working area

Figure 49 shows the normal working area and the storage area on the work-bench for the average operative. As far as possible, materials should not be stored in the area directly in front of him, as stretching forwards involve the use of the back muscles, thereby causing fatigue. This has been demonstrated by physiological research.

## 11.3 Classification of movements

The fourth 'rule' of motion economy in the use of the human body calls for movements to be of the lowest classification possible. This classification is built up on the pivots around which the body members must move, as shown in Table 10.

*Table 10. Classification of movements.*

| Class | Pivot | Body member(s) moved |
| --- | --- | --- |
| 1 | Knuckle | Finger |
| 2 | Wrist | Hand and fingers |
| 3 | Elbow | Forearm, hand and fingers |
| 4 | Shoulder | Upper arm, forearm, hand and fingers |
| 5 | Trunk | Torso, Upper arm, forearm, hand and fingers |

It is obvious that each movement above Class 1 will involve movements of all classes below it. Thus the saving in effort resulting from using the lowest class possible is obvious. If, in laying out the workplace, everything needed is placed within easy reach, this will minimise the class of movement which the work itself requires from the operative.

## 11.4 Further notes on workplace layout

A few general notes on laying out the workplace may be useful.

1) If similar work is being done by each hand, there should be a separate supply of materials or parts for each hand.

2) If the eyes are used to select material, as far as possible the material should be kept in an area where the eyes can locate it without there being any need to turn the head.

3) The nature and the shape of the material influence its position in the layout.

4) Hand tools should be picked up with the least possible disturbance to the rhythm and symmetry of movements. As far as possible the operator should be able to pick up or put down a tool as the hand moves from one part of the work to the next, without making a special journey. Natural movements are curved, not straight; tools should be placed on the arc of movements, but clear of the path of movement of any material which has to be slid along the surface of the bench.

5) Tools should be easy to pick up and replace: as far as possible they should have an automatic return, or the location of the next piece of material to be moved should allow the tool to be returned as the hand travels to pick it up.

6) Finished work should be—

    a) dropped down a hole or a chute;

    b) dropped through a chute when the hand is starting the first motion of the next cycle;

    c) put in a container placed so that hand movements are kept to a minimum; and,

    d) if the operation is an intermediate one, placed in a container in such a way that the next operative can pick it up easily.

7) Always look into the possibility of using pedals or knee-operated levers for locking or indexing devices on fixtures or devices for disposing of finished work.

## AN EXAMPLE OF A WORKPLACE LAYOUT

Let us now look at a typical workplace with the principles of motion economy and the notes in the previous section in mind.

Figure 50 shows a typical example of the layout of a workplace for the assembly of small electrical equipment (in this case electric meters). Certain points will be noticed at once:

1) A fixture has been provided for holding the workpiece (here the chassis of the meter), leaving both the operative's hands free for assembly work. The use of one hand purely for holding the part being worked on should always be avoided, except for operations so short that a fixture would not be justified.

2) The power screwdriver and box spanner are suspended in front of the operative so that she has to make only a very short and easy movement to grasp them and bring them to the work. They are, however, clear of the surface of the table and of the work. The hammer and hand screwdriver for use with the left hand are within easy reach, so that the operative can pick them up without searching, although picking up the screwdriver might involve a little fumbling. They are in line with the trays of parts but below them, and so do not get in the way.

3) All the small parts are close to the operative, well within the "maximum working area". Each part has a definite location, and the trays are designed with 'scoop' fronts for easy withdrawal, parts being drawn forward with the tips of the fingers and grasped as they come over the rounded edge. They are arranged for symmetrical movements of the arms, so that parts which are assembled simultaneously are picked up from trays in the same relative position to the operative, on either side of her. It will be noted that the trays come almost in front of the operative, but this is not very important in this case as the length of reach is not excessive and will not involve much play of the shoulder and back muscles.

4) The operative has taken a small number of the formed wire parts normally kept in a tray to her left front and placed them conveniently in front of and to the side of the workpiece, in order to make a shorter reach.

5) The backrest of the operative's chair is an interesting and ingenious improvisation. Special chairs with this type of backrest were not produced locally.

*Figure 50. Assembling an electric meter*

## 11.5   Notes on the design of jigs, tools and fixtures

A jig holds parts in an exact position and guides the tool that
works on them.

> **A fixture is a less accurate device for holding parts which would otherwise have to be held in one hand while the other worked on them.**

The designer's object in providing jigs and fixtures is primarily accuracy in machining or assembly. Often, opening and closing them or positioning the workpiece calls for more movements on the part of the operative than are strictly necessary. For example, a spanner may have to be used to tighten a nut when a wing nut would be more suitable; or the top of the jig may have to be lifted off when the part might be slid in.

Cooperation between the work study man and the jig and tool designers, in industries where they are employed (principally the engineering industry), should start in the early stages of designing, and tool designers should be among the first people to take appreciation courses in method study. Some points worth noting are—

1) Clamps should be as simple to operate as possible and should not have to be screwed unless it is essential for accuracy of positioning. If two clamps are required, they should be designed for use by the right and left hands at the same time.

2) The design of the jig should be such that both hands can load parts into it with a minimum of obstruction. There should be no obstruction between the point of entry and the point from which the material is obtained.

3) The action of unclamping a jig should at the same time eject the part, so that additional movements are not required to take the part out of the jig.

4) Where possible on small assembly work, fixtures for a part which does not require both hands to work on it at once should be made to take two parts, with sufficient space between them to allow both hands to work easily.

5) In some cases jigs are made to take several small parts. It will save loading time if several parts can be clamped in position as quickly as one.

6) The work study man should not ignore machine jigs and fixtures such as milling jigs. A great deal of time and power is often wasted on milling machines owing to the fact that parts are milled one at a time, when it may be quite feasible to mill two or more at once.

7) If spring-loaded disappearing pins are used to position components, attention should be paid to their strength of construction. Unless the design is robust, such devices tend to function well for a while but then have to be repaired or redesigned.

8) In introducing a component into a jig it is important to ensure that the operative should be able to see what he is doing at all stages; this should be checked before any design is accepted.


## 11.6   Machine controls and displays of dials

Until recently, machinery and plant of all kinds were designed with very little thought being given to the convenience of the operative. In short cycle work especially, the manipulation of the controls (changing speeds on a capstan lathe, for example) often involves awkward movements. There is not much that the user can do about the controls of a machine after he has bought it; but he can draw the attention of the makers to inconvenient controls so that they can make improvements in later models. There is some evidence that machinery makers generally are beginning to be more conscious of this problem, but a great deal remains to be done. In the few companies that make their own machinery or plant, the work study department should be called in at the earliest possible stage of the design process, to give assistance and advice.

Physiologists and psychologists have given some thought to the arrangement of dials with a view to minimising the fatigue to people who have to watch them. The arrangement of the control panels for chemical processes and similar types of process is often made at the works installing them, and the work study man should be consulted when this is done.

There is a good deal of published literature on the subject, and this can be consulted in order to arrive at an easily readable 'display' or arrangement of dials or visual indicators.

The growing awareness of the importance of arranging machine controls and workplaces so that they are convenient for the people who have to do the work has led in recent years to the development of a new field of scientific study which is concerned entirely with such matters. This is ergonomics:[3] the study of the relationship between a worker and the environment in which he works, particularly the application of anatomical, physiological and psychological knowledge to the problems arising therefrom. Ergonomists have carried out many experiments to decide on matters such as the best layout for machine controls, the best dimensions for seats and worktops, the most convenient pedal pressures, and so on. It may be expected that their findings will gradually be incorporated in the designs of new machines and equipment over the next few years, and will eventually form the basis of standard practice.

## 11.7   The two-handed process chart

The study of the work of an operative at the bench starts, as does method study over the wider field, with a process chart. In this case the chart used is the fifth of the charts indicating process sequence (Table 9), the one known as the two-handed process chart.

> **The two-handed process chart is a process chart in which the activities of a worker's hands (or limbs) are recorded in their relationship to one another.**

The two-handed process chart is a specialised form of process chart because it shows the two hands (and sometimes the feet) of the operative moving or static in relation to one another, usually in relation to a time scale. One advantage of incorporating a time scale in the chart form is that the symbols for what the two hands are doing at any given moment are brought opposite each other.

The two-handed process chart is generally used for repetitive operations, when one complete cycle of the work is to be recorded. Recording is carried out in more detail than is normal on flow process charts. What may be shown as a single operation of a flow process chart may be broken down into a number of elemental activities which together make up the operation. The two-handed process chart generally employs the same symbols as the other process charts; however, because of the greater detail covered, the symbols are accorded slightly different meanings—

| Symbol | Name | Description |
|---|---|---|
| ○ | OPERATION | is used for the activities of grasp, position, use, release, etc. of a tool, component or material. |
| ⇨ | TRANSPORT | is used to represent the movement of the hand (or limb) to or from the work, or a tool, or material. |
| D | DELAY | is used to denote time during which the hand or limb being charted is idle (although the others may be in use). |
| ▽ | HOLD "Storage" | The term storage is not used in connection with the two-handed process chart. Instead, the symbol is redesignated as hold and is used to represent the activity of holding the work, a tool or material—that is, when the hand being charted is holding something. |

The symbol for inspection is not much used because the hand movements when an operative is inspecting an article (holding it and examining it visually or gauging it) may be classified as 'operations' on the two-handed chart. It may, however, sometimes be useful to employ the 'inspection' symbol to draw attention to the examination of a piece[4].

The very act of making the chart enables the work study man to gain an intimate knowledge of the details of the job, and the chart itself enables him to study each element of the job by itself and in relation to other elements. From this study ideas for improvements are developed. These ideas should be written down in chart form when they occur, just as in all other process charting. It may be that different ways of simplifying the work can be found; if they are all charted, they can be compared easily. The best method is generally that which requires the fewest movements.

The two-handed process chart can be applied to a great variety of assembly, machining and clerical jobs. In assembly operations, tight fits and awkward positioning present certain problems. In the assembly of small parts with close fits, "positioning before assembly" may be the longest element in the cycle. In such cases 'positioning' should be shown as a separate movement ('Operation') apart from the actual movement of assembly (e.g. fitting a screwdriver in the head of a small screw). Attention can thus be focussed on it and, if it is shown against a time scale, its relative importance can be assessed. Major savings can be made if the number of such positionings can be reduced, as for example by slightly countersinking the mouth of a hole and putting a chamfer on the end of the shaft fitting in it, or by using a screwdriver with a self-centring bit.

## NOTES ON COMPILING TWO-HANDED PROCESS CHARTS

The chart form should include—

Spaces at the top for the usual information.

Adequate space for a sketch of the layout of the workplace (corresponding to the flow diagram used in association with the flow process chart), or sketch of jigs, etc.

Spaces for the movements of right and left hands.

Space for a summary of movements and analysis of idle time.

Examples are given in the following pages.

Some points on compiling charts are worth mentioning—

1) Study the operation cycle a few times before starting to record.

2) Chart one hand at a time.

3) Do not record more than a few symbols at a time.

4) The action of picking up or grasping a fresh part at the beginning of a cycle of work is a good point at which to start the record. Start with the hand that handles the part first or the hand that does the most work. The exact point of starting is not really important, as the complete cycle will eventually come round to it again, but the point chosen must be definite. Add in the second column the kinds of work done by other hand.

5) Only record actions on the same level when they occur at the same moment.

6) Actions which occur in sequence must be recorded on the chart at different horizontal levels. Check the chart for the time relation of the hands.

7) Care must be taken to list everything the operative does and to avoid combining operations and transports or positionings, unless they actually occur at the same time.


## EXAMPLE OF A TWO-HANDED PROCESS CHART: CUTTING GLASS TUBES

This very simple example describes how a two-handed process chart was constructed for cutting off short lengths of glass tube with the aid of a jig. This is illustrated on the form; the operations involved are self-explanatory (Figure 51).

- *RECORD*

In the original method the tube was pressed to the stop at the end of the jig marked with the file and then eased back for notching. It was then taken out of the jig for breaking. The chart goes into great detail in recording the movements of the hands, because in short cycle work of this kind fractions of seconds, when added together, may represent a large proportion of the total time needed for the job.

- *EXAMINE critically*

An examination of the details of the original method, using the questioning technique, at once raises certain points. (It is not considered necessary to go through the questions in sequence at this stage in the book: it is assumed that the reader will always do so.)

1) Why is it necessary to hold the tube in the jig?

Figure 51. Two-handed process chart cutting glass tubes (original method)

## TWO-HANDED PROCESS CHART

| CHART No.1   SHEET No.1   OF 1, | WORKPLACE LAYOUT |
|---|---|

DRAWING AND PART: *Glass tube 3 mm dia.,*
*1 metre original length*
OPERATION: *Cut to lengths of 1.5 cm*

LOCATION: *General shop*
OPERATIVE:
CHARTED BY:                    DATE:

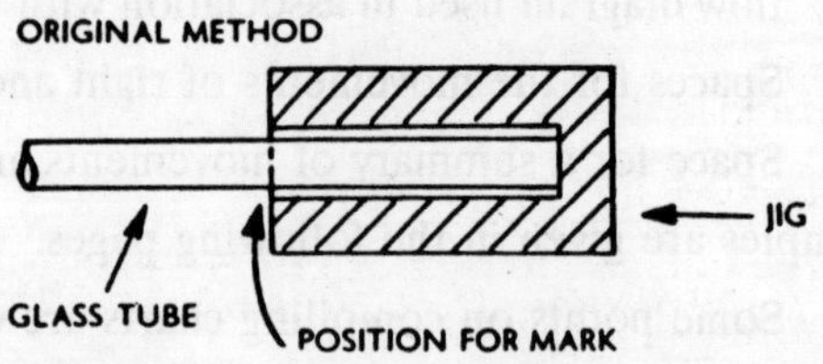

ORIGINAL METHOD

GLASS TUBE
POSITION FOR MARK
JIG

| LEFT-HAND DESCRIPTION | O ⇨ D ▽ | O ⇨ D ▽ | RIGHT-HAND DESCRIPTION |
|---|---|---|---|
| *Holds tube* | | | *Picks up file* |
| *To jig* | | | *Holds file* |
| *Inserts tube to jig* | | | *File to tube* |
| *Presses to end* | | | *Holds file* |
| *Holds tube* | | | *Notches tube with file* |
| *Withdraws tube slightly* | | | *Holds file* |
| *Rotates tube 120°/180°* | | | *Holds file* |
| *Pushes to end jig* | | | *Moves file to tube* |
| *Holds tube* | | | *Notches tube* |
| *Withdraws tube* | | | *Places file on table* |
| *Moves tube to R.H.* | | | *Moves to tube* |
| *Bends tube to break* | | | *Bends tube* |
| *Holds tube* | | | *Releases cut piece* |
| *Changes grasp on tube* | | | *To file* |

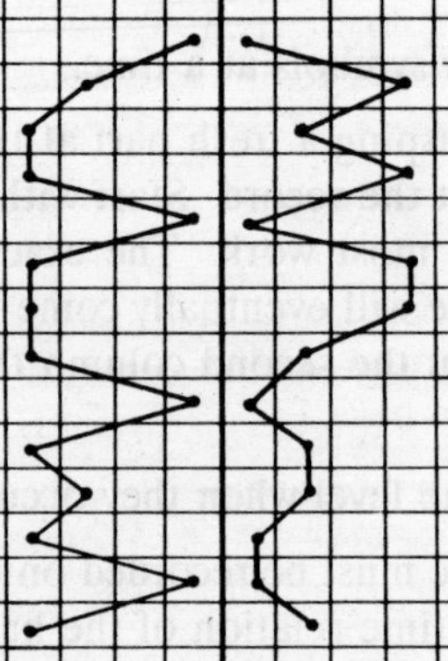

### SUMMARY

| METHOD | PRESENT | | PROPOSED | |
|---|---|---|---|---|
| | L.H. | R.H. | L.H. | R.H. |
| *Operations* | 8 | 5 | | |
| *Transports* | 2 | 5 | | |
| *Delays* | – | – | | |
| *Holds* | 4 | 4 | | |
| *Inspections* | – | – | | |
| *Totals* | 14 | 14 | | |

2) Why cannot the tube be notched while it is being rotated instead of the right hand having to wait?

3) Why does the tube have to be taken out of the jig to break it?

4) Why pick up and put down the file at the end of each cycle? Can it not be held?

A study of the sketch will make the answers to the first three questions plain.

1) The tube will always have to be held because the length supported by the jig is short compared with the total length of the tube.

2) There is no reason why the tube cannot be rotated and notched at the same time.

3) The tube has to be taken out of the jig to be broken because, if the tube were broken by bending against the face of the jig, the short end would then have to be picked out—an awkward operation if very little were sticking out. If a jig were so designed that the short end would fall out when broken, it would not then be necessary to withdraw the tube.

The answer to the fourth question is also obvious.

4) Both hands are needed to break the tube using the old method. This might not be necessary if a new jig could be devised.

● *DEVELOP the new method*

Once these questions have been asked and answered, it is fairly easy to find a satisfactory solution to the problem. Figure 52 shows one possible solution. It will be seen that, in redesigning the jig, the study man has arranged it in such a way that the notch is cut on the right-hand side of the supporting pieces, so that the short end will break away when given a sharp tap and it will no longer be necessary to withdraw the tube and use both hands to break off the end. The number of operations and movements has been reduced from 28 to six, as a result of which an increase in productivity of 133% was expected. In fact this was exceeded, because the job is now more satisfactory following the elimination of irritating work such as "position tube in jig". The new method can be carried out without looking closely at the work, so that workers can be trained more easily and become less fatigued.

## 11.8   Reorganisation of a workplace by means of a two-handed process chart

### ASSEMBLY OF POWER MOTOR STARTING WINDING TO CORE[5]

Figure 53 shows the workplace before reorganisation. Some thought has evidently been given to the operation, since a fixture has been provided for holding the assembly. Apart from this, the organisation of the workplace appears to have been left to the worker. The various tools and the ring gauge are placed quite conveniently at her right hand, although a study of the 'Before' process chart shows that she always has to pick up the tamping tools with her right hand and pass them to her left. This occurs seven times in the course of one assembly. The handles of the tools are awkward to grasp since they lie flat on the bench. Lengths of systoflex tubing are upright in a tin in front of the fixture (a long reach for the worker). The prepared coils (not visible in this figure but seen in

Figure 52. Two-handed process chart cutting glass tubes (improved method)

| TWO-HANDED PROCESS CHART | | | | | | | | |
|---|---|---|---|---|---|---|---|---|
| CHART No. 2   SHEET No. 1   OF 1 | | | | WORKPLACE LAYOUT | | | | |
| DRAWING AND PART: *Glass tube 3 mm dia.,* *1 metre original length* | | | | | | | | |
| OPERATION: *Cut to lengths of 1.5 cm* | | | | | | | | |
| LOCATION: *General shop* | | | | | | | | |
| OPERATIVE: | | | | | | | | |
| CHARTED BY:          DATE: | | | | | | | | |

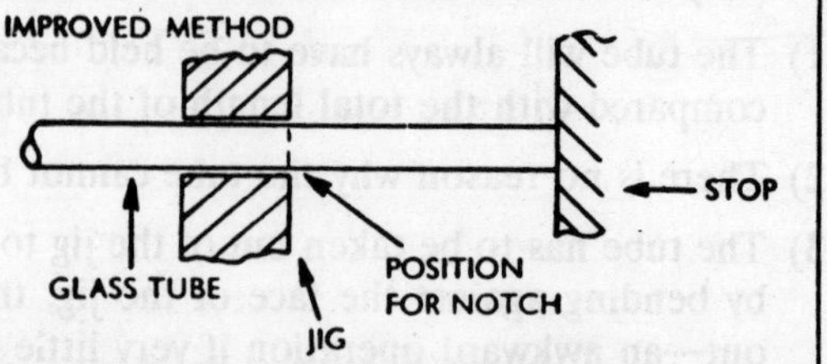

IMPROVED METHOD

STOP

GLASS TUBE

POSITION FOR NOTCH

JIG

| LEFT-HAND DESCRIPTION | ○ | ⇨ | □ | ▽ | ○ | ⇨ | □ | ▽ | RIGHT-HAND DESCRIPTION |
|---|---|---|---|---|---|---|---|---|---|
| *Pushes tube to stop* | | | | | | | | | *Holds file* |
| *Rotates tube* | | | | | | | | | *Notches with file* |
| *Holds tube* | | | | | | | | | *Taps with file:* |
| | | | | | | | | | *end drops to box* |

| SUMMARY | | | | |
|---|---|---|---|---|
| | PRESENT | | PROPOSED | |
| METHOD | L. H. | R. H. | L. H. | R. H. |
| *Operations* | 8 | 5 | 2 | 2 |
| *Transports* | 2 | 5 | – | – |
| *Delays* | – | – | | |
| *Holds* | 4 | 4 | 1 | 1 |
| *Inspections* | – | – | – | – |
| *Totals* | 14 | 14 | 3 | 3 |

*Figure 53. Example of workplace layout (original method)*

Figure 54. Example of workplace layout (improved method)

# FIGURE 56

## TWO-HANDED PROCESS CHARTS:

## ASSEMBLY OF POWER MOTOR STARTING WINDING TO CORE

**BEFORE**

| Left hand | Op. | Op. | Right hand |
|---|---|---|---|
| Transport stock of systoflex tubing to bench. | | H | ▽ |
| Transport stock of wedges to bench. | | H | ▽ |
| Transport stator to bench and place in position in fixture. | | H | ▽ |
| Transport set of coils to bench and place on shelf. | | H | ▽ |
| | 1 | 1 | Remove order card from stator bore & P.A. |
| | 2 | 2 | P.U. set of starting coils from shelf and place on bench. |
| Hold coils to assist. | 3 | 3 | P.U. scissors, cut tape. |
| | 4 | 4 | P.A. scissors. |
| | 5 | 5 | P.A. 3 groups of coils on shelf. |
| P.U. group of coils. | 6 | 6 | P.U. wire cutters. |
| Hold coils. | 7 | 7 | Cut off surplus wire from coils. |
| Hold coils. | 8 | 8 | P.A. wire cutters. |
| Hold coils. | 9 | 9 | Slide tapes to end of coils opposite to leads. |
| Hold coils. | 10 | 10 | Rotate stator to locate starting position. |
| Pass coils into stator bore. | 11 | 11 | |
| Aid R.H. | 12 | 12 | Insert 1st side of first (smallest) coil in slot. |
| | 13 | 13 | P.U. one slot wedge from bench. |
| Aid R.H. | 14 | 14 | Fold and bend to shape. |
| | 15 | 15 | Insert wedge in slot. |
| | 16 | 16 | P.U. tamping tool. |
| | 17 | 17 | Transfer to L.H. |
| Hold tamping tool. | 18 | 18 | P.U. hammer. |
| Hold tamping tool against wedge. | 19 | 19 | Hammer wedge into slot. |
| P.A. tamping tool. | 20 | 20 | P.A. hammer. |
| Repeat (12–20) for 1st side of second coil. | 21 to 29 | 21 to 29 | Repeat (12–20) for 1st side of second coil. |
| Repeat (12–20) for 1st side of third coil. | 30 to 38 | 30 to 38 | Repeat (12–20) for 1st side of third coil. |
| Rotate fixture. | 39 | 39 | |
| | 40 | 40 | P.U. scissors. |
| | 41 | 41 | Cut off two coil securing tapes. |
| | 42 | 42 | P.A. scissors. |
| Assist R.H. | 43 | 43 | Insert 2nd side of first coil into slot. |
| | | | P.U. one slot wedge from bench. |

| Left hand | | Right hand |
|---|---|---|
| Aid R.H. | (45) (45) | Fold and bend to shape. |
| | (46) (46) | Insert wedge in slot. |
| | (47) (47) | P.U. tamping tool. |
| | (48) (48) | Transfer to L.H. |
| Hold tamping tool. | (49) (49) | P.U. hammer. |
| Hold tamping tool against wedge. | (50) (50) | Hammer wedge into slot. |
| P.A. tamping tool. | (51) (51) | P.A. hammer. |
| Repeat (40–51) for 2nd side of second coil. | (52) to (63) | Repeat (40–51) for 2nd side of second coil. |
| Repeat (40–51) for 2nd side of third coil. | (64) to (75) | Repeat (40–51) for 2nd side of third coil. |
| Repeat (6–75) for 2nd group of coils. | (76) to (145) | Repeat (6–75) for 2nd group of coils. |
| Repeat (6–75) for 3rd group of coils. | (146) to (215) | Repeat (6–75) for 3rd group of coils. |
| Repeat (6–75) for 4th group of coils. | (216) to (285) | Repeat (6–75) for 4th group of coils. |
| | (286) (286) | P.U. mallet. |
| Position stator to suit. | (287) (287) | Knock back overhang. |
| Rotate fixture. | (288) (288) | |
| Position stator to suit. | (289) (289) | Knock back overhang. |
| | (290) (290) | P.A. mallet. |
| | (291) (291) | P.U. one length of 1 mm systoflex tubing and transfer to L.H. |
| Hold tubing. | (292) (292) | P.U. scissors. |
| Hold tubing | (293) (293) | Cut off 8 pieces 1½" long. |
| P.A. remaining systoflex. | (294) (294) | P.A. scissors. |
| | (295) (295) | Gather up 8 pieces of tubing. |
| Assist. R.H. | (296) (296) | Slide one tube over each lead. |
| | (297) (297) | P.U. one length of 3 mm tubing and transfer to L.H. |
| Hold tubing. | (298) (298) | P.U. scissors. |
| Hold tubing. | (299) (299) | Cut off 3 pieces about 1" long each. |
| P.A. remaining tubing. | (300) (300) | P.A. scissors. |
| | (301) (301) | Gather up three pieces of tubing. |
| Aid R.H. | (302) (302) | Slide one over one of each pair of leads to be connected for intercoil connection. |
| Aid R.H. | (303) (303) | Twist together 3 pairs of leads (6 turns per pair) to make intercoil connections. |
| | (304) (304) | P.U. wire cutters. |
| Aid R.H. | (305) (305) | Trim connections to length. |
| | (306) (306) | P.A. wire cutters. |
| Aid R.H. | (307) (307) | Fold all leads into bore of stator. |
| | (308) (308) | P.U. order card and place in bore of stator. |
| Aid R.H | (309) (309) | Transport stator to temporary storage. |

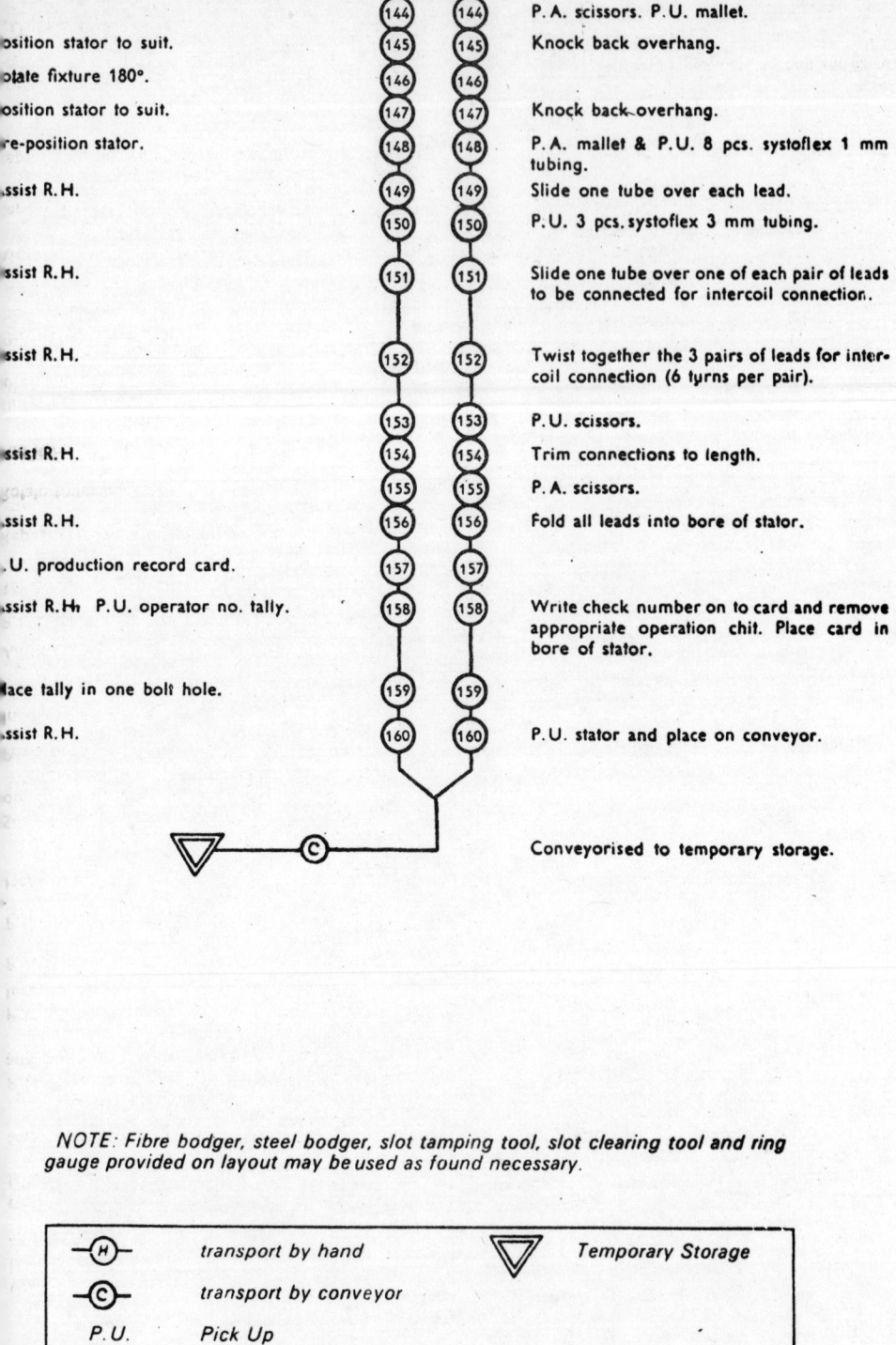

| Left | Op | | Description |
|---|---|---|---|
| epeat (5–38) with 4th group of coils. | 110 to 143 | | Repeat (5–38) with 4th group of coils. |
| | 144 | 144 | P.A. scissors. P.U. mallet. |
| osition stator to suit. | 145 | 145 | Knock back overhang. |
| otate fixture 180°. | 146 | 146 | |
| osition stator to suit. | 147 | 147 | Knock back overhang. |
| re-position stator. | 148 | 148 | P.A. mallet & P.U. 8 pcs. systoflex 1 mm tubing. |
| ssist R.H. | 149 | 149 | Slide one tube over each lead. |
| | 150 | 150 | P.U. 3 pcs. systoflex 3 mm tubing. |
| ssist R.H. | 151 | 151 | Slide one tube over one of each pair of leads to be connected for intercoil connection. |
| ssist R.H. | 152 | 152 | Twist together the 3 pairs of leads for intercoil connection (6 turns per pair). |
| | 153 | 153 | P.U. scissors. |
| ssist R.H. | 154 | 154 | Trim connections to length. |
| | 155 | 155 | P.A. scissors. |
| ssist R.H. | 156 | 156 | Fold all leads into bore of stator. |
| U. production record card. | 157 | 157 | |
| ssist R.H. P.U. operator no. tally. | 158 | 158 | Write check number on to card and remove appropriate operation chit. Place card in bore of stator. |
| lace tally in one bolt hole. | 159 | 159 | |
| ssist R.H. | 160 | 160 | P.U. stator and place on conveyor. |
| | | | Conveyorised to temporary storage. |

NOTE: Fibre bodger, steel bodger, slot tamping tool, slot clearing tool and ring gauge provided on layout may be used as found necessary.

|  |  | |  |  |
|---|---|---|---|---|
| —(H)— | transport by hand | | ▽ | Temporary Storage |
| —(C)— | transport by conveyor | | | |
| P.U. | Pick Up | | | |
| P.A. | Place Aside | | | |

**AFTER**

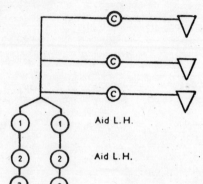

Wedges and systoflex tubing conveyorised from stores in bins for self service from bins as needed by operator.

Coils conveyorised from stores in bins containing 5 sets each.

Stators conveyorised from stores.

P.U. box of starting coils from conveyor and position on bench. — ① ① — Aid L.H.

P.U. stator from conveyor and place on fixture. — ② ② — Aid L.H.

P.A. production record card from stator. — ③ ③

P.U. one group of coils. — ④ ④

Hold coils. — ⑤ ⑤ — Untwine leads and lodge two coils between stator and fixture.

Slide 1st side securing tape to end of coil opposite to leads. — ⑥ ⑥ — Hold coil.

Position stator. — ⑦ ⑦ — Pass coil into stator.

Insert 1st side of coil into its slot. — ⑧ ⑧ — Aid L.H.

— ⑨ ⑨ — P.U. one wedge.

Aid R.H. — ⑩ ⑩ — Fold and insert in slot.

P.U. second coil. — ⑪ ⑪

Repeat (6–10) with second coil. — ⑫ to ⑯ — Repeat (6–10) with second coil.

Repeat (11–16) with third coil. — ⑰ to ㉒ — Repeat (11–16) with third coil.

Rotate fixture 180°. — ㉓ ㉓

Aid R.H. — ㉔ ㉔ — Fold back 2nd and 3rd coils in bore of stator.

Slide securing tape on 2nd side of first coil to end opposite leads. — ㉕ ㉕ — Aid L.H.

Insert 2nd side of first coil into slot. — ㉖ ㉖ — Aid L.H.

— ㉗ ㉗ — P.U. one wedge.

Aid R.H. — ㉘ ㉘ — Fold and insert in slot.

Repeat (25–28) with 2nd side of second coil. — ㉙ to ㉜ — Repeat (25–28) with 2nd side of second coil.

Repeat (25–28) with 2nd side of third coil. — ㉝ to ㊱ — Repeat (25–28) with 2nd side of third coil.

— ㊲ ㊲ — P.U. scissors.

Position stator to suit R.H. — ㊳ ㊳ — Cut off securing tapes, first, second and third coils.

P.U. 2nd group of coils. — ㊴ ㊴ — P.A. scissors.

Repeat (5–39) with 2nd group of coils. — ㊵ to ㊼ — Repeat (5–39) with 2nd group of coils.

Repeat (5–39) with 3rd group of coils. — ㊺ to ⑩⑨ — Repeat (5–39) with 3rd group of coils.

a tray in Figure 54) are stated in the process chart to be placed on the shelf in front of the worker (another long reach).

Figure 56 shows the two-handed process charts before and after the alteration in method and re-laying out the workplace, in the original form in which they were drawn at the time. The process charts are accompanied (Figure 55) by right- and left-handed activity charts (not described in this book) which show the relative activity of the individual hands. From these it will be seen that under the original method the left hand is idle during a considerable part of the cycle: the right hand performs nearly twice as many operations. Reference to the 'Before' process chart shows that the left hand is used very largely either to hold components or to assist the right hand.

The 'After' activity chart shows that the activities of the two hands are more nearly balanced. The number of operations performed by the right hand has been reduced to 143, although the number of delays has increased from nine to 16. This, however, is more than compensated by the reduction in both the number and the duration of the delays of the left hand, whose operations have been reduced to 129. It will be seen also in Figure 61 that transport by hand (H) has been eliminated by the use of a conveyor (C).

The 'After' process chart shows that the left hand is now much more usefully employed. There is only one 'hold' for each hand; although the left hand is still used to some extent to assist the right, it is also used to carry out a number of operations of its own.

The process chart, although it gives details of the change in method, does not give any indication of the changes in the workplace layout. These may be seen in Figure 54.

The workplace has been laid out according to the principles of motion economy and the working areas shown in Figure 49. The workpiece and all the components and tools are well within the maximum working area. The fixture is the same, but it has been placed nearer the edge of the bench, where it is more convenient to the worker. The systoflex, wedges and other components are conveniently located in standard trays; the coils (a larger item) are in the large tray within easy reach of the worker's left hand. Special note should be taken of the positioning of the tools. These are located for the use of the appropriate hand with the handles in a position that is easy to grasp: even the scissors are tucked between the trays with their handles upwards. The ring gauge, which in Figure 53, is to be seen lying flat on the bench (a difficult position from which to pick it up), is now upright in a specially shaped tin on the right-hand side of the bench where it is very simple to grasp: the operative need not look up from her work.

Figure 54 requires careful study. The compactness of the workplace encourages the operative to keep things in their proper places: a large amount of bench space is an invitation to scatter tools and components on it. As regards economy of factory space, this new layout will pay for itself in two ways: first by making it possible to establish more workplaces in a given area: and second, by providing greater output from a given workplace. The operative will also find the work much less tiring because she no longer has to stretch and search.

*Figure 55. Right- and left-handed activity charts: assembly of power motor starting winding to core*

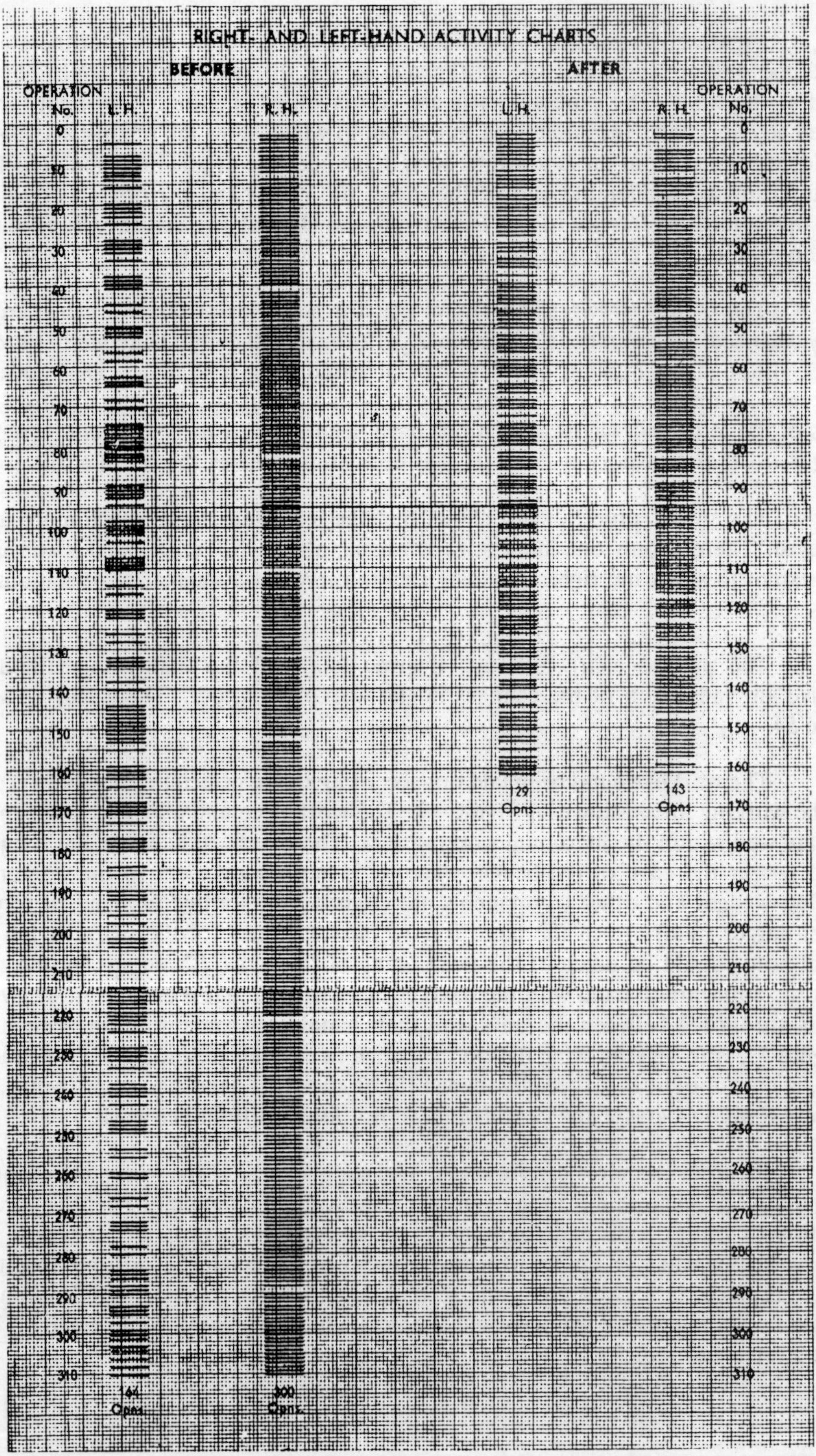

## 11.9   Micromotion study

In certain types of operation, and particularly those with very short cycles which are repeated thousands of times (such as the packing of sweets into boxes or food cans into cartons), it is worth while going into much greater detail to determine where movements and effort can be saved and to develop the best possible pattern of movement, thus enabling the operative to perform the operation repeatedly with a minimum of effort and fatigue. The techniques used for this purpose frequently make use of filming, and are known collectively as micromotion study.

The micromotion group of techniques is based on the idea of dividing human activity into divisions of movements or groups of movements (known as therbligs) according to the purpose for which they are made.

The divisions were devised by Frank B. Gilbreth, the founder of motion study; the word 'therblig' is an anagram of his name. Gilbreth differentiated 17 fundamental hand or hand and eye motions, to which an eighteenth has subsequently been added. The therbligs cover movements or reasons for the absence of movement. Each therblig has a specific colour, symbol and letter for recording purposes. These are shown in Table 11.

Therbligs refer primarily to motions of the human body at the workplace and to the mental activities associated with them. They permit a much more precise and detailed description of the work than any other method so far described in this book. On the other hand, considerable practice is required before they can be used for analysis with any degree of assurance.

It is not felt necessary in an introductory book of this kind to go deeply into these techniques, because so much can be done to improve productivity by using the simpler ones already described, before it becomes necessary to use such refinements. They are used much less than the simpler techniques, even in the highly industrialised countries, and then mainly in connection with mass-production operations, and they are more preached about than practised. They are, however, techniques for the expert, and in any case it would be imprudent for the trainee or comparatively inexperienced work study man to waste his time trying to save split seconds when there are sure to be plenty of jobs where productivity can be doubled and even trebled by using the more general methods.

## 11.10   The simo chart

Only one recording technique of micromotion study will be described here, namely the simultaneous motion cycle chart, known as the simo chart for short.

> **A simo chart is a chart, often based on firm analysis, used to record simultaneously on a common time scale the therbligs or groups of therbligs performed by different parts of the body of one or more workers.**

The simo chart is the micromotion form of the man type flow process chart. Because simo charts are used primarily for operations of short duration, often performed

*Table 11. Therbligs*

| Symbol | Name | Abbreviation | Colour |
|--------|------|--------------|--------|
| | Search | Sh | Black |
| | Find | F | Grey |
| | Select | St | Light Grey |
| | Grasp | G | Red |
| | Hold | H | Gold ochre |
| | Transport Load | TL | Green |
| | Position | P | Blue |
| | Assemble | A | Violet |
| | Use | U | Purple |
| | Disassemble | DA | Light violet |
| | Inspect | I | Burnt ochre |
| | Pre-position | PP | Pale blue |
| | Release load | RL | Carmine red |
| | Transport Empty | TE | Olive green |
| | Rest for overcoming fatigue | R | Orange |
| | Unavoidable delay | UD | Yellow |
| | Avoidable delay | AD | Lemon yellow |
| | Plan | Pn | Brown |

with extreme rapidity, it is generally necessary to compile them from films made of the operation which can be stopped at any point or projected in slow motion. It will be seen from the chart illustrated in Figure 57 that the movements are recorded against time measured in 'winks' (1 wink = 1/2000 minute). These are recorded by a "wind counter" placed in such a position that it can be seen rotating during the filming.

Figure 57. A simo chart [1]

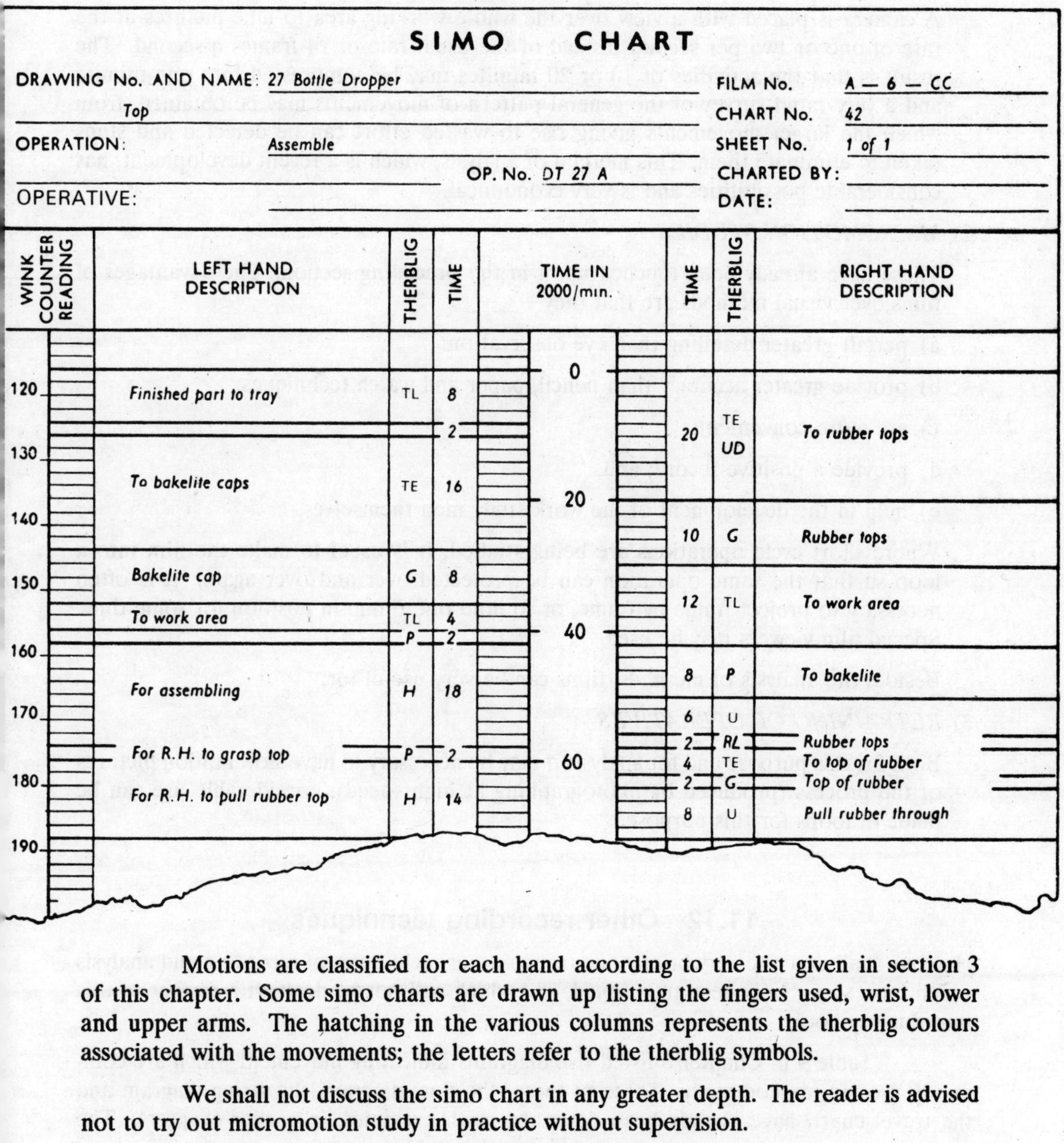

| WINK COUNTER READING | LEFT HAND DESCRIPTION | THERBLIG | TIME | TIME IN 2000/min. | TIME | THERBLIG | RIGHT HAND DESCRIPTION |
|---|---|---|---|---|---|---|---|
| | | | | 0 | | | |
| 120 | Finished part to tray | TL | 8 | | | | |
| | | | 2 | | 20 | TE UD | To rubber tops |
| 130 | To bakelite caps | TE | 16 | 20 | | | |
| 140 | | | | | 10 | G | Rubber tops |
| 150 | Bakelite cap | G | 8 | | 12 | TL | To work area |
| | To work area | TL | 4 | 40 | | | |
| | | P | 2 | | | | |
| 160 | | | | | 8 | P | To bakelite |
| | For assembling | H | 18 | | | | |
| 170 | | | | | 6 | U | |
| | | | | | 2 | RL | Rubber tops |
| | For R.H. to grasp top | P | 2 | 60 | 4 | TE | To top of rubber |
| 180 | | | | | 2 | G | Top of rubber |
| | For R.H. to pull rubber top | H | 14 | | 8 | U | Pull rubber through |
| 190 | | | | | | | |

Motions are classified for each hand according to the list given in section 3 of this chapter. Some simo charts are drawn up listing the fingers used, wrist, lower and upper arms. The hatching in the various columns represents the therblig colours associated with the movements; the letters refer to the therblig symbols.

We shall not discuss the simo chart in any greater depth. The reader is advised not to try out micromotion study in practice without supervision.

## 11.11  The use of films in methods analysis

In methods analysis, films may be used for the following purposes:

1) *MEMOMOTION PHOTOGRAPHY* (A form of time-lapse photography which records

activity by the use of a cine camera adapted to take pictures at longer intervals than normal. The time intervals usually lie between 1/2 second and 4 seconds).

A camera is placed with a view over the whole working area to take pictures at the rate of one or two per second instead of the usual rate of 24 frames a second. The result is that the activities of 10 or 20 minutes may be compressed into one minute and a very rapid survey of the general pattern of movements may be obtained, from which the larger movements giving rise to wasted effort can be detected and steps taken to eliminate them. This method of analysis, which is a recent development, has considerable possibilities and is very economical.

2) *MICROMOTION STUDIES*

These have already been touched upon in the preceding section. The advantages of films over visual methods are that they—

a) permit greater detailing than eye observation:

b) provide greater accuracy than pencil, paper and watch techniques;

c) are more convenient;

d) provide a positive record; and,

e) help in the development of the work study men themselves.

Where short cycle operations are being studied, it is usual to make the film into a loop so that the same operation can be projected over and over again. It is often necessary to project frame by frame, or to hold one frame in position for some time. Special film viewers may be used.

Besides the analysis of methods, films can be very useful for:

3) *RETRAINING OF OPERATIVES*

Both for this purpose and for analysis it may be necessary to have slow motion pictures of the process (produced by photographing at high speed); considerable use can be made of loops for this purpose.

## 11.12   Other recording techniques

Here we shall describe very briefly one or two other techniques of recording and analysis which have so far only been mentioned, and which will not be dealt with further in this introductory book.

Table 9 in Chapter 8 listed five diagrams indicating movement which are commonly used in method study. Three of these, the flow diagram, the string diagram and the travel chart, have already been described, with examples, in earlier chapters. The other two are the cyclegraph and the chronocyclegraph.

The cyclegraph is a record of a path of movement, usually traced by a continuous source of light on a photograph, preferably stereoscopic. The path of movement of a hand, for instance, may be recorded on a photograph in this way if the worker is asked to wear a ring carrying a small light which will make the trace on the photograph. Alternatively, such a light may be attached to a worker's helmet if the purpose is to obtain a record of the path over which he moves during the performance of a task.

The chronocyclegraph is a special form of cyclegraph in which the light source is suitably interrupted so that the path appears as a series of pear-shaped dots, the pointed end indicating the direction of movement and the spacing indicating the speed of movement.

In comparison with the other recording techniques outlined in this book, the cyclegraph and chronocyclegraph are of limited application, but there are occasions on which photographic traces of this sort can be useful.

## 11.13   The development of improved methods

In each of the examples of the different method study techniques given so far, our discussion has covered the three stages of *RECORD, EXAMINE* and *DEVELOP*, but has been focussed primarily on the first two, the development stage being discussed only as far as was necessary to draw attention to the improvements made in method as a result of using the particular diagram or form being demonstrated.

It will now be appropriate to study a little more closely the manner in which improved methods can be developed.

One of the rewards of method study is the large saving which can often be made from quite small changes and inexpensive devices, such as chutes or suitable jigs.

An example of this is a small spring-loaded table, very cheaply made in plywood, for removing the tiles from an automatic tile-making machine. The spring was so calibrated that, each time a tile was pushed on to it by the machine, it was compressed until the top of the tile dropped to the level of the machine platform so that the table was ready to receive the next tile. This enabled the girl operating the machine to concentrate on loading the finished tiles on to a rack ready for firing while the new stack was piling up. When about a dozen tiles were in place, she was able to lift them off the table, which immediately sprang up to the level of the machine platform ready to receive the first tile of the next stack. This very simple device enabled the second operative formerly employed on this operation to be released for other work, an important feature in an area where skilled tile-pressers were difficult to obtain.

In many manufacturing plants the work study man may have to go beyond the study of the movements of materials and workers if he is going to make the most effective contribution to increased productivity. He must be prepared to discuss with the designers the possibility of using alternative materials which would make the product easier and quicker to manufacture. Even if he is not an expert in design—and, indeed, he cannot be expected to be—drawing attention to the possibilities of an alternative may put ideas into the minds of the designers themselves which they had previously overlooked. After all, like everyone else, they are human and often hard-worked, and there is a strong temptation to specify a given material for a given product or component simply because it has always been used in the past.

Apart from the elimination of obviously wasteful movements—which can be done from the flow diagram or process chart—the development of improved methods calls for skill and ingenuity. It is likely to be more successful if the work study man is also well acquainted with the industry with which he is concerned. In any but the simplest manual operations, he will have to consult with the technical or supervisory staff and, even if he does know the right answer, it is better that he should do so, since a method which they have taken part in developing is likely to be accepted so, since a method which they have taken part in developing is likely to be accepted more readily than one which is introduced as someone else's idea. The same is true of the operatives. Let everyone put forward his ideas—two heads are better than one!

The fact that really successful methods improvement is a combined operation is being increasingly recognised. Many organisations, large and small, have set up groups for the improvement of manufacturing and operating methods. These groups may be permanent or set up for some particular job such as the re-laying out of a shop or factory, or the organisation of work. Such groups often decide on the division and allocation of work as well as other related functions such as the control of quality.

In the United Kingdom Joseph Lucas Ltd, manufacturers of electrical equipment and motor car accessories, have developed similar groups at various levels which consider every aspect of manufacturing efficiency from designing the product for more economic production onwards through all the processes and methods.

## 11.14   The methods laboratory

There is great value in having a small room or shop where the work study men can develop and try out new methods. It need not be elaborate or expensive; many devices can be tried out in wood before they are manufactured in metal. If the scale of the work study activities justifies it, one or two good all-round craftsmen can be seconded to this laboratory with some simple tools, such as a drill press and sheet-metal equipment, together with a good operative from the production shops who will try out the different 'gadgets' in collaboration with the work study staff until the best method has been found. Having such a place saves interfering with the production shops or the plant engineer's department when things are wanted in a hurry, and the work study staff feel much freer to try out revolutionary ideas. New methods can be demonstrated to the management, foremen and operatives, who can be encouraged to try them out and make suggestions to be incorporated in the final version.

Do not let the work study shop become the place where everyone in the works comes when they "want a little job done quickly" or private repairs executed. There is a real danger of this, as more than one company has discovered.

On no account may the operative attached to the work study shop be used for the setting of time standards. It is quite acceptable to time him in order to compare the effectiveness of different methods, but time studies for standard setting must always be made in the shops under production conditions with regular operatives.

## NOTES

1. In the B.S. *Glossary*, op. cit., the term "characteristics of easy movement" is preferred, rather than "principles of motion economy". The earlier term has been retained here, however, as being more descriptive of the rest of this section of the chapter.

2. See Ralph M. Barnes: *Motion and time study: Design and measurement of work* (New York and London, John Wiley, 6th ed., 1969), Chapters 17–19.

3. See Chapter 6.

4. Some authorities feel that the standard process-chart symbols are not entirely suitable for recording hand and body movements and have adopted variants, such as—

   O: Operation          H: Hold.
   TL: Transport Loaded   R: Rest
   TE: Transport Empty.

5. This example from industrial practice was provided by the General Electric Company Ltd, Witton (United Kingdom), through whose courtesy the photographs and process charts have been made available. The process charts are reproduced in their original form. It will be seen that they differ somewhat from the newer practice recommended in this book, but careful examination will make them perfectly clear to the reader.

6. Adapted from Marvin E. Mundel: *Motion and time study: Principles and practice* (Englewood Cliffs NJ, and Hemel Hempstead, UK, Prentice-Hall, 4th ed., 1970).

# Chapter 12
# Define, install, maintain

## 12.1 Introduction

To establish and maintain the new improved method in the system is one of the most important task because the new time standard provides the basic data for manpower control, production control, machine utilisation, standard cost, manpower requirement, budgets, etc.

Once the new method is developed it is necessary to obtain the approval of works management for the establishment and maintenance of it in the system.

The basic administrative steps required to establish and maintain the new methods are the following—

DEFINE        the new method and the related time so that it can always be identified.

INSTALL       the new method as agreed standard practice with the time allowed.

MAINTAIN      the new standard practice by proper control procedure.

## 12.2 Obtaining approval for the improved method

Once a complete study of the job has been made, and the preferred new method developed, it is generally necessary to obtain the approval of the works management before proceeding to install it. The work study man should prepare a report giving details of the existing and proposed methods and should give his reasons for the changes suggested.

The report should show—

a) Relative costs in material, labour and overheads of the two methods, and the savings expected.

b) the cost of installing the new method, including the cost of any new equipment and of re-laying out shops or working areas, if this is required.

c) Executive actions required to implement the new method.

Before it is finally submitted, the report should be discussed with the departmental supervision or management; if the costs of the change are small and all are agreed that it is a useful change, the work may proceed on the authority of the departmental manager or foreman.

If capital expenditure is involved, such as the purchase of material-handling equipment, or if complete agreement cannot be obtained from everyone concerned on the desirability of the change, the matter may have to be decided on by the management. In this case, it is almost certain that the work study man will be called upon to justify his estimates. If capital investment is involved to any extent, he will have to be able to convince doubting people, often non-technical, that it will really be justified. Great care must therefore be taken in preparing such estimates, since a failure to live up to them may damage both the work study man's own reputation and that of work study itself.

## 12.3   Defining the improved method

### THE WRITTEN STANDARD PRACTICE

For all jobs other than those performed on standard machine tools or specialised machines where the process and methods are virtually controlled by the machine, it is desirable to prepare a written standard practice, also known as an operative instruction sheet. This serves several purposes:

a) It records the improved method for future reference, in as much detail as may be necessary.

b) It can be used to explain the new method to the management, foremen and operatives. It also advises all concerned, including the works engineers, of any new equipment required or of changes needed in the layout of machines or workplaces.

c) It is an aid to training or retraining operatives and can be used by them for reference until they are fully conversant with the new method.

d) It forms the basis on which time studies may be taken for setting standards, although the element breakdown (see Chapter 16, section 6) will not necessarily be the same as the breakdown of motions.

The written standard practice outlines in simple terms the methods to be used by the operative. Therbligs and other method study symbols should not be used. Three sorts of information will normally be required—

a) The tools and equipment to be used and the general operating conditions.

b) A description of the method. The amount of detail required will depend on the nature of the job and the probable volume of production. For a job which will occupy several operatives for several months, the written standard practice may have to be very detailed, going into finger movements.

c) A diagram of the workplace layout and, possibly, sketches of special tools, jigs or fixtures.

A very simple written standard practice for the operation studied in Chapter 11, section 7 (cutting glass tubes to length), is illustrated in Figure 58. The same principle is followed in more complex cases. In some of these the description may run into several pages. The workplace layout and other diagrams may have to be put on a separate sheet. With the more widespread use in recent years of standardised printed sheets for process charts, it is becoming common practice to attach a fair copy of the appropriate process chart to the written standard practice, whenever the simple description entered thereon does not constitute a complete definition of the method.

## 12.4   Installing the improved method

The final stages in the basic procedure are perhaps the most difficult of all. It is at this point that active support is required from the management and trade unions alike. It is here that the personal qualities of the work study man, his ability to explain clearly and simply what he is trying to do and his gift for getting along with other people and winning their trust become of the greatest importance.

Installation can be divided into five stages, namely—

a) Gaining acceptance of the change by the departmental supervision.

b) Gaining approval of the change by the management.

These two steps have already been discussed. There is little point in trying to go any further unless they have been successful.

c) Gaining acceptance of the change by the workers involved and their representatives.

d) Restrain the workers to operate the new methods.

e) Maintaining close contact with the progress of the job until satisfied that it is running as intended.

If any changes are proposed which affect the number of workers employed in the operation—as is often the case—the workers' representatives should be consulted as early as possible. Plans for any displacement of labour must be very carefully worked out so that the least possible hardship or inconvenience is caused. Remember, even on a one-man operation, the workers in a workshop or any other organisation does not work in isolation. If he is not a member of a team for the specific purpose of his job, he is a member of his job, he is a member of a section or department; he gets used to having the same people working around him, to spending his meal breaks with the same gang. Even if he is too far away from them to carry on a conversation during his work, he can see them; he can, perhaps, exchange a joke with them from time to time or grouse at the management or the foreman. He adjusts his personality to them and they to him. If he is suddenly moved, even if it is only to the other end of the shop, his social circle is broken up, he feels slightly lost without them and they without him.

In the case of a team or gang working together, the bonds are far stronger; and breaking up such a team may have serious effects on productivity, in spite of improved methods. It is only since the 1930s that the importance of group behaviour in a working area has come to be recognised. Failure to take this into account may lead the workers to resist changes which they would otherwise accept.

It is in carrying out the first three steps of installation that the importance of preliminary education and training in work study for all those likely to be concerned with it—management, supervisors and workers' representatives—becomes evident. People are much more likely to be receptive to the idea of change if they know and understand what

*Figure 58. Standard practice sheet*

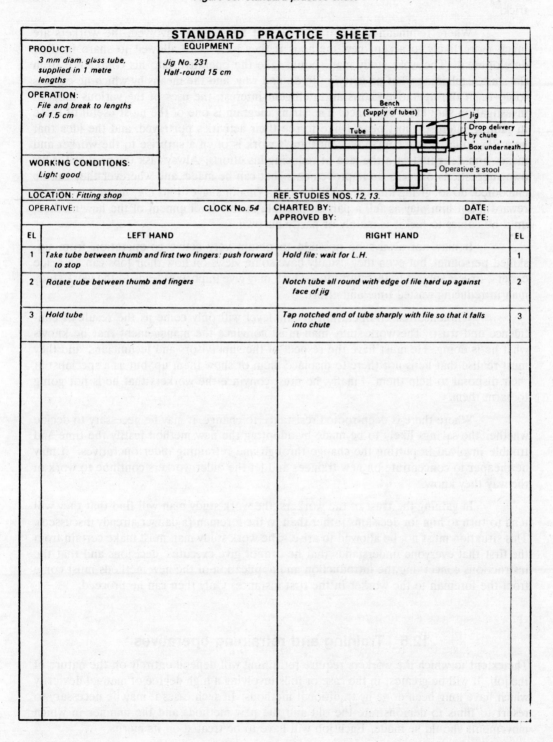

is happening than if they are merely presented with the results of a sort of conjuring trick.

Where redundancy or a transfer is not likely to be involved, the workers are much more likely to accept new methods if they have been allowed to share in their development. The work study man should take the operative into his confidence from the start, explaining what he is trying to do and why, and the means by which he expects to do it. If the operative shows an intelligent interest, the uses of the various tools of investigation should be explained. The string diagram is one of the most useful of these in gaining interest: most people like to see their activities portrayed, and the idea that he walks so far in the course of a morning's work is often a surprise to the worker and makes him delighted with the idea of reducing his efforts. Always ask the worker for his own suggestions or ideas on improvements that can be made, and wherever they can be embodied, do so, giving him the credit for them (major suggestions may merit a monetary reward). Let him play as full a part as possible in the development of the new method, until he comes to feel it is mainly or partly his own.

It may not always be possible to obtain very active co-operation from un-skilled personnel, but even they usually have some views on how their jobs can be made easier—or less subject to interruption—which may give important leads to the work study man in reducing wasted time and effort.

Wholehearted co-operation at any level will only come as the result of con-fidence and trust. The work study man must convince the management that he knows what he is doing. He must have the respect of the supervisors and technicians, and they must realise that he is not there to displace them or show them up, but as a specialist at their disposal to help them. Finally, he must convince the workers that he is not going to harm them.

Where there is deep-rooted resistance to change, it may be necessary to decide whether the savings likely to be made by adopting the new method justify the time and trouble involved in putting the change through and retraining older operatives. It may be cheaper to concentrate on new trainees and let the older workers continue to work in the way they know.

In gaining the trust of the workers, the work study man will find that they will tend to turn to him for decisions rather than to the foreman (a danger already discussed). This situation must not be allowed to arise. The work study man must make certain from the first that everyone understands that he cannot give executive decisions and that the instructions concerning the introduction and application of the new methods must come from the foreman to the worker in the first instance. Only then can he proceed.

## 12.5  Training and retraining operatives

The extent to which the workers require retraining will depend entirely on the nature of the job. It will be greatest in the case or jobs involving a high degree of manual dexterity which have long been done by traditional methods. In such cases it may be necessary to resort to films to demonstrate the old and the new methods and the manner in which movements should be made. Each job will have to be treated on its merits.

In the training or retraining of operatives, the important thing is to develop the habit of doing the job in the correct way. Habit is a valuable aid to increased productivity as it reduces the need for conscious thought. Good habits can be formed just as easily as bad ones.

Beginners can be taught to follow a numbered sequence illustrated on a chart or they may be taught on the machine itself. Either way, they must be made to understand the reason for every movement. Still pictures together with instruction sheets have proved very successful. Film strips can also be used.

Films are particularly valuable when retraining. When old habits have to be broken, it may be found that the operative is quite unaware of what he is doing. A film in slow motion will enable him to see his exact movements and, once he knows, he can start to learn the new method. It is important that the new method should be really different from the old, otherwise the operative will tend to slip back into his old ways, especially if he is not young and has spent many years doing the job.

In learning a new series of movements, the operative gathers speed and reduces the time required to perform them very quickly at first. The rate of improvement soon begins to slow up, however, and it often requires long practice to achieve really high and consistent speed, although the adoption of modern accelerated training methods will considerably shorten the time needed. A typical learning curve is shown in Figure 59.

*Figure 59. A typical learning curve*

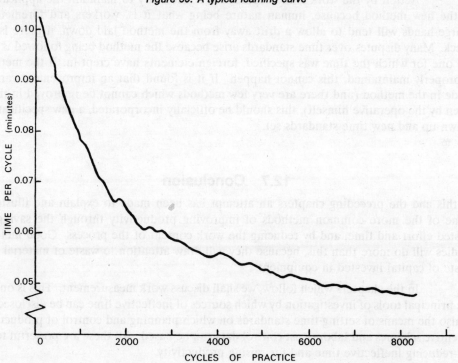

Experiments have shown that in the first stages of learning, to obtain the best results, rests between periods of practice should be longer than the periods of practice themselves. This situation alters rapidly, however, and when the operative has begun to grasp the new method and to pick up speed, rest periods can be very much shorter.

As part of the process of installation it is essential to keep in close touch with the job, once it has been started, to ensure that the operative is developing speed and skill and that there are no unforeseen snags. This activity is often known as nursing the new method, and the term is an apt one. Only when the work study man is satisfied that the productivity of the job is at least at the level he estimated and that the operative has settled down to it can it be left—for a time.

## 12.6   Maintaining the new method

It is important that, when a method is installed, it should be maintained in its specified form, and that workers should not be allowed to slip back into old methods, or introduce elements not allowed for, unless there is very good reason for doing so.

To be maintained, a method must first be very clearly defined and specified. This is especially important where it is to be used for setting time standards for incentive or other purposes. Tools, layout and elements of movement must be specified beyond any risk of misinterpretation. The extent to which it is necessary to go into minute details will be determined by the job itself.

Action by the work study department is necessary to maintain the application of the new method because, human nature being what it is, workers and foremen or charge-hands will tend to allow a drift away from the method laid down, if there is no check. Many disputes over time standards arise because the method being followed is not the one for which the time was specified; foreign elements have crept in. If the method is properly maintained, this cannot happen. If it is found that an improvement can be made in the method (and there are very few methods which cannot be improved in time, often by the operative himself), this should be officially incorporated, a new specification drawn up and new time standards set.

## 12.7   Conclusion

In this and the preceding chapters an attempt has been made to explain and illustrate some of the more common methods of improving productivity through the saving of wasted effort and time, and by reducing the work content of the process. Good method studies will do more than this, because they will draw attention to waste of material and waste of capital invested in equipment.

In the chapters which follow, we shall discuss work measurement. This is one of the principal tools of investigation by which sources of ineffective time can be disclosed. It is also the means of setting time standards on which planning and control of production, incentive schemes and labour cost control data can be based. All these are powerful tools for reducing ineffective time and for raising productivity.

# Part C
# Work measurement

# Part C

## Work measurement

# Chapter 13
# General remarks on work measurement

## 13.1 Introduction

Method study technique reveals the realistic and most constructive proposals for method improvement. After reducing the work by eliminating redundant elements on the part of the material or operatives and substituting improved methods for poor ones by method study, work measurement starts investigating, and eliminating ineffective time in the improved method, ultimately suggested by method study, and measures time taken in performance of the same identifying and then eliminating ineffective time.

## 13.2 Definition

In Chapter 4 it was said that work study consists of two complementary techniques—method study and work measurement. In that chapter both were defined, and before we go on to discuss work measurement it is worthwhile repeating the definition of work measurement given there.

---

**Work measurement is the application of techniques designed to establish the time for a qualified worker to carry out a specific job at a defined level of performance.**

---

We shall have occasion to examine several factors of this carefully throughout definition in more detail in later chapters. For instance, the reader will have noted the reference to "a qualified worker", and to "a defined level of performance". We need not concern ourselves with the exact meaning of these terms for the moment. It is worth noting, however, that the term "work measurement", which we have referred to hitherto as a technique is really a term used to describe a family of techniques, any one of which can be used to measure work, rather than a single technique by itself. The principal techniques which are classed as work measurement techniques are listed in section 13.6 of this chapter.

## 13.3   Purpose of work measurement

Work measurement provides the management with a means of measuring the time taken in the performance of an operation or series of operations in such a way that ineffective time is shown up and can be separated from effective time. In this way its existence, nature and extent become known where previously they were concealed within the total. One of the surprising things about factories where work measurement has never been employed is the amount of ineffective time whose very existence is unsuspected or which is accepted as "the usual thing" and something inevitable that no one can do much about—that is built into the process. Once the existence of ineffective time has been revealed and the reasons for it tracked down, steps can usually be taken to reduce it.

Not only can work measurement reveal the existence of ineffective time; it can also be used to set standard times for carrying out the work, and standard output level.

Work measurement technique can also be used for the following purposes:

a) To evaluate a worker's performance: This can be done by comparing actual output over a given period of time with standard output level determined from work measurement.

b) To plan work-face needs: Work measurement can be used to determine the labour input required for any given output level.

c) To determine available capacity: Work measurement standards can be used to estimate available capacity for a given level of work-force and equipment availability.

d) To determine price or cost of a product: Labour standard is one of the most important parameters of a costing or pricing system which is crucial to the survival and growth of the business.

e) To compare work methods: When alternative methods for a job are being considered, work measurement can provide the basis for economic comparison of the methods.

f) To facilitate operations scheduling: To all scheduling systems one of the data inputs is time estimates for work activities which are derived from work measurement.

g) To establish wage incentive schemes: Workers receive more wage for more output under wage incentive schemes. Underlying these schemes is a time standard which defines the reference output.

It is clear that work measurement provides the basic information necessary for all the activities of organising and controlling the work of an enterprise in which the time element plays a part.

Work measurement may start a chain reaction throughout the organisation. How does this come about?

The first thing to realise is that breakdowns and stoppages taking effect at the shop floor level are generally only the end results of a series of management actions or failures to act.

Let us take an example of excessive idle time of an expensive machine, revealed by a study taken over several days. This piece of plant is very productive when operating but takes a long time to set up. It is found that a great deal of the idle time is due to the fact that the batches of work being put on this machine are very small, so that almost as much time is spent in resetting it to do new operations as is spent in actual production. The chain of reactions resulting from this discovery may be something like this:

**The work study department**

reports that work measurement reveals that the machine is idle for excessively long periods because of small orders coming from the planning office. This is substantially increasing the cost of manufacture. It suggests that the planning office should do some proper planning and either combine several orders for the same product into one large order or make more for stock.

**The planning office**

complains that it has to work on the instructions of the sales office which never seems to sell enough of any one product to make up a decent-sized batch and cannot give any forecast of future orders so that more can be made for stock.

**The sales office**

says that it cannot possibly make forecasts or provide large orders of any one product as long as it remains the policy of top management to accept every variation that customers like to ask for. Already the catalogue is becoming too large: almost every job is now a 'special'.

**The managing director**

when the effect of his marketing policy (or lack of it) on the production costs is brought to his attention, is surprised and says that he never thought of it like that; all he was trying to do was to prevent orders going to his competitors by being as obliging to his customers as possible.

One of the principal purposes of Work Study will have been served if the original investigation leads the managing director to think again about his marketing policy. Enthusiasts work study men may, however, find it well to pause a moment and think about the fact that such chains of reaction tend to make someone ask: "Who started this, anyway?" People do not like being "shown up". This is one of the situations in which a good deal of tact may have to be used. It is not the work study man's job to dictate marketing policy, but merely to bring to the attention of the management the effect of that policy on the company's costs and hence on its competitive position.

Thus it can be seen that the purposes of work measurement are to reveal the nature and extent of ineffective time, from whatever cause, so that action can be taken to eliminate it; and then to set standards of performance of such a kind that they will be attainable only if all avoidable ineffective time is eliminated and the work is performed by the best available method and by personnel suitable in training and ability to their tasks.

## 13.4   Where work measurement can be applied

Work measurement is widely used in factories where work or at least some parts of the work are repetitive. There are three general criteria for measurable jobs. These are:

i) The work should be identifiable in terms of the number of units a worker performed.

ii) The work should be performed in a reasonably consistent manner.

iii) There should be a considerable volume of the work to justify performing a study and keeping counts and records.

## 13.5  The basic procedure

In section 3 of Chapter 4 we described the basic steps of Work Study, embracing both method study and work measurement. The basic procedure of method study has been described separately in Figure 60. We shall now isolate those steps which are necessary for the systematic carrying out of work measurement. These steps and the techniques necessary for achieving them are shown diagrammatically in Figure 61. They are—

SELECT the work to be studied.

RECORD all the relevant data relating to the circumstances in which the work is being done, the methods and the elements of activity in them.

EXAMINE the recorded data and the detailed breakdown critically to ensure that the most effective method and motions are being used and that unproductive and foreign elements are separated from productive elements.

MEASURE the quantity of work involved in each element, in terms of time, using the appropriate work measurement technique.

COMPILE the standard time for the operation, which in the case of stop-watch time study will include time allowances to cover relaxation, personal needs, etc.

DEFINE precisely the series of activities and method of operation for which the time has been compiled and issue the time as standard for the activities and methods specified

It will be necessary to take the full range of steps listed above only if a time is to be published as a standard. When work measurement is being used only as a tool of investigation of ineffective time before or during a method study, or to compare the effectiveness of alternative methods, only the first four steps are likely to be needed.

## 13.6  The techniques of work measurement

The following are the principal techniques by which work measurement is carried out:

  work sampling;
  stop-watch time study;
  predetermined time standards (PTS); and,
  standard date.

In the next few chapters we shall describe each of these techniques in some detail.

## Figure 60. Method study

*(Reproduction and adapted by permission of Imperial Chemical Industries Ltd, London)*

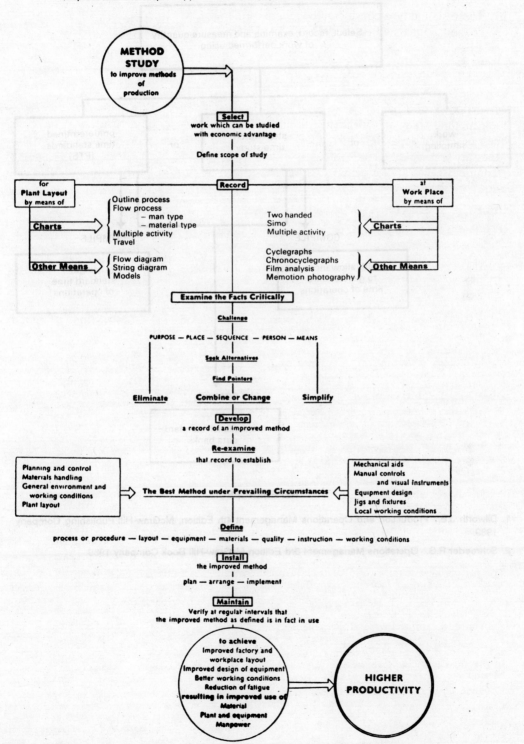

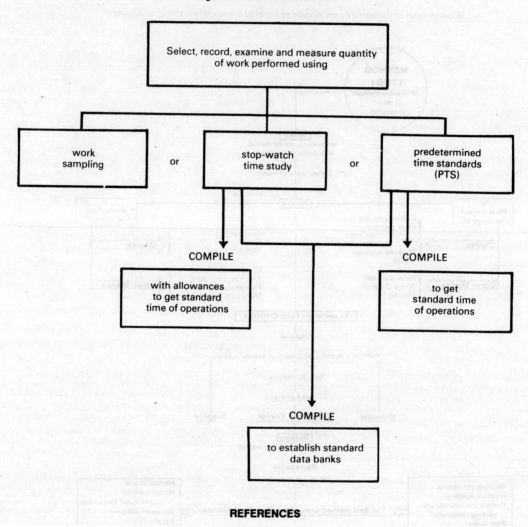

Figure 61. *Work measurement*

**REFERENCES**

1. Dilworth J.B.: Production and Operations Management 4th Edition, McGraw-Hill Publishing Company 1989.

2. Schroeder R.G.: Operations Management 3rd Edition McGraw-Hill Book Company 1989.

# Chapter 14
# Work sampling

## 14.1 Introduction

Before setting the standard time for an activity it is necessary to make allowances for delays. Continuous stop-watch time study is generally used for obtaining complete information about the activities of a man, machine or process equipment particularly delays and interruptions. These studies are made to minimise chance fluctuations on a number of operators, machines or process equipment doing the same or similar work which is a time consuming and costly affair and more so when the work follows long and irregular cycle. To get rid of this difficulty L.H.C. Tippet of British Cotton Industry Research Association in 1934 developed the idea of work sampling (also known as "activity sampling", "ratio-delay study", "random observation method", "snap-reading method", and "observation ratio study") technique which enables the analyst to obtain the information about an activity, machine or process equipment by simply making random observations of the worker, the machine or the process without the use of a stop-watch. Observations are made over a more extended time so that they are more likely to take variations into account and produce fewer complaints from individuals under study or unions than continuous observation.

## 14.2 Basic concepts and definition

Work sampling is a work measurement technique in which a large number of instantaneous observations are made at random intervals over a specified period of time of a group of workers, machines and process. Each observation records the state of the system observed, the percentage of observations recorded for a particular activity or delay over the specified period is a measure (estimate) of the percentage of time during which that activity or delay occurs. This estimate resembles closely to the actual situation if the specified time interval is taken to be very long. Thus, work sampling is defined as—

---

**Work sampling is a method of finding the percentage occurrence of a certain activity by statistical sampling and random observations.**

---

Unlike the costly and tedious method of continuous observations, work sampling is based on the law of probability. Probability has been defined as "the extent to which an event is likely to occur". A simple and oft-mentioned example that illustrates the point is that of tossing a coin. When we toss an ideal coin there are two possibilities; that it will come down 'heads' or that it will come down 'tails'. Suppose we want the probability of the event that the coin will turn up heads. Since, the coin can fall only in two ways, heads or tails, and since the coin is well balanced, we would expect it to fall heads and tails with about equal frequency; hence in the long run would expect it to fall heads about one-half of the time, and so the probability in the event of a head will be given by the value 1/2. The law of probability says that we are likely to have 50 heads and 50 tails in every 100 tosses of the coin. Note that we use the term "likely to have". In fact we might have a score of 55–45 or 48–52, or some other ratio. But it has been proved that the law becomes increasingly accurate as the number of tosses increase. In other words, the greater the number of tosses, the more chance we have of arriving at a ratio of 50 heads to 50 tails. This suggests that the larger the size of the sample, the more accurate or representative it becomes with respect to the original 'population' or group of terms under consideration.

We can therefore visualise a scale; at one end, we can have the complete accuracy achieved by the continuous observation and, at the other end very doubtful results derived from a few sample observations only. The size of the sample is therefore very important, and we can express our confidence in whether or not the sample is representative by using a certain confidence level.

To have an idea about the confidence level let us consider the example of tossing five ideal coins at a time and then record the number of times we have tails per each toss of these coins. Let us then repeat this operation 100 times. The results could be presented as in Table 12, or graphically as in Figure 62.

*Figure 62. Proportional distribution of 'heads' and 'tails'*
*(100 tosses of five coins at a time)*

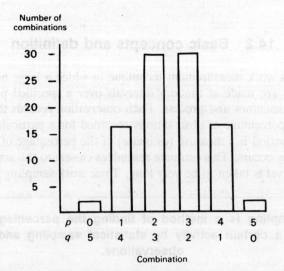

*Table 12. Proportional distribution of 'heads' and 'tails'*
*(100 tosses of five coins at a time).*

| Combination | | No. of combinations |
|---|---|---|
| Heads (p) | Tails (q) | |
| 5 | 0 | 3 |
| 4 | 1 | 17 |
| 3 | 2 | 30 |
| 2 | 3 | 30 |
| 1 | 4 | 17 |
| 0 | 5 | 3 |
| | | 100 |

If we considerably increase the number of tosses and in each case toss a large number of coins at a time, we can obtain a smoother curve, such as that shown in Figure 63.

*Figure 63. Distribution curve showing probabilities of combinations when large samples are used*

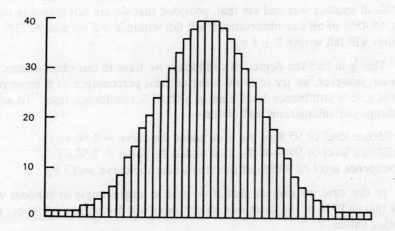

This curve, called the curve of normal distribution, may also be depicted as in Figure 64. Basically, this curve tells us that, in the majority of cases, the tendency is for the number of heads to equal the number of tails in any one series of tosses (when $p = q$ the number of tosses is maximum). In few cases, however, $p$ is markedly different from $q$ due to mere chance.

Curves of normal distribution may be of many shapes. They may be flatter, or more rounded. To describe these curves we use two attributes: $\bar{x}$, which is the average or measure of central dispersion; and $\sigma$, which is the deviation from the average, referred

Figure 64. Curve of normal distribution

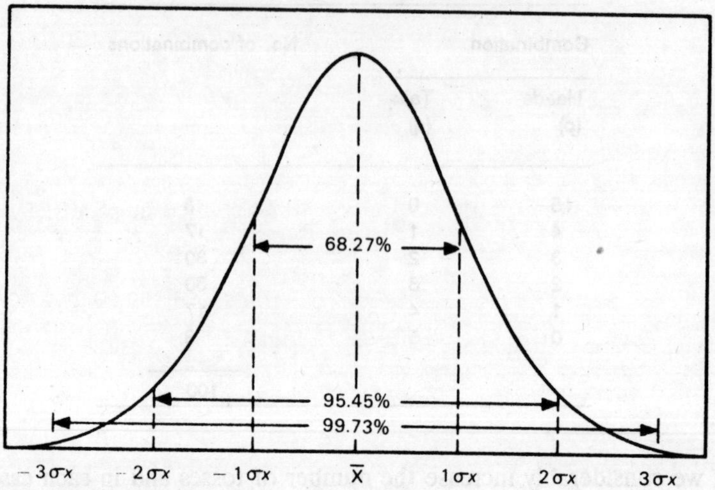

to as standard deviation. Since in this case we are dealing with a proportion, we use $\sigma p$ to denote the standard error of the proportion.

The area under the curve of normal distribution can be calculated. In Figure 68 on $\sigma p$ on both sides of $\bar{x}$ gives an area of 68.27% of the total area; two $\sigma p$ on both sides of $\bar{x}$ gives an area of 95.45% and three $\sigma p$ on both sides of $\bar{x}$ an area of 99.73%. We can put this in another way and say that, provided that we are not biased in our random sampling, 95.45% of all our observations will fall within $\bar{x} \pm 2\, \sigma p$ and 99.73% of all our observations will fall within $\bar{x} \pm 3\, \sigma p$.

This is in fact the degree of confidence we have in our observations. To make things easier, however, we try to avoid using decimal percentages: it is more convenient to speak of a 95% confidence level than of a 95.45% confidence level. To achieve this we can change our calculations and obtain—

95% confidence level or 95% of the area under the curve = 1.96 $\sigma p$
99% confidence level or 99% of the area under the curve = 2.58 $\sigma p$
99.9% confidence level or 99.9% of the area under the curve = 3.3 $\sigma p$

In this case we can say that if we take a large sample at random we can be confident that in 95% of the cases our observations will fall within $\pm 1.96\, \sigma p$, and so on for the other values.

In work sampling the most commonly used level is the 95% confidence level.

## 14.3   Procedure

The work sampling procedure can be divided into the following three phases:

a) Preparing for work sampling.

   i) Statement of the main objective of the study.

ii) Obtain the approval of the supervisor of the department in which work sampling is to be performed.

iii) Establish quantitative measure of activity.

iv) Selection of training of personnel.

v) Making a detail plan for taking observations.

b) Performing work sampling.

i) Describing and classifying the elements to be studied in details.

ii) Design the observation form.

iii) Determine the number of days or shifts required for the study.

iv) Develop properly randomised times of observations.

v) Observing activity and recording data.

vi) Summarising the data at the end of each day.

c) Evaluating and presenting results of work sampling.

i) Evaluate the validity and reliability of data.

ii) Presenting and analysing data.

iii) Planning for future studies.

## 14.4  Determination of sample size

The work sampling technique is similar to the sampling technique used in statistical quality control where conclusion about the proportion defective items 'p' in a large number of items is to be drawn after observing the proportion defective '$p^1$' in a suitably selected sample of size 'n' from the large number of items.

In work sampling the instants of random observations during a specified period of time may be considered to constitute the sample from large time horizon. Then from the ratio '$p^1$' of the number of non-working (say) period observations to total number of random observations during the specified time interval, conclusion about the proportion 'p' of activity (non-working) with respect to total time is to be made.

The question—how far the actual proportion, 'p' may '$p^1$' be found in a sample size 'n' depart can be answered by finding a suitable indicator for measuring the error which will be a measure of sampling variations.

This indicator is the standard deviation of '$p^1$' for sample size 'n' which is also termed as standard error of proportion and is given by

$$\sigma p = \sqrt{\frac{p(1-p)}{n-1}}$$

For large 'n' (as a rule of thumb if $n \geq 30$) the above expression may be approximated without too much error as

$$\sigma p = \sqrt{\frac{p(1-p)}{n}}$$

and normal distribution of the error of proportion may be assumed. When 'p' is unknown, the average value of 'p¹' computed from previous samples may be used in place of 'p' and the following confidence level statements can be made:

i) 68% of the time $p$ will lie within $p^1 \pm \sigma p$.

ii) 95% of the time $p$ will lie within $p^1 \pm 1.96\,\sigma p$.

iii) 99% of the time $p$ will lie within $p^1 \pm 2.58\,\sigma p$.

As well as defining the confidence level for our observations we have to decide on the margin of error that we can allow for these observations so that we must be able to say that: "We are confident that say, for 95% of the time this particular observation is correct within ± 5%" or 10%, or whatever other range of accuracy we may decide on.

Let us now return to our example about the working time and non-working time of the machine in a factory. The formula for determining the number of observations required is given by

$$P.\ S = K\,\sigma p$$

$$= K\sqrt{\frac{p(1-p)}{n}}$$

When, $p$ = percentage occurrence of the activity (working or non-working) being measured in fraction.

$S$ = error (accuracy required) in fraction.

$K$ = a factor, the value of which depends on the desired confidence level. For example, for 95% confidence level, $K = 1.96$.

$n$ = number of observations required for the desired confidence level and margin of error.

In order to find out the value of '$n$', '$p$' is initially assumed or estimated taking some trial observations and then estimate of '$p$' is revised from time to time after regular intervals of observations during the study. Using the estimated value of '$p$', value of '$n$' is determined satisfying the confidence level and accuracy constraint. The sequential process of finding out the value of '$n$' can be explained through the following example:

## EXAMPLE

Consider the determination of the percentage idle time of a turner by work sampling assuming the desired confidence level to be 95% and the error (accuracy required) be ± 5%.

Before using the above formula we have to get a first estimate of '$p$', the percentage idle time measured in fraction. Let a total of 100 observations be made out of which 25 observations show that the turner is idle. We are now ready to calculate the initial value of '$n$' from

$$P.\ S = K\sqrt{\frac{p(1-p)}{n}}$$

for $p = 0.25$, $S = 0.05$, and $R = 1.96$ we get $0.25 \times 0.05 = 1.96 \sqrt{\dfrac{0.25(1 - 0.25)}{n}}$ or, $n = 4610$

After 500 observations are taken the revision value of '$n$' is done as illustrated below:

Number of times the turner is observed to be idle  = 150
Total number of observations  = 500
So, the new estimate of '$p$' would be $\dfrac{150}{500}$  = 0.30

This new information would enable us to recalculate the number of observations needed as follows:

$$0.3 \times 0.05 = 1.96 \sqrt{\frac{0.30(1 - 0.30)}{n}}$$

or, $n = 4774$

It is advisable to recalculate '$n$' at regular intervals. Let us assume that at certain stages when 4000 observations are taken, the result of the study is as follows:

Number of times the turner is observed to be idle  = 1400
Total number of observations  = 4000

So, the new estimate of '$p$' becomes $\dfrac{1400}{4000} = 0.35$, and to get a new value of '$n$' we have $0.35 \times 0.35 = 1.96 \sqrt{\dfrac{0.35(1 - 0.35)}{n}}$ to yield $n = 2854$.

As $n = 2854$ which is less than 4000, the total number of observations already taken, we stop and recalculate $S$, as follows:

$$0.35 \times S = 1.96 \sqrt{\frac{0.35(1 - 0.35)}{4000}}$$

or, $S = 0.0422$ or 4.22%

In this case, the inference could be made that we are 95% confident that the turner is idle between $(35 \pm 0.0422 \times 35)$% i.e. 33.523% to 36.477%.

## 14.5  Procedure for selecting random observations

In designing the work sampling study a time schedule should be prepared for taking observations. Our previous conclusions are valid provided that we can make the number of observations needed to attain the confidence level and accuracy required, and also provided that these observations are made at random i.e. each individual moment has an equal opportunity of being chosen, and also they must be unbiased and independent.

To ensure that our observations are in fact made at random, we can use a random table such as the one in Table 13. Various types of random tables exist, and these can be used in different ways. In our case let us assume that we shall carry out our observations during a day shift of eight hours, from 7 a.m. to 3 p.m. An eight-hour day has 480 minutes. These may be divided into 48 ten-minute periods.

*Table 13.   Table of random numbers*

| | | | |
|---|---|---|---|
| 49 54 43 54 82 | 17 37 93 23 78 | 87 35 20 96 43 | 84 26 34 91 64 |
| 57 24 55 06 88 | 77 04 74 47 67 | 21 76 33 50 25 | 83 92 12 06 76 |
| 16 95 55 67 19 | 98 10 50 71 75 | 12 86 73 58 07 | 44 39 52 38 79 |
| 78 64 56 07 82 | 52 42 07 44 38 | 15 51 00 13 42 | 99 66 02 79 54 |
| 09 47 27 96 54 | 49 17 46 09 62 | 90 52 84 77 27 | 08 02 73 43 28 |
| | | | |
| 44 17 16 58 09 | 79 83 86 19 62 | 06 76 50 03 10 | 55 23 64 05 05 |
| 84 16 07 44 99 | 83 11 46 32 24 | 20 14 85 88 45 | 10 93 72 88 71 |
| 82 97 77 77 81 | 07 45 32 14 08 | 32 98 94 07 72 | 93 85 79 10 75 |
| 50 92 26 ⑪ 97 | 00 56 76 31 38 | 80 22 02 53 53 | 86 60 42 04 53 |
| 83 39 50 08 30 | 42 34 07 96 88 | 54 42 06 87 98 | 35 85 29 48 39 |
| | | | |
| 40 33 20 38 26 | 13 89 51 03 74 | 17 76 37 13 04 | 07 74 21 19 30 |
| 96 83 50 87 75 | 97 12 25 93 47 | 70 33 24 03 54 | 97 77 46 44 80 |
| 88 42 95 45 72 | 16 64 36 16 00 | 04 43 18 66 79 | 94 77 24 21 90 |
| 33 27 14 34 09 | 45 59 34 68 49 | 12 72 07 34 45 | 99 27 72 95 14 |
| 50 27 89 87 19 | 20 15 37 00 49 | 52 85 66 60 44 | 38 68 88 11 80 |
| | | | |
| 55 74 30 77 40 | 44 22 78 84 26 | 04 33 46 09 52 | 68 07 97 06 57 |
| 59 29 97 68 60 | 71 91 38 67 54 | 13 58 18 24 76 | 15 54 55 95 52 |
| 48 55 90 65 72 | 96 57 69 36 10 | 96 46 92 42 45 | 97 60 49 04 91 |
| 66 37 32 20 30 | 77 84 57 03 29 | 10 45 65 04 26 | 11 04 96 67 24 |
| 68 49 69 10 82 | 53 75 91 93 30 | 34 25 20 57 27 | 40 48 73 51 92 |
| | | | |
| 83 62 64 11 12 | 67 19 00 71 74 | 60 47 21 29 68 | 02 02 37 03 31 |
| 06 09 19 74 66 | 02 94 37 34 02 | 76 70 90 30 86 | 38 45 94 30 38 |
| 33 32 51 26 38 | 79 78 45 04 91 | 16 92 53 56 16 | 02 75 50 95 98 |
| 42 38 97 01 50 | 87 75 66 81 41 | 40 01 74 91 62 | 48 51 84 08 32 |
| 96 44 33 49 13 | 34 86 82 53 91 | 00 52 43 48 85 | 27 55 26 89 62 |
| | | | |
| 64 05 71 95 86 | 11 05 65 09 68 | 76 83 20 37 90 | 57 16 00 11 66 |
| 75 73 88 05 90 | 52 27 41 14 86 | 22 98 12 22 08 | 07 52 74 95 80 |
| 33 96 02 75 19 | 07 60 62 93 55 | 59 33 82 43 90 | 49 37 38 44 59 |
| 97 51 40 14 02 | 04 02 33 31 08 | 39 54 16 49 36 | 47 95 93 13 30 |
| 15 06 15 93 20 | 01 90 10 75 06 | 40 78 78 89 62 | 02 67 74 17 33 |
| | | | |
| 22 35 85 15 33 | 92 03 51 59 77 | 59 56 78 06 83 | 52 91 05 70 74 |
| 09 98 42 99 64 | 61 71 62 99 15 | 06 51 29 16 93 | 58 05 77 09 51 |
| 54 87 66 47 54 | 73 32 08 11 12 | 44 95 92 63 16 | 29 56 24 29 48 |
| 58 37 78 80 70 | 42 10 50 67 42 | 32 17 55 85 74 | 94 44 67 16 94 |
| 87 59 36 22 41 | 26 78 63 06 55 | 13 08 27 01 50 | 15 29 39 39 43 |
| | | | |
| 71 41 61 50 72 | 12 41 94 96 26 | 44 95 27 36 99 | 02 96 74 30 83 |
| 23 52 23 33 12 | 96 93 02 18 39 | 07 02 18 36 07 | 25 99 32 70 23 |
| 31 04 49 69 96 | 10 47 48 45 88 | 13 41 43 89 20 | 97 17 14 49 17 |
| 31 99 73 68 68 | 35 81 33 03 76 | 24 30 12 48 60 | 18 99 10 72 34 |
| 94 58 28 41 36 | 45 37 59 03 09 | 90 35 57 29 12 | 82 62 54 65 60 |

We can start by choosing any number at random from our table, for example by closing our eyes and placing a pencil point somewhere on the table. Let us assume that in this case we pick, by mere chance, the number 11 which is in the second block, fourth column, fourth row (see Table 13). We now choose any number between 1 and 10. Assume that we choose the number 2; we now go down the column picking out every second reading and noting it down, as shown below (if we had chosen the number 3, we should pick out every third figure, and so on).

11  38  45  87  68  20  11  26  49  05

Looking at these numbers, we find that we have to discard 87, 68 and 49 because they are too high (since we have only 48 ten-minute periods, any number above 48 has to be discarded). Similarly, the second 11 will also have to be discarded since it is a number that has already been picked out.

We therefore have to continue with our readings to replace the four numbers we have discarded. Using the same method, that is, choosing every second number after the last one (05), we now have

14  15  47  22

These four numbers are within the desired range and have not appeared before. Our final selection may now be arranged numerically and the times of observation throughout the eight-hour day worked out. Thus our smallest number (05) represents the fifth ten-minute period after the work began at 7 a.m. and thus our first observation will be at 7.50 a.m., and so on (see Table 14).

Table 14.   Determining the sequence of time for random observations

| Usable number as selected from the random table | Arranged in numerical order | Time of observation[1] |
|---|---|---|
| 11 | 05 | 7.50 a.m. |
| 38 | 11 | 8.50 a.m. |
| 45 | 14 | 9.20 a.m. |
| 20 | 15 | 9.30 a.m. |
| 26 | 20 | 10.20 a.m. |
| 05 | 22 | 10.40 a.m. |
| 14 | 26 | 11.20 a.m. |
| 15 | 38 | 1.20 a.m. |
| 47 | 45 | 2.30 a.m. |
| 22 | 47 | 2.50 a.m. |

[1]Multiply each number by ten minutes and start from 7 a.m.

Random number tables may also be used to determine the time of the days when observations should be made, to indicate the order in which the workers should be observed, to determine the specific location in the plant where an observation should be taken, etc.

An example is considered in which the flash observations of average duration of half a minute each are to be made on two operators A and B during 8 a.m. to 5 p.m. in a day with a lunch break from 12 noon to 1 p.m. The plan for determination of a flash observation is given in the Table 15 using five digit random number. This can be carried out for the number of observations needed.

**Table 15.**

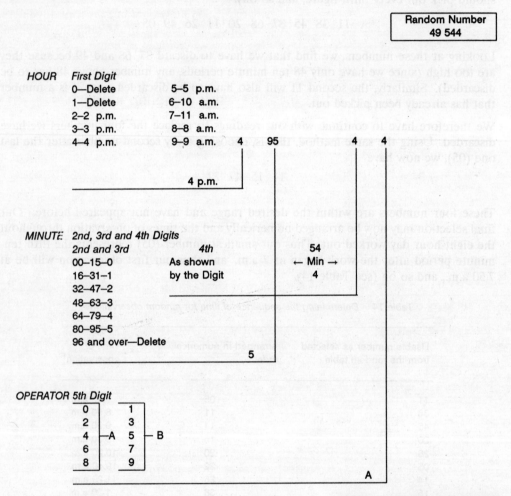

| | Random Number |
|---|---|
| | 49 544 |

**HOUR    First Digit**

| | |
|---|---|
| 0—Delete | 5—5 p.m. |
| 1—Delete | 6—10 a.m. |
| 2—2 p.m. | 7—11 a.m. |
| 3—3 p.m. | 8—8 a.m. |
| 4—4 p.m. | 9—9 a.m. |

4 p.m.

**MINUTE    2nd, 3rd and 4th Digits**

| 2nd and 3rd | 4th |
|---|---|
| 00—15—0 | As shown |
| 16—31—1 | by the Digit |
| 32—47—2 | |
| 48—63—3 | |
| 64—79—4 | |
| 80—95—5 | |
| 96 and over—Delete | |

*OPERATOR 5th Digit*

```
0     1
2     3
4 —A  5 — B
6     7
8     9
```

| HOUR | MINUTE | OPERATOR |
|---|---|---|
| 4 p.m. | 54 | A |

## 14.6   Error in work sampling

Work sampling is subjected to the two kinds of error as follows:

i) *Observational error*: This is due to the presence and behaviour of the observer, the observed and/or the environment in which the observations are made. Further, the study is made on a finite period of time, and the period may not be representative.

ii) *Experimental error*: Due to finite number of random observations on a specified activity, an experimental error is introduced which is a measure of the sampling variations in terms of the standard error of proportion as mentioned earlier.

## 14.7   Conducting the work sampling study

As has been mentioned in section 14.3 of this chapter, it is important in the outset that we decide on the objective of our work sampling. The simplest objective is that of determining whether a given machine is idle or working, our observations then aim at detecting one of two possibilities only:

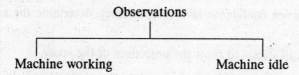

We can, however, extend this simple model to try and find out the cause of the stoppage of the machine:

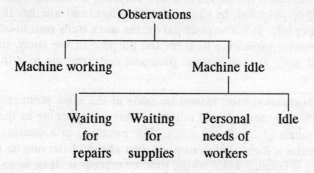

Again, we may be interested in determining the percentage of time spent on each activity while the machine is working:

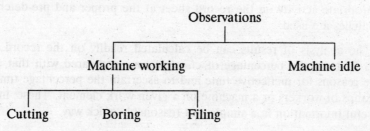

Or perhaps we may wish to get an idea of the percentage distribution of time when the machine is working and when it is idle, in which case we combine the last two models.

We may also be interested in the percentage time spent by a worker or groups of workers on a given element of work. If a certain job consists of ten different elements, by observing a worker at the defined points in time we can record on which element he is working and therefore arrive at a percentage distribution of the time he has been spending on each element.

The objectives to be reached by the study will therefore determine the design of the recording sheet used in work sampling, as can be seen from Figures 65, 66 and 67.

## MAKING THE OBSERVATIONS

So far we have taken the following logical steps in performing work sampling—

1) Selecting the job to be studied and maintaining the scope of the study.

2) Making preliminary observations to determine the approximate value of $p$ for the activity.

3) In terms of chosen confidence level and accuracy, determine the number of observations needed '$n$'.

4) Designing record sheets to meet the objectives of the study.

The next step is making and recording the observations. In making the observations, it is essential from the outset that the work study man is clear in his own mind about what he wants to achieve and why. He should avoid ambiguity when classifying activities. For example, if the engine of a fork-lift truck is running while the truck is waiting to be loaded or unloaded, he should decide beforehand whether this means that the truck is working or idle. It is also essential for the work study man to contact the persons he wishes to observe, explaining to them the purpose of the study, indicating to them that they should work at their normal pace and endeavouring to gain their confidence and co-operation.

The observation itself should be made at the same point relative to each machine. The work study man should not note what is happening at the machines ahead of him, as this tends to falsify the study. For example, in a weaving department, the observer may notice a loom that is stopped, just ahead of the one he is observing. The weaver may have it running again by the time he reaches it. If he were to note it as idle, he would be giving an untrue picture.

The recording itself, as can be seen, consists simply of making a stroke in front of the appropriate activity on the record sheet at the proper and pre-determined time. No stop-watches are used.

The analysis of results can be calculated readily on the record sheet. It is possible to find out the percentage of effective time compared with that of delays, to analyse the reasons for ineffective time and to ascertain the percentage time spent by a worker, groups of workers or a machine on a given work element. These, in themselves, provide useful information in a simple and reasonably quick way.

*Figure 65. Example of a simple work sampling record sheet*

| Date: | Observer: | | Study No : | |
|---|---|---|---|---|
| Number of observations: 75 | | | Total | Percentage |
| Machine running | ЖЖ ЖЖ ЖЖ ЖЖ ЖЖ ЖЖ ЖЖ ЖЖ ЖЖ ЖЖ ЖЖ ЖЖ II | | 62 | 82.7 |
| Machine idle | ЖЖ ЖЖ III | | 13 | 17.3 |

*Figure 66. Work sampling record sheet showing machine utilisation and distribution of idle time*

| Date: | | Observer: | | Study No : | |
|---|---|---|---|---|---|
| Number of observations: 75 | | | | Total | Percentage |
| Machine running | | ЖЖ ЖЖ ЖЖ ЖЖ ЖЖ ЖЖ ЖЖ ЖЖ ЖЖ ЖЖ ЖЖ ЖЖ II | | 62 | 82.7 |
| Machine idle | Repairs | II | | 2 | 2.7 |
| | Supplies | ЖЖ I | | 6 | 8.0 |
| | Personal | I | | 1 | 1.3 |
| | Idle | IIII | | 4 | 5.3 |

*Figure 67. Work sampling record sheet showing distribution of time on ten elements of work performed by a group of four workers*

| Date: | Observer: | | | | | | Study No.: | | | |
|---|---|---|---|---|---|---|---|---|---|---|
| Number of observations: | | | | | | | | | | |
| | Elements of work | | | | | | | | | |
| | 1 | 2 | 3 | 4 | 5 | 6 | 7 | 8 | 9 | 10 |
| Worker No. 1 | | | | | | | | | | |
| Worker No. 2 | | | | | | | | | | |
| Worker No. 3 | | | | | | | | | | |
| Worker No. 4 | | | | | | | | | | |

## 14.8   Uses of work sampling

Work sampling is widely used. It is a relatively simple technique that can be used advantageously in a wide variety of situations, such as manufacturing, servicing and office operations. It is, moreover, a relatively low-cost method and one that is less controversial than stop-watch time study. The information derived from work sampling can be used to compare the efficiency of two departments, to provide for a more equitable distribution of work in a group and, in general, to provide the management with an appreciation of the percentage of and reasons behind ineffective time. As a result, it may indicate where method study needs to be applied, material handling improved or better production planning methods introduced, as may be the case if work sampling shows that a considerable percentage of machine time is spent idle, waiting for supplies to arrive.

Some of the uses of work sampling may be stated as follows:-

i) To aid in determination of time standards and delay allowances.

ii) To aid in the measurement of overall performances.

iii) To determine the nature and extent of cycles and 'peak load' Variations in observable activity.

iv) To study the time utilisation by supervisors and establishing goals for supervision.

v) To aid in job evaluation.

vi) To assist in engineering economy studies.

vii) To aid in manpower planning.

viii) For appraisal of safety performance.

ix) For appraisal of organisational efficiency.

# Chapter 15

# Time study:
# The equipment

## 15.1   Introduction

In a previous chapter (i.e. Chapter 13) the main techniques of work measurement have been stated. In the next few chapters details of the most important of such techniques "time study" will be discussed.

> Time study is defined as a work measurement technique for recording the times and rates of working for the elements of a specified job carried out under specified conditions, and for analysing the data so as to obtain the time necessary for carrying out the job at a defined level of performance.

## 15.2   Time study equipment

The equipment logistic support required to carry out time study activity basically belongs to two groups. The first group represent those which are to be used at site i.e., during the data collection. Such a set of equipment which is essential, are: (i) a stop-watch (ii) a study board, and (iii) time study forms. While the second group of equipment, representing those which are used as supplementary ones, usually not carried to the site always, includes: (i) a small calculator, (ii) a reliable clock with a seconds hand, (iii) set of measuring instruments such as tape, steel rule, micrometer, spring balance and tachometer (to measure rotational speed) and such other items relevant to and useful for the particular type of work being measured.

### THE STOP-WATCH

Usually three types of stop-watches are used for performing time study:

> (i) The flyback type, (ii) The non-flyback type, and (iii) the split hand stop-watch type. However the first two types are used for a large majority of cases.

All the above types may have any one of the following three graduations in them:

i) Recording one minute per revolution with the smallest graduation of 1/5th of a second with a small hand recording 30 minutes.

ii) Recording one minute per revolution with the smallest graduation of 1/100th of a minute, with a small hand recording 30 minutes.

iii) Recording 1/100th of an hour per revolution with the smallest graduation of 1/10,000th of an hour, with a small hand recording 1 hour.

The versatility of the second type of stop-watches can further be extended by incorporating an optional scale, usually in red, providing graduations in seconds and fifths of a second, outside the main scale.

A flyback decimal-minute stop-watch having the smallest graduation representing 1/100th of a minute (i.e., the second type) is the most extensively used one. A typical stop-watch of this type is shown in Figure 68. Such a watch is operated with the help of the slide (1) which when slided anti-clockwise starts the recording while when slided clockwise ceases the recording. A subsequent sliding in the anti-clockwise direction initiates further starting of the measurement in continuance with the previous recording unless the winding knob (2) is used to restore zero setting of both the measuring indices (hands). Such facilities allow recording of time in either 'flyback' or 'cumulative' mode depending on whether or not the zeros are restored before every individual observation.

However, stop-watches of only non-flyback type are also available. Such watches are put into operation by applying pressure on the winding knob (the only knob) and is stopped by applying pressure on the same knob for the second time. The zero setting is obtained pressing the knob for the third time. Such a device would however give 'cumulative' timing.

In split hand type stop-watches there exists two sets of indices (hands) instead of one set as in other types. Pressing a secondary knob makes one of the indices (hands) to come to a still condition while the other continues to move. A second press of the same knob put back the indices (hands) along with the other set. So when a particular reading is being recorded with one of the indices (hands) in still condition, the measurement of subsequent elements goes on with the second index (hands) which continues to move. Such a system because of its versatility facilitates more accurate recording but at the same time becomes more expensive and complex insofar as maintenance is concerned because of the additional features.

The type of stop-watches put to use primarily depends on the experience and expertise of the observer. But for convenience of calculation and subsequent analysis the decimal-minute stop-watches with flyback facilities has gained more acceptance than the other types. Irrespective of the type of stop-watch ultimately chosen or preferred to regular winding, periodic calibration and routine cleaning and overhauling are imperatives to keep such delicate and costly equipment in proper working condition.

Keeping in tune with the development of technology taking place several cost-effective alternative electronic stop-watches are available presently. Even wrist watches do have stop-watch facilities with accuracy of 0.1 seconds. These electronic alternatives

Figure 68. Decimal-minute stop-watch

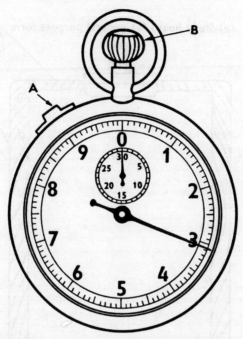

A = Slide for stopping and starting the movement.
B = Winding knob. Pressure on this knob returns both
    the hands to zero.

are lighter, less expensive, relatively insensitive to environmental condition and certainly more reliable and easy to operate.

## TIME STUDY BOARD

A light flat board is usually of plywood or plastic sheet having a size larger than the document i.e., the time study form that it has to hold properly. It should normally have provisions for holding the stop-watch in a convenient orientation and location to facilitate easy operation and clear reading. The time study form is held in position with the help of a spring-loaded clip. A typical Time Study board is depicted in Figure 69.

During the observation for a right-handed observer the board is held against the body of the observer with the upper left arm in such a way that the watch can be freely operated by the thumb and index finger of the same left hand. The right hand of the observer, thus kept free, in such a process can be used to record the data. To enable the use of the same board by a left-handed observer, facility should be there to keep the stop-watch properly located towards the left hand top corner of the board instead of the right hand top corner in case of a right-handed observer.

The overall size of the board should be such that not too much static strain gets developed in the body support systems that is required to keep the board in proper position all throughout the observation phase. To allow maximum comfort to the ob-

**Figure 69. Time study boards**

*(a)* Study board for general purpose form

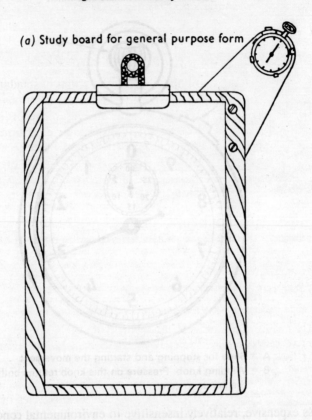

*(b)* Study board for short cycle form

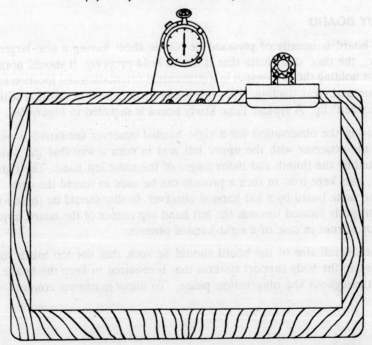

server the overall size of the boards should be allowed to vary marginally to match with the relevant anthropometric parameters of the observer.

## 15.3  Time study forms

These are predesigned printed/cyclostyled/photocopied forms of standard size which allow the observer to record the relevant observations in present locations in the form. The design of the form should be such that it automatically ensures recording of all pertinent data. The standard size also enables subsequent filing of individual forms without affecting the readability of such data even after being filed.

The design of such forms, particularly so far as its layout is concerned is found to vary rather widely.

Two types of forms are used: those used to record the observation with such overall size that they can be fitted on to the time-study board conveniently; and those which are to be used in the time study office after the observations have been recorded in the first type of forms.

### FORMS USED ON THE TIME-STUDY BOARD

Top or Introductory sheet: This sheet contains the provision to record information pertaining to the study like the elements into which the operation under study has been decomposed and the break points that have been used. Besides such information, the recording of the first few cycles of the study can also be done. A typical layout of a general purpose top sheet is given in Figure 70.

Continuation Sheet: The recording of subsequent cycles of study is made in these sheets. The layout of a typical general purpose continuation sheet is given in Figure 71. The reverse side is similar except for the top line of heading which is not there.

Short cycle study form: A typical layout for a most used simple short cycle study form is given in Figure 72.

### FORMS USED IN THE TIME STUDY OFFICE

Working Sheet: This sheet is used for analysing the data obtained during the study to derive representative times for the elemental operations identified. A typical example of such a sheet is given in Figure 92 in Chapter 20. However, since an analysis could be made in different ways with the same set of data, the layout would also require corresponding change to facilitate easy analysis. So to facilitate maximum versatility simple ruled sheets of the same size as that of the observation sheets are used.

Study summary sheet: This sheet is used for recording the selected or derived times for the elemental activities along with their frequency of occurrence. This summarises the information recorded during the course of study. The headings include all the pertinent details recorded in the time study top sheet. The summary is placed on top of the time study sheets providing the data base which are of the same size. A typical layout of such a sheet is given in Figure 73.

*Figure 70. Layout of a typical general purpose TIME STUDY TOP SHEET*

| TIME STUDY TOP SHEET | | | | | | | | | | |
|---|---|---|---|---|---|---|---|---|---|---|

| DEPARTMENT: | | | | STUDY No.: | | | | | |
|---|---|---|---|---|---|---|---|---|---|
| OPERATION: | M.S. No.: | | | SHEET No.:          OF | | | | | |
| | | | | TIME OFF: | | | | | |
| PLANT/MACHINE: | No.: | | | TIME ON: | | | | | |
| TOOLS AND GAUGES: | | | | ELAPSED TIME: | | | | | |
| | | | | OPERATIVE: | | | | | |
| | No.: | | | CLOCK No.: | | | | | |
| PRODUCT/PART: | | | | STUDIED BY: | | | | | |
| DWG. No.: | MATERIAL | | | DATE: | | | | | |
| QUALITY: | | | | CHECKED: | | | | | |

*N.B.* Sketch the WORKPLACE LAYOUT/SET-UP/PART on the reverse, or on a separate sheet and attach

| ELEMENT DESCRIPTION | R. | W.R. | S.T. | B.T. | ELEMENT DESCRIPTION | R. | W.R. | S.T. | B.T. |
|---|---|---|---|---|---|---|---|---|---|
| | | | | | | | | | |
| | | | | | | | | | |
| | | | | | | | | | |
| | | | | | | | | | |
| | | | | | | | | | |
| | | | | | | | | | |
| | | | | | | | | | |
| | | | | | | | | | |
| | | | | | | | | | |
| | | | | | | | | | |
| | | | | | | | | | |
| | | | | | | | | | |
| | | | | | | | | | |
| | | | | | | | | | |
| | | | | | | | | | |
| | | | | | | | | | |
| | | | | | | | | | |
| | | | | | | | | | |
| | | | | | | | | | |
| | | | | | | | | | |
| | | | | | | | | | |
| | | | | | | | | | |
| | | | | | | | | | |
| | | | | | | | | | |
| | | | | | | | | | |
| | | | | | | | | | |
| | | | | | | | | | |
| | | | | | | | | | |

*N.B.*   R. = Rating.   W.R. = Watch Reading.   S.T. = Subtracted Time.   B.T. = Basic Time.

Figure 71. Layout of a typical general purpose continuation sheet

| STUDY No.: | **TIME STUDY CONTINUATION SHEET** | | | | SHEET No. | | OF | |
|---|---|---|---|---|---|---|---|---|
| ELEMENT DESCRIPTION | R. | W.R. | S.T. | B.T. | ELEMENT DESCRIPTION | R. | W.R. | S.T. | B.T. |
| | | | | | | | | | |
| | | | | | | | | | |
| | | | | | | | | | |
| | | | | | | | | | |
| | | | | | | | | | |
| | | | | | | | | | |
| | | | | | | | | | |
| | | | | | | | | | |
| | | | | | | | | | |
| | | | | | | | | | |
| | | | | | | | | | |
| | | | | | | | | | |
| | | | | | | | | | |
| | | | | | | | | | |
| | | | | | | | | | |
| | | | | | | | | | |
| | | | | | | | | | |
| | | | | | | | | | |
| | | | | | | | | | |
| | | | | | | | | | |
| | | | | | | | | | |
| | | | | | | | | | |
| | | | | | | | | | |
| | | | | | | | | | |
| | | | | | | | | | |
| | | | | | | | | | |
| | | | | | | | | | |
| | | | | | | | | | |
| | | | | | | | | | |
| | | | | | | | | | |
| | | | | | | | | | |
| | | | | | | | | | |

*N.B.* Reverse side similar, but without upper line of heading.

*Figure 72. Layout of a typical Short Cycle Time Study Form*

# SHORT CYCLE STUDY FORM

| DEPARTMENT: | SECTION: | STUDY No. SHEET No.: OF: |
|---|---|---|
| OPERATION: | M. S. No.: | TIME OFF: TIME ON: |
| PLANT/MACHINE: | No. | ELAPSED TIME: |
| TOOLS AND GAUGES: | | OPERATIVE: |
| PRODUCT/PART: | No. | CLOCK No.: |
| DWG. No. | MATERIAL: | STUDIED BY: |
| QUALITY: | WORKING CONDITIONS: | DATE: CHECKED: |

N.B.   Sketch the workplace overleaf.

| El. No. | ELEMENT DESCRIPTION | Observed Time | | | | | | | | | | Total O.T. | Average O.T. | R | B.T. |
|---|---|---|---|---|---|---|---|---|---|---|---|---|---|---|---|
| | | 1 | 2 | 3 | 4 | 5 | 6 | 7 | 8 | 9 | 10 | | | | |
| | | | | | | | | | | | | | | | |
| | | | | | | | | | | | | | | | |
| | | | | | | | | | | | | | | | |
| | | | | | | | | | | | | | | | |
| | | | | | | | | | | | | | | | |
| | | | | | | | | | | | | | | | |
| | | | | | | | | | | | | | | | |
| | | | | | | | | | | | | | | | |
| | | | | | | | | | | | | | | | |
| | | | | | | | | | | | | | | | |
| | | | | | | | | | | | | | | | |
| | | | | | | | | | | | | | | | |
| | | | | | | | | | | | | | | | |
| | | | | | | | | | | | | | | | |
| | | | | | | | | | | | | | | | |
| | | | | | | | | | | | | | | | |
| | | | | | | | | | | | | | | | |
| | | | | | | | | | | | | | | | |
| | | | | | | | | | | | | | | | |

*N.B.*   R = Rating.     O.T. = Observed Time.     B.T. = Basic Time.

Figure 73. A typical Time Study summary sheet layout

| Summary Sheet | | |
|---|---|---|
| Department | Section | Study no |
| | | Sheet no of |
| Operation | M.S. No | Date |
| Plant/Machine | No | Time off |
| Facility | | Time on |
| Tools/Gauges | | Time elapsed |
| Aids | | Checked time |
| | | net time |
| Product/Part | No | OBS. Time |
| Service | | UNACC. time |
| | | U.T.A.S. % |
| Drawing No | Material | Studied by |
| Quality | Working conditions: | Checked |
| Operative: | Watch no | |
| Sketch and notes on reverse side | | |
| El.No. | Element description        B.T. | OBS |
| B.T. = Basic Times      F = Frequency of occurence | | OBS = No of observation |

Study analysis sheet: This sheet is used for recording the results obtained through analysis of the summarised data. The analysis sheet records the results of *all* the studies made on any particular operation, no matter when they were made or by whom. The basic times for the elements of any operation are finally compiled from this sheet. A typical layout is given in Figure 74 while Figure 94 in Chapter 20 represents another form.

Relaxation Allowance sheet: often used for compiling of Relaxation allowances.

Other Equipment: Stop-watches give reasonably accurate recording of times for most of the real life operations that are expected to be undertaken by a professional during the first few years of his career.

However, the Motion picture camera running at constant speed can produce a photographic record of the operation at that particular speed. The series of photographs when projected at the same speed enable one to derive the times taken by different elements constituting the operation. The time taken is observed by either counting the number of picture frames through which that particular element proceeds or by actually observing the time during which the particular element remains on the screen. The particular advantage of such devices are basically three; the first being that the time study professional does not have to confront the operator being observed directly, secondly the derivation of time taken by the elements can be made in a much better working environment at least insofar as the time study professional is concerned while lastly the film can be stored as a documentary evidence.

Time study machines produce marks, on a paper tape running at constant speed, by means of pressure applied by fingers on the keys. Such a facility allows the professional to observe the operation continuously instead of simultaneous observation and recording as in stop-watch based system.

Figure 74. A typical layout of study sheet

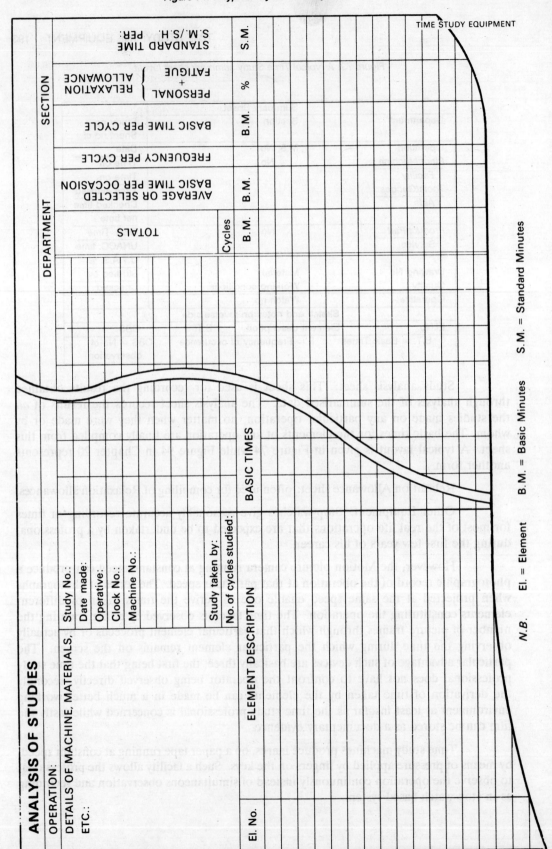

N.B.  El. = Element    B.M. = Basic Minutes    S.M. = Standard Minutes

Besides the sets of equipment and document already discussed, the time study man should preferably have a reliable wrist watch to keep a record of the time taken for each study. Sets of simple standard measuring tools are also useful to generate the data regarding the dimensions of the job and measurement of the workplace.

# Chapter 16

# Time study: Selecting and timing the job

## 16.1  Introduction

The types and nature of different equipment and documentation which are commonly used have been discussed in the previous chapter. The present chapter will deal with two other basic components of the system while the job is being performed by a particular worker in the context of the considerations to be made in identifying them. Since the primary object is to set performance standards for a particular job the selection of the job and worker both assume significant importance. Any job in its turn is composed of several discrete components which can be identified and measured as distinct identities. The method of breakdown and identification of elements will also be discussed. Similarly the number of observations to be obtained so that a statistically reliable estimate can be arrived at is another important aspect of discussion made in this chapter.

## 16.2  Selecting the job

Selection of the job is the first step in time study as in the use of method study. Almost always the time study man goes to a shop or a department after having selected the job. Some of the possible criteria for any job being selected *a priori* are as follows:

a) Recent introduction.

b) Change in material and method.

c) Counter-claims put by the workers or their representative informs of the time standard.

d) Appearance of a 'bottleneck' in a particular operation holding up subsequent operations and even previous operations (through accumulation of work in progress in excess of planned capacity).

e) Need for preparation of an Incentive scheme for the job.

f) Any equipment involved in the production process of the job for having excessive idle time or a low output.

g) For undertaking a method study of the job or comparing alternative process options in terms of time involved.

h) Excessive cost associated with the job.

Any one or more of the above-listed criteria make(s) one job eligible for selection as a job to be studied. However, before finally selecting any job, it must be ascertained that the best method of doing the job has already been identified (defined, standardised) and is being applied while carrying out the job. In cases where the best method has not been identified it becomes imperative to specify one time for each probable best method and never any attempt should be made to introduce any single time estimate to be valid irrespective of the method. However, there are other problems which have nothing to do with the importance of the jobs or the abilities of the worker. One such possible reason exists in situations where a rather liberal work incentive system for the particular job is already in existence allowing high incentives being received by the workers. Attempts to alter the method, likely affecting the already set time standards resulting in the reassessment of the incentive schemes, is expected to meet with the resistance. To accept such a situation, relatively less important jobs, which when improved by application of time study increase the earnings of the workers, should be taken at first, thereby ensuring an acceptance of the analyst and also the technique in general. Subsequently, the previously mentioned problem jobs could be taken up after undertaking necessary negotiating and conciliatory steps which may even include the offer of compensation, in some appropriate cases, to the workers whose earnings will be adversely affected.

## 16.3   Selecting the worker

The selection of the worker to be studied is also very important. After the job being done in the best method has been identified, the next step is obviously to select the particular worker to be studied while doing the job. Since the primary objective is to derive a time standard for the job under study and the capability of worker affects the time taken by him to do the job in the method indicated, an improper selection of worker would lead to derivation of improper time standards.

If a choice of workers is feasible, the suggestion of the foreman, supervisor and workers representation in the matter of identifying the most suitable worker, must be considered first. The rate of the worker so chosen should preferably be either average or little above average. However, there are also certain persons who despite being average or even later being more than average are not temperamentally suited to being studied and should be avoided at all cost. It is, however, always advisable in situations where a large number of workers are individually and independently involved in doing the job, already identified, to select a number of qualified workers instead of only one.

At this point, the point of distinction between a 'representative' and a 'qualified' worker needs to be made out explicitly in the context of time study practice. A representative worker is one who has skill and performance level equalling to the respective averages of the group of workers performing the same job. Such a worker need not be a qualified worker.

A qualified worker on the other hand is one who is accepted as having the necessary physical attributes, who possesses the required intelligence and education, who has acquired the necessary skill and knowledge to carry out the work in hand to satisfactory standards of safety, quantity and quality.

Selection of such qualified workers, for being studied, ensures the setting of time standards which are attainable without any undue fatigue being caused. The very fact that workers work at different speeds, observed times are required to be adjusted by factors to give such a standard. These factors are dependant on the judgment of the time study man, the accuracy of which can only be ensured if the observed times are within a fairly narrow range of speeds, close to that of the qualified workers.

Unduly large and hence uneconomic, or, unduly short and hence unfair to a worker and even unattainable by the worker, time standards may be obtained if slow unskilled or exceptionally fast workers are observed. Such a standard is most likely to generate future complaints at later dates.

After the worker to be studied has been selected, he along with the foreman and workers representative should be approached. They should be explained carefully and in as much details as possible the purpose and process of the study. The worker should be asked to work at the usual pace, taking the normal customary rests and also to point out and explain any difficulties encountered. However, in situations where the practice of Work Study is firmly established beyond doubt, the set of procedures made out in this paragraph should be bypassed. But in no case the immediate supervisor should be present nor should he be directly observing the worker under study.

In situations where a new method has been installed, the learning curve phenomenon must be given due consideration so that the workers get adapted to it by working on the method for a sufficiently long duration and reaches the maximum steady speed. Similarly, a new worker should be selected for being observed, even while working with relatively well-established methods, until he has grown thoroughly accustomed with the job.

The location of the time study man relative to the worker being observed is also a very ticklish issue since to perform his functions efficiently the work study man should be able to see everything the worker does including the movements of his hands without interfering with the movement of the worker as well as not exerting any undue psychological burden on the worker like the feeling of "having somebody standing over him". Though the best relative location depends on the nature of the job being studied, yet as a general rule a relative location on one side of the worker, slightly towards the rear and about 2 metres away is recommended. In such a position the time study man becomes visible to the worker when the worker tilts his head slightly and any conversation between them becomes less effort consuming and easily audible. The time study board and the watch should be held well up in line with the job, to make reading the watch and residing easy while maintaining continuous observation.

On no account under any pretext no attempt should ever be made to time the worker without his knowledge, from a concealed position with the watch in pocket. It is dishonest and in any case, some are sure to see and the news will spread like wildfire. Work study should have nothing to hide.

It is also imperative that the work study man should be in a comfortable standing posture, which he can maintain over a sufficiently long period. Otherwise the work study man is likely to lose the respect and in turn the confidence of the worker who is more inclined to believe that he has to do all the work while the work study man has just to stand around and watch him. The intense concentration and alertness called for from the work study man, specially while timing very short elements or cycles can best be ensured by the standing posture.

Most workers get quickly settled down to the normal work pace but nervous workers and women have a tendency to work fast and commit mistakes. In case if it is found that the workers are fumbling too much then the work study man should take appropriate action to relieve the tension in the worker.

The situation in which the work study man has to confront young clever workers who are inclined to get the better of the work study man by working at unnaturally slow speed and or by intermingling unnecessary movements, a looser standard is set. This tendency is seen to increase if it is known that the observations being made will be used for preparing incentive schemes.

For repetitive work the identification of workers, who intentionally resort to working at slower pace, can be made by observing the variation in timing of the cycles with the machine, material and method remaining unchanged. Since if the worker follows his own natural pace the variation of cycle timings are bound to be rather low, whereas for situations wherein slower pace is intentionally maintained, the variation of cycle time is seen to be wider.

The work study man with the help of his personal qualities should try to tackle the situation in the best possible way so that there is no loss of his popularity.

In case of jobs involving craft skill it may not be worthwhile for the work study man to stretch the time of a job unless he himself is an expert in the process. This problem can also be bypassed by establishing precisely the method and conditions of an operation before attempts are even initiated to time the same.

Some of the practical problems faced by the work study man in real life have been highlighted as also certain recommended ways of tackling them or bypassing them. However, the learnings of the problems and their solution or avoidance cannot be complete without gaining first-hand experience in the field.

## 16.4    Basic steps in time study

The following eight steps constitute the time study process excluding the selections of the job for the worker which have to be done before the steps in the list are taken up:

a) Obtaining and recording all the available information about the job, operator and the surrounding conditions likely to affect the execution of the work.

b) Recording the complete description of the method, breaking down the operation into 'elements'.

c) Examining the detailed breakdown to ensure the most effective method and motions are being used, and determining sample size.

d) Measuring with a time device (most commonly a stop-watch) and recording the time taken by the operator to perform each 'element' of the operation.

e) At the same time, assessing the effective speed of working the operator relative to the observer's concept of the rate corresponding to standard rating.

f) Extending observed time to "basic times".

g) Determining the allowances to be made over and above the basic time for the operation.

h) Determining the "standard time" for the operation.

## 16.5   Recording the information

Correctly designed time study top sheet headings ensures that no vital information gets overlooked. The exact number of items listed below which may have to be included when a time study form is designed depends on the type of work carried out in the organisation in which it has to be used. As for example it is not necessary to provide any space for 'product', its non-manufacturing industries like transport, catering etc.

The availability of instant print type camera with flash attachment and automatic exposure control has enabled a fast and accurate recording of the workplace details.

The filling-in of all the relevant information from direct observations is important in case the time study has to be referred later; incomplete information may make a study practically useless a few months after it has been made. The following groups of information are often required to be recorded:

a) Information to enable the study to be found and identified quickly when needed i.e.
(i) Study number, (ii) Sheet number and number of sheets, (iii) Name and initials of the work study man making the study, (iv) Date of study, (v) Name of the person approving the Study (head of the work study department, production management or other appropriate executive).

b) Information to enable the product or past being processed to be identified accurately, i.e.,
(i) Name of product or past, (ii) Drawing or specification number, (iii) Past number, (iv) Material, (v) Quality requirements.

c) Information to enable the process, method, plant or machine to be accurately identified i.e.
(i) Department or location where the operation is taking place, (ii) Description of the operation or activity, (iii) Method study sheet numbers, (iv) Plant or machine, (v) Tools, jigs, fixtures and gauges used, (vi) Sketch of the workplace layout, machine set up and/or part showing surfaces worked recorded, (the reverse side of the time study top sheet or on a separate sheet attached), (vii) Machine speeds and feeds and other setting information governing the rate of production of the machine or process.

d) Information to enable identification of the workers, i.e.
(i) Name, (ii) Clock Number.

e) Duration of the study i.e.
(i) The start of the study ("Time on"), (ii) The finish of the study ("Time off"), (iii) Elapsed time.

f) Working conditions i.e.
   (i) Temperature, (ii) Humidity, (iii) Adequacy of lighting.

## 16.6    Classification of elements

The elements found in general can be classified in the following eight broad groups:

i) A repetitive element—an element which occurs in every work cycle of the job like the element of picking up a past prior to assembly operation, the element of locating a workpiece in a holding divide; and putting aside a finished component for assembly.

ii) An occasional element—an element which does not occur in every work cycle of the job but which may occur at regular or irregular intervals, line adjusting the tension or machine setting; receiving instruction from a foreman. Such an element is useful work and part of the job. It is incorporated in the final standard time for the job.

iii) A constant element—an element for which the basic time remains constant whenever it is performed, like switching on a machine; gauging diameter; screwing and tightening a nut; inserting a particular cutting tool into the machine.

iv) A variable element—an element for which the basic time varies in relation to some characteristics of the product, equipment or process, e.g. dimensions, weight and quality like screwing legs with a handsaw (time varies with softness and diameter); sweeping the floor (varies with area); or, transferring material from one shop to the other (varies with distance).

v) A manual element—an element performed by the worker.

vi) A machine element—an element automatically performed by a power driven machine (or process) like forming glass bottles; machining with automatic feed etc.

vii) A governing element—an element which occupies a longer time than that of any other element which is performed concurrently on the same job like turning to certain diameter in a lathe while gauging/measuring from time to time; matching jobs while removing the chips from the tool tip etc.

viii) A foreign element—an element, observed during a study, which after analysis is not found to be an essential part of the job like degreasing a part which is to be machined and removed subsequently etc.

Examples of element description and of various types of element are given in Figures 95 and 97.

## 16.7    Breaking the job into elements

After recording of the previously mentioned information relating to the job and worker is over and the work study man is ensured that the best method is being adopted, then as the next step the job should be broken down into elements.

An element being defined as a distinct part of a specified job selected for convenience of observation, measurement and analysis.

A detailed systematic breakdown of job into elements is necessary—

a) To ensure that productive work (or effective time) is separated from unproductive activity (or ineffective time).

b) To permit the rate of working to be assessed more accurately than would be possible if the assessment was made over a complete cycle. The worker may not work at the same pace throughout the cycle and may find to perform some elements quicker than others.

c) To enable the different types of element to be identified and distinguished, so that each may be accorded the treatment appropriate to its type.

d) To enable elements involving a high degree of fatigue to be isolated and to make the allocation of fatigue allowances more accurate.

e) To facilitate checking the method so that the subsequent omission or insertion of elements may be detected quickly. This may become necessary if at a future date the time standard for the job is queried.

f) To enable a detailed work specification to be produced.

g) To enable time values for frequently recurring elements, such as the operation of machine controls or loading and unloading workpieces from fixtures to be extracted and used in the compilation of standard data.

The set of general rules concerning the way in which a job should be broken down into elements are as follows:

a) Elements should be easily identifiable having a definite beginning and ending ensuring their subsequent repeated recognition. These beginnings and ends can often be recognised by a sound or by a change of direction of the hand or arm. They are known as "break points" and should be clearly described on a study sheet. A break point thus is the instance when one element in a work cycle and another begins excepting in situations where overlapping elements are involved.

b) Elements should be as short as possible so that they can be measured reliably by a trained observer. Though a difference of opinion still exists on the smallest practical unit that can be timed with a stop-watch, yet it is generally accepted that elements having a duration equal to or more than about 0.04 minutes (2.4 seconds) can be reliably measured. However, the lowest limit is recommended to be raised to 0.07 to 0.10 minutes for trained observers. Very short elements (as described), should be next to longer elements, if possible, to facilitate accurate timing and reading. Long manual elements should be rated about every 0.33 minutes (20 seconds).

c) As far as practicable, elements—particularly manual ones—should be chosen so that they represent naturally unified and recognisably distinct segments of the operation. An analysis of the action of reaching for a wrench, moving it to the work and positioning it to tighten a nut, leads to the identification of the actions reaching, grasping, moving to the workpiece, shifting the wrench in the hand to the position giving the best grip for turning it, and positioning it. The worker will probably perform all these actions as one natural set of motions rather than as a series of independent acts. It is better to treat the group as a whole, defining the element as "get wrench" or "get and position wrench" and to time the whole set of motions which make up the group, than to select a break point at say, the instant the fingers first touch the wrench, which would result in the natural group of motions being divided between two elements.

d) Manual elements should be separated from the machine elements. Theoretically calculated times of machine elements like automatic feed incorporated machining at fixed speeds can be used as a check on stop-watch data. The manual time elements are normally completely within the control of the worker.

e) Constant elements should be separated from variable elements.

f) Occasional and foreign elements which do not occur in every cycle should be timed separately.

The type of manufacturing, the nature of operation and end use of work study result, dictates the level of details to which any operation has to be broken down, for instance—

1) Small batch jobs which occur infrequently require less detailed element descriptions than long running high-output lines.

2) Movement from place to place generally requires less description than hand and arm movements.

## 16.8   Determination of sample size

The sample size of observations i.e., the number of observations required for each element, is taken, so that a predetermined confidence level and accuracy margin are achieved. This can be done by applying either a statistical method or a conventional method. While applying the statistical method, a number of preliminary readings ($n'$) are first taken. Subsequently, the number of observations (i.e. the sample size) required to reach the set confidence and accuracy level can be calculated.

For example for a 95% confidence level and an accuracy level of $\pm 10\%$ the following formula is used:

$$n = \left[\frac{20\sqrt{n'\Sigma x^2 - (\Sigma x)^2}}{\Sigma x}\right]^2$$

where, $n'$ is the determined sample size;

$\Sigma x$   is the sum of the preliminary set of individual observations;

$\Sigma x^2$ is the sum of the squares of the individual observations of the preliminary set of readings.

Let the following example be considered:

At the preliminary level 5 observations are made with the individual observations being 7.6.7.7. and 6 units of time (1 unit of time = 0.01 minute).

$$n' = 5, \Sigma x = 7 + 6 + 7 + 7 + 6 = 33$$
$$\Sigma x^2 = 49 + 36 + 49 + 49 + 36 = 219$$
$$n = \left[\frac{20\sqrt{5 \times 219 - 33^2}}{33}\right]^2 = 2.4$$

Hence, a sample size of 3 observations is enough to attain the confidence and accuracy level. If however, the accuracy level is increased to ± 2.5% then 'n' becomes

$$n = \left( \frac{80\sqrt{5 \times 219 - 33^2}}{33} \right)^2 = 35.28$$

i.e., at least 36 observations are required. It is a fact to be taken note of that a fresh set of 36 observations should be made to arrive at the average value of the time estimate corresponding to a 95% confidence level and an accuracy level of ± 2.5%.

The statistical method however poses some practical difficulties in its application acceptance; like

i) The method is cumbersome.

ii) It is difficult to determine the best sample size for an operation, since the operation usually consists of several elements each of which may be required to be observed for different number of times for any set of confidence and accuracy level.

iii) Ensuring the holding of the assumptions inherent to the statistical method formulation.

These limitations have led to a conventional guide of a number of cycles to be timed based on the total number of minutes per cycle (Table 15a).

In conducting this study the Table of random numbers may be used to determine the times at which the readings are to be taken to ensure the statistical independence.

Table 15a. Conventional standard of number of recommended cycles for time study

| Minutes per cycle | To 0.10 | To 0.25 | To 0.50 | To 0.75 | To 1.0 | To 2.0 | To 5.0 | To 10.0 | To 20.0 | To 40.0 | To 40.0 |
|---|---|---|---|---|---|---|---|---|---|---|---|
| Number of cycles to be observed | 200 | 100 | 60 | 40 | 30 | 20 | 15 | 10 | 8 | 5 | 3 |

Sources: A.E. Shaw: "Stop-watch time study, in HB Manyasd (ed):
          Industrial Engineering Handbook, op.

## 16.9   Timing elements by stop-watch

After the elements to be timed have been identified, selected and listed as well as the number of times each of such elements to be timed is ascertained to ensure certain confidence and accuracy level; timing can start.

There are two principal methods of timing with the stop-watch.

(a) Cumulative timing. (b) Flyback timing.

In cumulative method the watch runs continuously throughout the study. It is started at the beginning of the first element of the first cycle to be timed and is stopped only after the study is completed. The purpose of this procedure is to ensure that all the time during which the job is observed is recorded in the study.

In flyback method the stop-watch is reset to zero reading, by returning the hands of the watch to zero, at the end of each element and the hands of the watch are allowed to start immediately at the beginning of the next element, the time for each element being observed directly. The mechanism of the water is never stopped.

In all time studies it is usual to make an independent check of the overall time of the study, using either a stop-watch or a clock in the study office. This also enables in performing retiming, if called for, exactly at the same time which eliminates the effect of location of the observation time in the working hour.

In case of flyback timing, the study man reaches the clock at an exact minute, preferably at the next major division such as the hour or one of the five minute points, and sets his stop-watch running, noting the exact time in the "time on" space. He reaches the location where the study is to be made with the water running and allows it to do so till he is ready to start timing. At the beginning of the first element of the first work cycle, as the hands are snapped back there is nothing in the first entry to show for the time that has elapsed. At the end of the study, the hand is snapped back to zero on completion of the last element of the last cycle and thereafter allowed to run continuously until he can again reach the clock and note the time of finishing when the watch is finally stopped. The final clock time is entered in the "time off" space on the form. The two times recorded before and after the study are known as "check times". The clock reading at the beginning of the study is subtracked from the clock reading at the end of the study yielding the elapsed time, to be entered in its appropriate location.

The recorded time is obtained as the aggregate of time of all the elements, i.e., other activities noted in the study and ineffective time and check time are also noted. This aggregate should ideally equal the elapsed time but in practice is found to be different from the elapsed time. The difference may be attributed to the cumulative loss of very small fractions of time at the return of the hand to zero and to bad reading of missed elements. The difference observed in case of cumulative (non-flyback) timing is less since there is no loss due to snapback effects.

Cumulative timing has the significant advantage that even in the event of missing an element or non-recording of some occasional element it does not have any effect on the overall time. However, cumulative timing calls for spending of more time in determining individual element timing which can be only obtained after performing a subtraction operation.

In the case of a work study man who has a fairly high level of experience and skill both the methods almost result accuracy at the same level. However, the ILO missions experiences in training in applying time study indicates that in general cumulative timing should be taught and used. Such an indication is based on the following reasons:

a) Reasonable accuracy level can be attained by the trainees more quickly if they use cumulative timing.

b) The overall time of the study does not get affected even if element times are occasionally missed by inexperienced observers.

c) Automatic inclusion of foreign elements and interruptions.

d) In assessing the working pace of the operator ('rating'), it is less easy to fall into the temptation to adjust the rating to the time taken by the element than with the flyback method, since watch readings and not actual times are recorded.

e) Workers and their representatives are likely to have greater faith in the fairness of time studies as a basis of incentive plans if they can see that no time could have been omitted.

Differential timing, a third type of timing, can provide possibly the only way of getting accurate times with a stop-watch for elements which are so very short that there is not enough time for the work study man to read his watch and make a record on the Study sheet. In differential timing, observations are made of a group of activities once including the particular element and then excluding the same element. For example, in an operation having five elements (1, 2, 3, 4 and 5) to be timed. In the first few cycles timings are observed for the group (1, 2 and 3) and (4 and 5) then subsequently a few more cycles are timed for the group (1, 2) and 4 and thereby marking it possible to determine the timing for the elements 3 and 5 individually even if they are so very short that they can be timed individually by a stop-watch by either cumulative timing or flyback timing technique.

In this chapter all the preliminaries for making a time study have been discussed. The next chapter will discuss the means of modifying these observed times to take into account the variations of work pace.

# Chapter 17

# Time study: Rating—Determination of basic time from observed time

## 17.1  Introduction

In section 4 of the previous chapter, the making of a time study was broken down into eight steps or stages, the first four of which were discussed in that chapter. We now come to the fifth step, namely "assessing the effective speed of working of the operative relative to the observer's concept of the rate corresponding to standard rating."

The treatment of rating which follows has been selected because experience in the use of this book for training purposes by ILO management and productivity missions suggests that this approach to the subject is best suited to the conditions in most of the countries for which the book is primarily intended.

Rating and 'allowances' (dealt with in the next chapter) are the two most controversial aspects of time study. Most time studies in industry are used to determine standard times for setting workloads and as a basis for incentive plans. The procedures employed have a bearing on the earnings of the workers as well as on the productivity and, possibly, the profits of the enterprise. Time study is not an exact science, although much research has been and continues to be undertaken to attempt to establish a scientific basis for it. Rating (the assessment of a worker's rate of working) and the allowances to be given for recovery from fatigue and other purposes are still, however, largely matters of judgment and therefore of bargaining between management and labour.

Various methods of assessing the rate of working, each of which has its good and bad points, have been developed. The procedures set out in this chapter represent sound current practice and, properly applied, should be acceptable to management and workers alike, particularly when used to determine standards for medium-batch production, which is the most common type in industry all over the world outside the United States and a few large or specialised undertakings elsewhere. They will certainly provide the reader with a sound basic system which will serve him well for most general applications, and one which can later be refined if the particular nature of certain special operations requires a modification of the system, so as to rate something other than effective speed.

## 17.2   The qualified worker

It has already been said that time studies should be made, as far as possible, on a number of qualified workers; and that very fast or very slow workers should be avoided, at least while making the first few studies of an operation. What is a "qualified worker"?

Different jobs require different human abilities. For example, some demand mental alertness, concentration, visual acuity; others, physical strength; most, some acquired skill or special knowledge. Not all workers will have the abilities required to perform a particular job, though if the management makes use of sound selection procedures and job training programmes, it should normally be possible to arrange that most of the workers engaged on it have the attributes needed to fit them for the task. The definition of a qualified worker given in the previous chapter is repeated here—

> **A qualified worker is one who is accepted as having the necessary physical attributes, who possesses the required intelligence and education, and who has acquired the necessary skill and knowledge to carry out the work in hand to satisfactory standards of safety, quantity and quality.**

The acquisition of skill is a complicated process. It has been observed [1] that among the attributes which differentiate the experienced worker from the inexperienced are the following. The experienced worker—

a)  achieves smooth and consistent movements;

b)  acquires rhythm;

c)  responds more rapidly to signals;

d)  anticipates difficulties and is more ready to overcome them; and,

e)  carries out the task without giving the appearance of conscious attention, and is therefore more relaxed.

It may take a good deal of time for a worker to become fully skilled in the performance of a job. In one study (see Figure 59) it was noted that it was only after some 8,000 cycles of practice that the times taken by workers began to approach a constant figure—which was itself half the time they took when they first essayed the operation. Thus time standards set on the basis of the rate of working of inexperienced workers could turn out to be quite badly wrong, if the job is one with a long learning period. Some jobs, of course, can be learned very quickly.

It would be ideal if the time study man could be sure that, whatever job he selected for study, he would find only properly qualified workers performing it. In practice this is too much to hope for. It may indeed be that none of the workers engaged on the task can really be said to be completely qualified to carry it out, though it may be possible to alter this in time, by training; or that, though some of the workers are qualified, these are so few in number that they cannot be considered to be average or representative of the group. A representative worker is defined as one whose skill and performance is the average of a group under consideration. He is not necessarily a qualified worker.

If the working group is made up wholly or mainly of qualified workers, there will be one—or perhaps several—of these qualified workers who can be considered as representative workers also. The concept of a standard time is, at root, that it is a time for a job or operation that should normally be attainable by the average qualified worker, working in his ordinary fashion, provided that he is sufficiently motivated to want to get on with the job. In theory, therefore, the time study man should be looking for the average qualified worker to study. In practice, this is not as easy as it might seem. It is worth looking more closely into what 'average' might mean in this context.

## 17.3   The 'average' worker

The truly average worker is no more than an idea. A completely average worker does not exist, any more than an "average family" or an "average man" exists. They are the inventions of statisticians. We are all individuals: no two of us are exactly alike. Nevertheless, among a large number of people from, for instance, the same country or area, variations in measurable characteristics such as height and weight tend to form a pattern which, when represented graphically, is called the "normal distribution curve". To take one characteristic, height: in many western European countries the average height for a man is about 5 ft. 8 in. (172 cm). If a crowd is a western European crowd, a large number of the men in it will be between 5 ft. 7 in. and 5 ft. 9 in. tall (170 to 175 cm). The number of men of heights greater or smaller than this will become fewer and fewer as those heights approach the extremes of tallness and shortness.

The case as regards the performance of operatives is exactly the same. This can be shown very conveniently in a diagram (Figure 75). If 500 qualified workers in a given factory were to do the same operation by the same methods and under the same conditions, the whole operation being within the control of the worker himself, the times taken to perform the operation would be distributed in the manner shown in the figure. To simplify the figure, the times have been divided into groups at intervals of four seconds. It will be seen that the workers fall into the groups shown in Table 16.

If the time groups are examined, it will be seen that 32.4% of the times are less than 46 seconds and 34.8% of the times are greater than 50 seconds. The largest single group of times (32.8%) lies between 46 and 50 seconds. We should therefore be justified in saying that for this group of 500 workers the average time taken to perform this operation was between 46 and 50 seconds (say, 48 seconds). We could call 48 seconds the time taken by the average qualified worker to do this job under these conditions. The time might not hold good for any other factory. Factories which are well run, where working conditions and pay are good, tend to attract and keep the best workers, so that in a better run factory the average worker's time might be less (say, 44 seconds), while in a poorly run factory with less able workers it might be more (say, 52 seconds).

If a curve is drawn to fit this distribution it will be found to assume the shape of the curve in the figure. This is known as the "normal distribution curve". In general, the larger the sample the more the curve will tend to be symmetrical about the peak value, but this can be altered if special conditions are introduced. For example, if the slower workers were to be transferred to other work, the right-hand side of the curve

*Figure 75. Distribution of times taken by workers to perform a given job.*

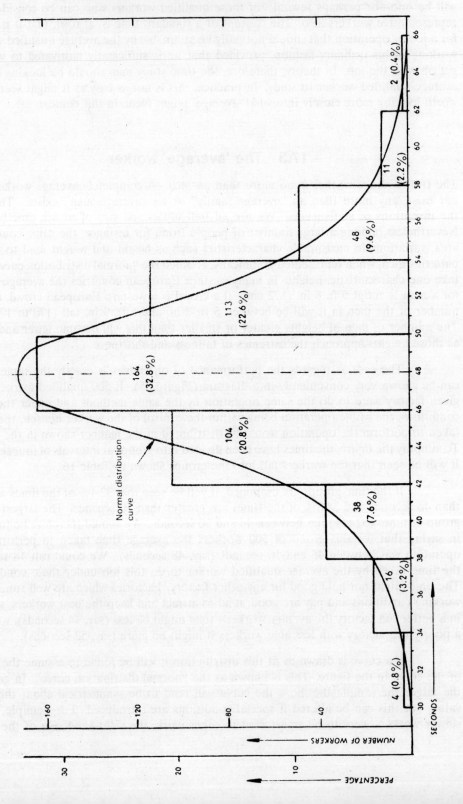

Table 16. Specimen performance distribution.

| Time group (sec) | Number of workers (out of 500) | Percentage of total workers | |
|---|---|---|---|
| 30–34 | 4 | 0.8 | |
| 34–38 | 16 | 3.2 | 32.4 |
| 38–42 | 38 | 7.6 | |
| 42–46 | 104 | 20.8 | |
| 46–50 | 164 | 32.8 | 32.8 |
| 50–54 | 113 | 22.6 | |
| 54–58 | 48 | 9.6 | 34.8 |
| 58–62 | 11 | 2.2 | |
| 62–66 | 2 | 0.4 | |
| | 500 | 100.0 | 100.0 |

of performances of the group would probably become foreshortened, for there would be fewer workers returning the very long times.

## 17.4   Standard rating and standard performance

In Chapter 13 it was said that the principal use of work measurement (and hence of time study) is to set time standards which can be used for a number of different purposes (including programme planning, estimating, and as a basis for incentives) for the various jobs carried out in the undertaking[2]. Obviously, if those time standards are to be of any value at all, their achievement must be within the capacity of the majority of workers in the enterprise. It would be no use setting standards so high that only the best could attain them, since programmes or estimates based on them would never be fulfilled. Equally, to set standards well within the achievement of the slowest workers would not be conductive to efficiency.

How does the work study man obtain such a fair time from time studies?

We have already said that, as far as possible, studies should be taken on qualified workers. If it were possible to obtain the times taken by 500 qualified operatives for a single operation and plot them in the manner shown in Figure 75, a reliable average time would be obtained. Unfortunately, this is hardly ever possible. It is not always possible to time a job on an average qualified worker; moreover, even if it were, people do not work consistently from day to day or even from minute to minute. The work study man has to have some means of assessing the rate of working of the operative he is observing and of relating it to standard pace. This process is known as rating.

> **Rating is the assessment of the worker's rate of working relative to the observer's concept of the rate corresponding to standard pace.**

By definition, rating is a comparison of the rate of working observed by the work study man with a picture of some standard level which he is holding in his mind. This standard level is the average rate at which qualified workers will naturally work at a job, when using the correct method and when motivated to apply themselves to their work. This rate of working corresponds to what is termed the standard rating [3], and is denoted by 100 on the rating scale recommended to readers of this book (see section 7 below). If the standard pace is maintained and the appropriate relaxation is taken, a worker will achieve standard performance over the working day or shift.

> **Standard performance is the rate of output which qualified workers will naturally achieve without over-exertion as an average over the working day or shift, provided that they know and adhere to the specified method and provided that they are motivated to apply themselves to their work.**
>
> **This performance is denoted as 100 on the standard rating and performance scales.**

The rate of working most generally accepted in the United Kingdom and the United States as corresponding to the standard rating is equivalent to the speed of motion of the limbs of a man of average physique walking without a load in a straight line on level ground at a speed of 4 miles an hour (6.4 kilometres per hour). This is a brisk, business-like rate of walking, which a man of the right physique and well accustomed to walking might be expected to maintain, provided that he takes appropriate rest pauses every so often. This pace has been selected, as a result of long experience, as providing a suitable benchmark to correspond to a rate of working which would enable the average qualified worker who is prepared to apply himself to his task to earn a fair bonus by working at that rate, without there being any risk of imposing on him any undue strain which would affect his health, even over a long period of time. (As a matter of interest, a man walking at 4 miles an hour (6.4 kilometres per hour) appears to be moving with some purpose or destination in mind: he is not sauntering, but on the other hand he is not hurrying. Men hurrying, to catch a train for instance, often walk at a considerably faster pace before breaking out into a trot or a run, but it is a pace which they would not wish to keep up for very long.)

It should be noted, however, that the "standard pace" applies to Europeans and North Americans working in temperate conditions; it may not be a proper pace to consider standard in other parts of the world. In general, however, given workers of proper physique, adequately nourished, fully trained and suitably motivated, there seems little evidence to suggest that different standards for rates of working are needed in different localities, though the periods of time over which workers may be expected to average the standard pace will vary very widely with the environmental conditions. At the very least, the standard rate as described above provides a theoretical datum line with which comparisons of performance in different parts of the world could be made in order to determine whether any adjustment may be necessary. Another accepted example of working at the standard rate is dealing a pack of 52 playing cards in 0.375 minutes.

Standard performance on the part of the average qualified worker (that is, one with sufficient intelligence and physique, adequately trained and experienced in the job he is doing) will probably show as such only over a period of several hours. Anyone doing manual work will generally carry out the motions directly concerned with his work at his own natural working rate, which may not be exactly the standard rate, since some men work faster than others. There will of course be different standard paces (or speeds of movement) for different activities, according to the complexity or arduousness of the element making up the activity (among other things), so that working at the standard rate will not always mean moving the hands or limbs at the same speed. And in any event, it is not uncommon for workers to work faster at some periods of the day than at others, so that the standard performance is rarely achieved as the result of working, without any deviation, at the standard rate throughout the working periods of the shift, but rather as the cumulative outcome of periods of work at varying paces.

When time standards are used as a basis for payment by results, many union-management agreements stipulate that the time standards should be such that a representative or average qualified worker on incentive pay can earn 20–35% above his time rate by achieving the standard performance. If the worker has no target to aim at and no incentive to make him desire a higher output, he will (apart from any time he may waste consciously) tolerate the intrusion of small pieces of ineffective time, often seconds or fractions of seconds between and within elements of work. In this way he may easily reduce his performance over an hour or so to a level much below that of the standard performance. If, however, he is given enough incentive to make him want to increase his output, he will get rid of these small periods of ineffective time, and the gaps between his productive movements will narrow. This may also alter the pattern of his movements [4]. The effect of the elimination of these small periods of ineffective time under the influence of an incentive can be illustrated diagrammatically (see Figure 76).

What happens may be seen in the case of a man working on a lathe who has to gauge his workpiece from time to time. His gauge is laid on the tool locker beside him. If he has no particular reason to hurry, he may turn his whole body round every time he wishes to pick up the gauge, turn back to the lathe, gauge the workpiece and turn again to put the gauge down, each of these movements being carried out at his natural pace. As soon as he has reason to speed up his rate of working, instead of turning his whole body he will merely stretch out his arm, perhaps glancing round to check the position of the gauge on the locker, pick up the gauge, gauge the workpiece and replace the gauge on the locker with a movement of his arm, without bothering to look. In neither case would there be a deliberate stopping of work, but in the second some movements—ineffective from the point of view of furthering the operation—would have been eliminated.

The effect of putting a whole shop or factory (such as the 500 workers in Figure 75) on an incentive is shown in Figure 77.

*Figure 76. Effect of ineffective time on performance.*

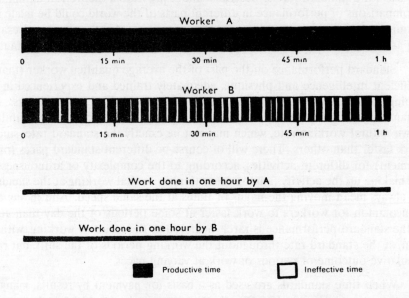

*Figure 77. Effect of a payment-by-results incentive on the time taken to perform an operation.*

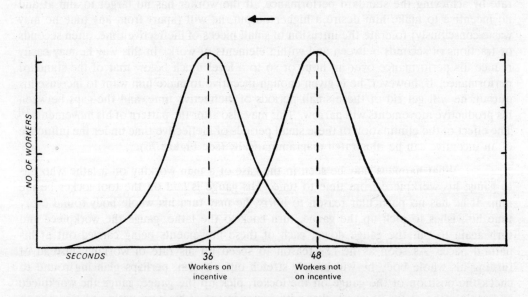

Offering an incentive in the form of payment in proportion to output will not make the unskilled or slow worker as fast or as skilled as the skilled or naturally fast worker; but if everyone in the shop is put on a well-designed incentive plan, other conditions remaining the same, the result will be that everyone will tend to work more consistently. The short periods of ineffective time discussed above will disappear, and everyone's average time for the job will be reduced. (This is an over-simplification but true enough for purposes of illustration.) The normal distribution curve shown in Figure 75 will move to the left while retaining approximately the same shape. This is quite clearly shown in Figure 77, where the peak of the curve (the average time) now comes at 36 seconds instead of 48—a reduction of 25%.

It should be added that, although the standard rate of working is that at which the average qualified worker will naturally perform his movements when motivated to apply himself to his task, it is of course quite possible and indeed normal for him to exceed this rate of working if he wishes to do so, just as a man can walk faster than 4 miles an hour if he wants to.  Men will be observed to be working, sometimes faster, sometimes slower than the standard rate, during short periods. Standard performance is achieved by working over the shift at paces which average the standard rate.

## 17.5  Comparing the observed rate of working with the standard

How is it possible accurately to compare the observed rate of working with the theoretical standard? By long practice.

Let us return once more to our man walking. Most people, if asked, would be able to judge the rate at which a man is walking. They would start by classifying rates of walking as slow, average or fast.  With a little practice they would be able to say: "About 3 miles an hour, about 4 miles an hour, or about 5 miles an hour" (or of course the equivalent rates in kilometres if they are more used to kilometres). If, however, a reasonably intelligent person were to spend all his time watching men walking at different speeds, he would soon reach the point where he could say: "That man is walking at $2\frac{1}{2}$ miles an hour and this one at $4\frac{1}{4}$ miles an hour", and he would be right, within close limits.  In order to achieve such accuracy, however, he would need to have in his mind some particular rate with which to compare those which he sees.

That is exactly what the work study man does in rating; but, since the operations which he has to observe are far more complex than the simple one of walking without load, his training takes very much longer. Judgment of walking pace is only used for training work study men in the first stages; it bears very little resemblance to most of the jobs that have to be rated. It has been found better to use films or live demonstrations of industrial operations.

Confidence in the accuracy of one's rating can be acquired only through long experience and practice on many types of operation—and confidence is essential to a work study man. It may be necessary for him to back his judgment in arguments with the management, foremen or workers' representatives; unless he can do so with assurance, the confidence of all parties in his ability will quickly disappear, and he might as well give up practising time study. This is one of the reasons why trainees may attempt method study after a comparatively short training but should on no account try to set time standards—except under expert guidance—without long practice, especially if the standards are to be used for incentive payments.

## 17.6  Rating of effort

The purpose of rating is to determine, from the time actually taken by the operative being observed, the standard time which can be maintained by the average qualified worker and which can be used as a realistic basis for planning, control and incentive schemes. What the study man is concerned with is therefore the speed with which the operative carries out the work, in relation to the study man's concept of a normal speed. In fact, speed of working as recorded by the time taken to carry out the elements of the operation is the only thing which can be measured with a stop-watch. Most authorities on time study agree on this point.

Speed of what? Certainly not merely speed for movement, because an unskilled operative may move extremely fast and yet take longer to perform an operation than a skilled operative who appears to be working quite slowly. The unskilled operative puts in a lot of unnecessary movements which the experienced operative has long since eliminated. The only thing that counts is the effective speed of the operation. Judgment of effective speed can only be acquired through experience and knowledge of the operations being observed. It is very easy for an inexperienced study man either to be fooled by a large number of rapid movements into believing that an operative is working at a high rate or to underestimate the rate of working of the skilled operative whose apparently slow movements are very economical of motion.

A constant source of discussion in time study is the rating of effort. Should effort be rated, and if so, how? The problem arises as soon as it becomes necessary to study jobs other than very light work where little muscular effort is required. Effort is very difficult to rate. The result of exerting effort is usually only seen in the speed.

The amount of effort which has to be exerted and the difficulty encountered by the operative is a matter for the study man to judge in the light of his experience with the type of job. For example, if an operative has to lift a heavy mould from the filling table, carry it across the shop and put it on the ground near the ladle, only experience will tell the observer whether the speed at which he is doing it is normal, above normal or sub-normal. Anyone who had never studied operations involving the carrying of heavy weights would have great difficulty in making an assessment the first time he saw such an operation.

Operations involving mental activities (judgment of finish, for example, in inspection of work) are most difficult to assess. Experience of the type of work is required before satisfactory assessments can be made. Inexperienced study men can be made to look very foolish in such cases, and moreover can be unjust to above-average and conscientious workers.

In any job the speed of accomplishment must be related to an idea of a normal speed for the same type of work. This is an important reason for doing a proper method study on a job before attempting to set a time standard. It enables the study man to gain a clear understanding of the nature of the work and often enables him to eliminate excessive effort or judgment and so bring his rating process nearer to a simple assessment of speed.

In the next section some of the factors affecting the rate of working of the operative will be discussed.

## 17.7   Factors affecting the rate of working

Variations in actual times for a particular element may be due to factors outside or within the control of the worker. Those outside his control may be—

1) Variations in the quality or other characteristics of the material used, although they may be within the prescribed tolerance limits.

2) Changes in the operating efficiency of tools or equipment within their useful life.

3) Minor and unavoidable changes in methods or conditions of operation.

4) Variations in the mental attention necessary for the performance of certain of the elements.

5) Changes in climatic and other surrounding conditions such as light and temperature.

These can generally be accounted for by taking a sufficient number of studies to ensure that a representative sample of times is obtained.

Factors within his control may be—

a) Acceptable variations in the quality of the product.

b) Variations due to his ability.

c) Variations due to his attitude of mind, especially his attitude to the organisation for which he works.

The factors within the worker's control can affect the times of similarly described elements of work by affecting—

i) The pattern of his movements.

ii) His working pace.

iii) Both, in varying proportions.

The study man must therefore have a clear idea of the pattern of movement which a qualified worker should follow, and of how this pattern may be varied to meet the range of conditions which that worker may encounter. Highly repetitive work likely to run for long periods should have been studied in detail through the use of refined method study techniques, and the worker should have been suitably trained in the patterns of movement appropriate to each element.

The optimum pace at which the worker will work depends on—

1) The physical effort demanded by the work.

2) The care required on the part of the worker.

3) His training and experience.

Greater physical effort will tend to slow up the pace. The ease with which the effort is made will also influence the pace. For example, an effort made in conditions where the operative cannot exert his strength in the most convenient way will be made much more slowly than one of the same magnitude in which he can exert his strength in a straightforward manner (for instance, pushing a car with one hand through the window on the steering wheel, as opposed to pushing it from behind). Care must be taken to distinguish between slowing up due to effort and slowing up due to fatigue.

When the element is one in which the worker is heavily loaded, so that he has to exert considerable physical effort throughout, it is unlikely that he will perform it at anything other than his natural best pace. In such circumstances rating may be superfluous: it may be sufficient to determine the average of the actual times taken during an adequate number of observations. This was very strikingly shown during an ILO study of manual earth-moving operations carried out in India. The workers—men, women and youths—carried loads of earth up to 84 lb (38 kg) in weight on their heads, in wicker baskets. A man with 84 lb on his head does not dawdle. He is anxious to get to the end of his walk and get rid of the load, and so performs the task at the best rate that he can naturally achieve. In doing so he shortens his stride, taking very short paces very quickly so that it looks almost as though he is going to break out into a trot at any moment. In point of fact, the stop-watch showed that the time taken for the loaded walk was a good deal longer than that needed for the apparently more leisurely return unloaded, so that the study man without experience of the effort involved in the operation could very easily be led into making false ratings. In fact, for the loaded walk ratings were not necessary, except when contingencies occurred. Similar heavily loaded elements occur in factories, as in carrying sacks, picking them up, or throwing them down on to stacks. These operations are most likely to be carried out at the best natural pace which the worker can manage.

An increased need for care in carrying out an element will reduce the pace. An example is placing a peg with parallel sides in a hole, which requires more care than if the peg is tapered.

Fumbling and hesitation on the part of the worker are factors which the study man must learn to recognise and cope with. A worker's natural skill and dexterity combined with training and experience will reduce the introduction of minor method variations (fumbling), and also the foreign element 'consider' (hesitation). Very slight deviations from the standard method can be taken into account by assigning a lower rating, but fumbling and hesitation usually signal a need for further training.

The study man should be careful not to rate too highly when—

a) The worker is worried or looks hurried.

b) The worker is obviously being over-careful.

c) The job looks difficult to the study man.

d) The study man himself is working very fast, as when recording a short-element study.

Conversely, there is a danger of rating too low when—

1) The worker makes the job look easy.

2) The worker is using smooth, rhythmic movements.

3) The worker does not pause to think when the study man expects him to do so.

4) The worker is performing heavy manual work.

5) The study man himself is tired.

The study man must take such factors into account. Rating is very much easier if a good method study has been made first, in which the activities calling for special skill or effort have been reduced to a minimum. The more the method has been simplified, the less the element of skill to be assessed and the more rating becomes a matter of simply judging pace.

## 17.8  Scales of rating

In order that a comparison between the observed rate of working and the standard rate may be made effectively, it is necessary to have a numerical scale against which to make the assessment. The rating can then be used as a factor by which the observed time can be multiplied to give the basic time, which is the time it would take the qualified worker, motivated to apply himself, to carry out the element at standard rating.

There are several scales of rating in use, the most common of which are those designated the 100–133 scale, the 60–80, the 75–100, and the British Standard scale used in this book (essentially a restatement of the 75–100 scale) which is the 0–100 scale.

Table 17 shows examples of various rates of working on the scales mentioned.

In the 100–133, 60–80 and 75–100 scales, the lower figure in each instance was defined as the rate of working of an operative on time rates of pay; and the higher, in each case one-third higher, corresponded to the rate of working we have called the standard rate, that of a qualified worker who is suitably motivated to apply himself to his work, as for instance by an incentive scheme. The underlying assumption was that workers on incentive perform, on average, about one-third more effectively than those who are not. This assumption has been well substantiated by practical experience over many years, but it is largely irrelevant in the construction of a rating scale. All the scales are linear. There is therefore no need to denote an intermediate point between zero and the figure chosen to represent the standard rating as we have defined it. Whichever scale is used, the final time standards derived should be equivalent, for the work itself does not change even though different scales are used to assess the rate at which it is being carried out.

The newer 0–100 scale has, however, certain important advantages which have led to its adoption as the British Standard. It is commended to readers of this book and is used in all the examples which follow. In the 0–100 scale, 0 represents zero activity and 100 the normal rate of working of the motivated qualified worker—that is, the standard rate.

## 17.9  Determination of basic time

The number 100 represents standard performance. If the study man decides that the operation he is observing is being performed with less effective speed than his concept of standard, he will use a factor of less than 100, say 90 or 75 or whatever he considers represents a proper assessment. If, on the other hand, he decides that the effective rate of working is above standard, he gives it a factor greater than 100—say, 110, 115 or 120.

It is the usual practice to round off ratings to the nearest multiple of five on the scale; that is to say, if the rate is judged to be 13% above standard, it would be put

*Table 17. Examples of various rates of working on the principal rating scales*

| Scales | | | | Description | Comparable walking speed[1] | |
|--------|--------|---------|-----------------|-------------|------|------|
| 60–80 | 75–100 | 100–133 | 0–100 Standard | | (mi/h) | (km/h) |
| 0 | 0 | 0 | 0 | No activity. | – | – |
| 40 | 50 | 67 | 50 | Very slow; clumsy, fumbling movements; operative appears half asleep, with no interest in the job. | 2 | 3.2 |
| 60 | 75 | 100 | 75 | Steady, deliberate, unhurried performance, as of a worker not on piecework but under proper supervision; looks slow, but time is not being intentionally wasted while under observation. | 3 | 4.8 |
| 80 | 100 | 133 | 100 (Standard Rating) | Brisk, business-like performance, as of an average qualified worker on piecework; necessary standard of quality and accuracy achieved with confidence. | 4 | 6.4 |
| 100 | 125 | 167 | 125 | Very fast; operative exhibits a high degree of assurance, dexterity and coordination of movement, well above that of an average trained worker. | 5 | 8.0 |
| 120 | 150 | 200 | 150 | Exceptionally fast; requires intense effort and concentration, and is unlikely to be kept up for long periods; a 'virtuoso' performance achieved only by a few outstanding workers. | 6 | 9.6 |

[1]Assuming an operative of average height and physique, unladen, walking in a straight line on a smooth level surface without obstructions.

*Source:* Freely adapted from a table issued by the Engineering and Allied Employers (West of England) Association, Department of Work Study.

down at 115. During the first weeks of their training, study men are unlikely to be able to rate more closely than the nearest ten.

If the study man's ratings were always impeccable, then however many times he rates and times an element the result should be that—

$$\text{Observed Time} \times \text{Rating} = \text{A Constant}$$

provided that the element is of the type described as a constant element in section 6 of the previous chapter, and that it is always performed in the same way.

An example, expressed numerically, might read as follows:

| Cycle | Observed time (decimal minutes) | | Rating | | Constant |
|---|---|---|---|---|---|
| 1 | 0.20 | × | 100 | = | 20 |
| 2 | 0.16 | × | 125 | = | 20 |
| 3 | 0.25 | × | 80 | = | 20 |

and so on.

It must be remembered, however, that rating does not stand by itself. It is always a comparison with the standard rating. So, if the standard rating is taken to be 100, then dividing the constant by the standard rating (100) will yield the constant known as the "basic time" for the element.

$$\text{Observed Time} \times \frac{\text{Rating}}{\text{Standard Rating}} = \text{Basic Time}$$

For example—

$$0.16 \text{ min} \times \frac{125}{100} = 0.20 \text{ min.}$$

This basic time (0.20 minutes in the example) represents the time the element would take to perform (in the judgment of the observer) if the operative were working at the standard rate, instead of the faster one actually observed.

If the operative was judged to be working more slowly than the standard, a basic time less than the observed time would be arrived at, for example—

$$0.25 \text{ min} \times \frac{80}{100} = 0.20 \text{ min.}$$

In actual practice, the multiple Observed Time × Rating is very rarely exactly constant when taken over a large number of readings, for various reasons such as—

a) Variations in the work content of the element.

b) Inaccuracies in noting and recording observed times.

c) Inaccuracies in rating.

d) Variations due to rating to the nearest five points.

## 17.10   Recording the rating

We have discussed the theory of rating at some length and are now in a position to undertake the complete study.

In general, each element of activity must be rated during its performance before the time is recorded, without regard to previous or succeeding elements. No consideration should be given to the aspect of fatigue, since the allowance for recovery from fatigue will be assessed separately (see Chapter 18).

In the case of very short elements and cycles this may be difficult. If the work is repetitive, it is possible to rate every cycle or possibly the complete study. This is done when the short cycle study form (Figure 72) is used.

It is most important that the rating should be made while the element is in progress and that it should be noted before the time is taken, as otherwise there is a very great risk that previous times and ratings for the same element will influence the assessment. For this reason the 'Rating' column on the time study sheet in Figures 75 and 76 is placed to the left of the "Watch Reading" column. It is, perhaps, a further advantage of the cumulative method of timing that the element time does not appear as a separate figure until the subtractions have been made later in the office. If it did, it might influence the rating or tempt the study man to "rate by the watch".

Since the rating of an element represents the assessment of the average rate of performance for that element, the longer the element the more difficult it is for the study man to adjust his judgment to that average. This is a strong argument in favour of making elements short, subject to the conditions discussed in Chapter 16. Long elements, though timed as a whole up to the break points, should be rated every half minute.

Rating to the nearest five is found to give sufficient accuracy in the final result. Greater accuracy than this can be attained only after very long training and practice.

We may now refer back to the time study form in Figures 70 and 71. We have discussed the filling-in of two columns, namely "Watch Reading" (WR) and the 'Rating' (R), both entries being made on the same line.

These readings are continued for a sufficient number of cycles, at the end of which the watch is allowed to run on until compared with the clock with which it was synchronised when started. The "time after" can then be noted and recorded. The study is then at an end. The next step, after thanking the operative for his cooperation, is to work out the basic time for each element. How to do this is described in the next chapter.

## REFERENCES

1. W.D. Seymour: *Industrial training for manual operations* (London, Pitman, 1966).

2. For details of various well-known types of incentive plans, see ILO: *Payment by results*, Studies and reports, New series, No. 27 (Geneva, 14th impr., 1977).

3. The definition given in the B.S. *Glossary*, op. cit., concludes with the words "standard rating", rather than "standard pace", as used here. It is considered that the word "pace" more exactly conveys the sense of a rate of working than "rating", which has connotations implying a factor, or ratio, which do not help clarity at this point in the explanation.

4. Research carried out under the late Professor T.U. Mathew at the University of Birmingham (United Kingdom) tended to confirm this.

# Chapter 18

# Time study: From study to standard time

## 18.1  Introduction

Once the observations on job timing and rating at the workplace are recorded, an analysis of these data is to be carried out first to obtain basic time for the job and then to determine work content and standard time for the job considering various allowances.

## 18.2  Summarising the study

At the stage we have now reached, the study man has completed his observations at the workplace and has returned to the work study office with his study. No doubt he will later be making further studies on the same job or operation as performed by different operatives, but for the moment we shall consider how he works up the study he has just taken and enters the results obtained on the analysis of studies sheet for the operation. Later in the chapter we shall see how standard times are compiled from the entries on the analysis of studies sheet.

All the entries made so far on the time study top sheet (Figure 70) and the continuation sheets (Figure 71) have been written in pencil. As well as the heading details shown in the data block on the top sheet, there will be the "time before", the first entry on the study proper; the "time after", which will be the last entry; and two entries for each watch reading made—the rating and the watch reading itself. The ratings will all be in the column headed 'R' and will consist of numbers such as 95, 115, 80, 100, 75, 105, etc., though until the study man has had considerable practice he should confine his ratings to steps of ten, such as 80, 90, 100, etc. In the next column, that headed 'WR', will be the watch readings in decimal minutes. Since watch readings will have been made at intervals of half-a-minute or less (long elements being rated and timed every half-minute during the element as well as at the break point which signals its end), most of the entries will consist of two figures only, with a three-figure entry occurring whenever a full minute has been crossed. It is usual to omit the decimal points. This saves the study man a certain amount of writing and in practice gives rise to no ambiguity.

Let us assume that the "time before" was 2.15 minutes. The first entry on the study proper will thus be 215. The next may be 27, indicating that the watch was read 2.27 minutes after it was started. If the next three entries are 39, 51 and 307, these will signify that the watch was read at 2.39, 2.51 and 3.07 minutes after it was started. Two- and three-figure entries will continue in this way down the sheet until ten minutes have elapsed, when the next entry will be a four-figure one. Most study men then revert to three-figure and four-figure entries again until another ten minutes have passed, using four figures only for the first entries after the ten-minute intervals. The study will close with the "time after" entry, at which time also the "time off" will be noted in the data panel on the study top sheet. Every now and then in the study there may be watch readings without accompanying ratings, when some delay or stoppage has occurred. These of course cannot be rated, for they are not work.

It should be made a working rule that none of these pencil entries may ever be erased and replaced. Occasionally a study may contain a very obvious error, of a sort which may be corrected without invalidating the study. If so, the correction should be made in ink, over the original pencil entry, so that it may always be seen later as a change made in the study office, not at the place where the study was made. Whenever there is an error about which there is doubt as to how it should be corrected, that part of the study should be ignored. It may be necessary to scrap the study and make another.

It is good practice to carry out all subsequent work on the study sheets either in ink or in pencil of a different colour from that used for the initial recordings. Many study departments make this a standing rule also. There is then no doubt whatever about what was actually recorded from direct observation and what represents subsequent calculation. Quite apart from its merits in obtaining orderly processing of the data recorded, the practice helps also to maintain the confidence of workers and their representatives that nothing improper is permitted in the working up of studies.

## 18.3  Preparing the study summary sheet

As will be seen a little later, much of the work necessary before the study summary sheet can be completed consists of quite simple routine calculations which may be done by a clerk while the study man gets on with something else. In the beginning, however, the study man should do everything himself, until he is so thoroughly familiar with all the procedures involved that he cannot only instruct the clerk on what has to be done but can also check the calculations easily and quickly. It is also a good idea to provide the clerk with a calculator to help to reduce the number of mistakes and increase the amount of useful information that can be extracted from the study.

The first step is to complete the data at the head of the study summary sheet (Figure 73), copying the details neatly, in ink, from the study sheets. From the time off and the time on, the elapsed time may be calculated and entered. When cumulative timing is being practised, the elapsed time should of course agree with the final watch reading. If it does not, there is an error which must at once be investigated. It is no use doing further work on the study until this is cleared up, for a serious error may be cause for scrapping the study and starting again. Deducting from the elapsed time the total "check time"—the sum of the "time before" and the "time after"—yields the net time.

This should agree with the sum of all the observed times when using flyback timing, or the sum of all the subtracted times with cumulative timing. If flyback timing has been used, this check should be made before proceeding further, by adding up all the element times recorded and seeing how the total compares with the net time. It is unlikely that there will be an exact agreement, for the reasons noted earlier, but the discrepancy should be within ± 2%. If it is greater than this, some departments make it their practice to ignore the study and make another.

When cumulative timing has been used, the check cannot be made until the subtracted times have been obtained and totalled. The comparison then serves as a check on the accuracy with which the subtractions have been made. Any error should be investigated and corrected before the work of extension is undertaken.

On the body of the study summary sheet the study man next lists in order all the repetitive elements observed, in order of their occurrence, noting the break points used on the reverse of the sheet.

Some of these repetitive elements may be variable elements, which will have to be treated in a different way from the constant elements. These variable elements are therefore listed again in a fresh tabulation below the full list of repetitive elements. Below the variable elements the study man next lists any occasional elements observed, including with them any contingency elements of work which actually occurred during the study. Below these again are listed any foreign elements and ineffective time. When these entries have been made, the sheet should provide for a summarised record of everything that has been observed during the study.

## ENTER FREQUENCIES

The next step is to enter against each element listed on the study summary sheet the frequency with which that element occurred. Repetitive elements, by definition, occur at least once in every cycle of the operation so the entry to be made against a repetitive element will read 1/1, or 2/1, etc., indicating that the element concerned occurs once in every cycle (1/1), twice (2/1), or whatever may be the case. Occasional elements (for example, the element "sharpen tools") may occur only once every 10 or 50 cycles, when the entry would be 1/10, 1/50, or as appropriate. The entries are made in the column headed 'F' on the study summary sheet.

## 18.4   Extension: The calculation of basic time

The study man has now completed the entries in the heading block of the study summary sheet, listed the elements, entered frequencies and (if necessary) made a clear sketch of the workplace layout on the reverse of the sheet (when appropriate, the use of a simple instant-print-type camera can save a great deal of time and money; it is usually necessary to include in the photograph a simple scale, such as a square rod painted in 1 cm bands). He must turn next to the calculations which have to be made on the time study sheet themselves before he can go any further with his study summary. The results of his calculations will be entered on the time study sheets in ink or pencil of a different colour from that used when recording observations at the workplace.

If flyback timing has been used, the study man may proceed direct to extension. When using cumulative timing, however, it is first necessary to subtract each watch reading from the one following it, in order to obtain the observed time for each element. The entries obtained in this way should properly be styled "subtracted times" rather than "observed times"; they are entered in the third column on the time study sheets, that are headed 'ST'. The subtracted times derived when using cumulative timing are of course exactly equivalent to the observed times entered directly at the workplace when using flyback timing; so for the sake of simplicity the single term "observed time" is used during the rest of this chapter to mean both directly observed and subtracted times.

The next step is to convert each observed time to a basic time, entering the result in the column headed 'BT' on the time study sheets.

---

**Basic Time is the time for carrying out an element of work at standard rating, i.e.**

$$\frac{\text{Observed Time} \times \text{Observed Rating}}{\text{Standard Rating}}.$$

**Extension is the calculation of basic time from observed time.**

---

The effect of extending an observed time for an element to the basic time is shown graphically in Figure 78.

## 18.5   The selected time

---

**The selected time is the time chosen as being representative of a group of times for an element or group of elements.  These times may be either observed or basic and should be denoted as selected observed or selected basic times.**

---

### CONSTANT ELEMENTS

In theory the results of all the calculations of the basic time for any single constant element should be the same, but for the reasons given in Chapter 17 this is rarely so. It is necessary to select from all the basic times which have been entered on the time study sheets a representative time for each element. This will be recorded against the element description on the study summary sheet and will later be transferred to the analysis of studies sheet as the end result of the study, at least insofar as that particular element is concerned.

The calculations necessary to arrive at the selected basic time are carried out

Figure 78. Effect of extension on the time of an element

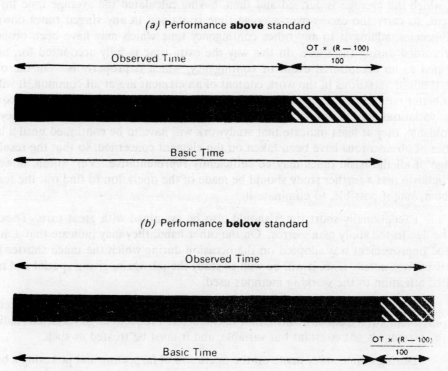

(a) Performance **above** standard

(b) Performance **below** standard

on the working sheet. As was noted in Chapter 15, it is quite common to use simple lined sheets for making the analysis (or, for variable elements, squared paper), without having any special forms printed. The working sheets, when completed, are stapled to the time study sheets and filed with them. Much time can be saved and accuracy can be greatly improved by using a small calculator or computing equipment.

There are various methods of examining and selecting the representative basic time for a constant element. Perhaps the most common, and in many ways often the most satisfactory, is by making a straight average of the element times arrived at, adding all the calculated basic times together and dividing the total by the number of occasions on which the element was recorded. Before doing this, however, it is usual to list all the basic times for the element and to scrutinise the list, ringing out any times which are excessively high or low, well outside the normal range. These ringed times are sometimes styled 'rogues'. They should be examined carefully.

An exceptionally high time may be due to an error in timing. If cumulative timing is being used, an error of this sort will be revealed by examining the study, because an excessively long time for one element will cause shortening of the recorded time for the next. A high time may also be due to an error having been made in extension. But perhaps the most common cause, apart from errors, is that there has been some variation in the material being worked on or in some other aspect of the working method, which has caused a higher work content on the particular occasion recorded. If so, it is necessary to establish the cause and to consider whether it is likely to recur frequently or only very rarely. If the latter, it is usual to exclude the element's basic time from the total from which the average is derived and then, having calculated the average time for the element, to carry the excess-over-average time contained in any ringed times down to contingencies, adding it to any other contingency time which may have been observed and recorded during the study. In this way the extra time is fully accounted for, but it is treated as an exceptional event or contingency, which it properly is. On the other hand, if minor variations in the work content of an element are at all common, it will be much better not to exclude any calculations at all when calculating the average. Frequent minor variations should always be treated as signals to alert the study man. If they are unavoidable, they at least indicate that studywork will have to be continued until a large number of observations have been taken on the element concerned, so that the resulting average of all the basic times may be sufficiently representative. Very often, however, they indicate that a further study should be made of the operation to find out the reason for them, and, if possible, to eliminate it.

Exceptionally short times should also be examined with great care. They too may be due to the study man's error. On the other hand, they may indicate that a minor method improvement was adopted on the occasion during which the much shorter time than usual was noted. If so, it will be well to study the job again, giving special and more detailed attention to the working methods used.

The approach outlined above is valid so long as the exceptional times are either very infrequent, or, if frequent, only minor in character. Frequent large variations indicate that the element is not constant but variable, and it must be treated as such.

During a time study made on the operation of inspecting and jacketing a book, one element was described as; "Pick up one book, inspect, initial back end paper (break point: book closed)". This element was observed 31 times, and the basic minutes calculated were as shown below:

*Basic minutes*

| | | |
|---|---|---|
| 27 | 26 | 28 |
| 26 | 25 | 25 |
| 27 | 29 | 27 |
| 27 | 28 | 27 |
| 26 | 28 | 26 |
| 27 | 27 | 25 |
| 26 | 27 | 26 |
| 25 | 26 | 26 |
| 26 | 27 | (49) (Faulty part) |
| 27 | 26 | 26 |
| | | 28 |

It will be seen that one figure has been ringed—the basic time of 0.49 minutes which arose when a faulty book was encountered examined and rejected. Excluding this figure, the total of the remaining 30 basic times is 7.97 minutes, which yields an average of 0.266 minutes per occasion. At this stage in the studywork the number 266 would be entered on the study summary sheet and be carried to the analysis of studies sheet; but at the end of the calculations for the element, the basic time finally selected would be rounded off to the nearest two figures—in this case 0.27 minutes. The excess work observed in the ringed observation $(0.49 - 0.27 = 0.22)$ would be carried down to the contingencies record.

Selection by averaging in this way is simple to teach and to understand, and is readily accepted by both study men and workers. When the total number of observations made on an element is relatively small, averaging usually gives a more accurate result than is obtainable with other methods of selection. It does, however, give rise to a great deal of clerical work when many observations have been recorded, particularly when short elements have been observed very many times. Consequently, other methods of selection have been devised to reduce the calculation effort required.

One method, which obviates the necessity for extending observed times to basic times, is to tabulate the observed times for the element under the ratings recorded as corresponding to each observation, so as to form a distribution table against ratings. The table can be compiled direct from the entries made on the time study sheets at the workplace. For the element is the example above, the distribution table would appear as follows:

| Rating: | 80 | 85 | 90 | 95 | 100 | 105 | |
|---|---|---|---|---|---|---|---|
| Observed | 31 | 32 | 30 | 28 | 28 | 27 | |
| times | | 31 | 30 | 30 | 27 | | |
| | | 30 | 30 | 27 | 27 | | |
| | | 31 | 26 | 28 | 26 | | |
| | | 31 | 27 | 27 | 27 | | |
| | | | 28 | 26 | 28 | | |
| | | | 29 | 29 | 27 | | |
| | | | 29 | | | | |
| | | | 29 | | | | |
| | 31 | 155 | 258 | 195 | 190 | 27 | Totals of observed times |
| **Basic times** | **25** | **132** | **232** | **185** | **190** | **28** | **Total = 792** |

In the tabulation above, all the 30 observed times from which the basic times shown in the earlier example were calculated are listed, the one ringed observation having been excluded. The observed times are then totalled under each rating, and these totals are then extended by multiplying by the corresponding ratings, to yield the basic times (totals) shown in the line below. The grand total of all these basic times comes to 7.92 minutes, which, when divided by 30 (the number of observations) gives the selected basic time for the element—0.264 minutes. This may be compared with the result of 0.266 minutes achieved by averaging the individual basic times.

A third method also avoids the need to extend each observed time, the selection being made by constructing a plot as shown in Figure 79. In this method there are two sections to the plot, and two entries are made for each observation, but the entries are crosses or dots. The left-hand axis contains the time scale and shows the range of times observed for the element, in this case from 26 to 32. The scale at the top of the right-hand part of the plot shows the ratings observed, from 80 to 105. To make the plot, the study man runs down his study, and each time the element is recorded he makes a cross against the time observed, and a second cross, also against the observed time but under the rating observed, on the right-hand side of the plot.

*Figure 79. A graphical method of selecting basic time*

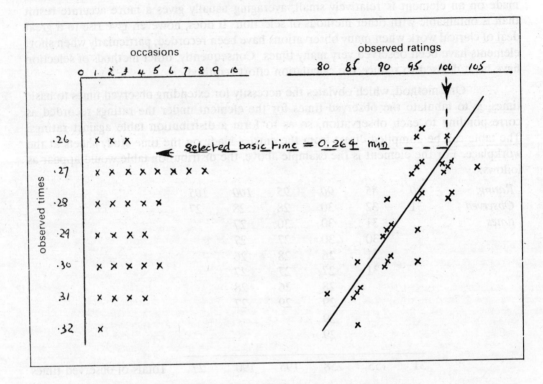

When all these entries are made, the left-hand side of the diagram will exhibit a frequency distribution of observed times. On the right-hand side, the best straight line is through the points plotted. The selected basic time for the element can then be read off by entering the right-hand plot under 100 rating, going vertically down until the line through the points is reached, and then reading on the scale at the left the time which corresponds to the intersection.

It is essential that the plot on the left-hand side be completed, in order to check whether the distribution follows the normal pattern. If it does not, the method should not be used. Distributions which are irregular—lopsided, skewed, or having two humps—should be treated as signals that the method will not be reliable, at any rate in the simple form described here. The different distribution patterns which can be produced, each have significant meanings, indicating different variations in the work itself, in the operative's rate of working, or in the study man's rating efficiency; but it will be better not to get involved in sophisticated analyses of this sort until considerable experience has been gained. The method is illustrated briefly here because it is typical of several which make use of graphical means to select representative basic times without extending each observation. Most of them are valid only when the distribution is normal or when the precise significance of any abnormality is thoroughly understood. It is recommended that the graphical methods be avoided unless expert guidance is available. The first two methods described will suffice for all normal needs, and have the merit that they are more easily understood by workers or their representatives.

Before leaving the subject of constant elements, the reader may like to refer again to the comments made in Chapter 17 about certain manual elements when the worker is heavily loaded, so that in all probability he normally performs the element at his best natural pace. Such elements are comparatively rare, but when they occur it may be sufficient to calculate the selected basic time by simply averaging observed times, without recourse to extension. It is essential, however, to have a large number of observations if this is to be done.

## VARIABLE ELEMENTS

The analysis of variable elements presents more difficulty. It is necessary to find out what it is that causes the basic time to vary, and quite often there may be several variables to take into account at once. For example, consider the operation of cross-cutting wooden planks with a handsaw. The basic time needed to make the cut will vary with the width of the plank, which establishes the length of cut that has to be made, and also with the thickness of the planks and the hardness of the wood being cut. If the saw needs sharpening, the cut will take longer; however, this would be considered to be the use of an incorrect method, and any observations made while the operative is using a blunt saw would therefore be disregarded.

The first step in the treatment of variable elements is almost always to extend observed times to basic times. The basic times will then be plotted on squared paper against the known variables. Thus for variable elements the analysis of studies sheet takes the form of graph paper, and the graph constructed at the time of summarising the study will probably be attached to the analysis of studies sheet, in place of the entries made on this sheet for constant elements.

Whenever possible, the basis chosen for the plot should be some variable which yields a straight line when the basic times are entered. Sometimes this can be done by using logarithmic paper, when an analysis of the operation suggests that the variability with time may not be arithmetically linear. Quite often, however, it is not possible to discover a straight-line relationship between time and the main variable, or with any combination of variables which is tried. In these cases the end product will be a curved line, drawn as smoothly as possible between all the plots made from all the studies on the element. Basic times for the element will then be selected by reading off the curve at the appropriate point on each occasion on which a standard time has to be compiled.

The treatment which the study man would accord to the times derived from studying the cross-cutting of planks would depend on whether the operation is an incidental one, performed only rarely, or whether it is an element performed many times each day, forming a substantial proportion of the total work done. In the latter case he will probably need to build up a series of graphs, each for a different hardness of wood, with each graph having a family of lines on it, one for each thickness of plank. Basic times would be plotted on these graphs against length of cut. The relationship should be linear, so that once it has been discovered the lines can be expressed as formulae, with factors to take into account the variables, thus dispensing with the graphs for the calculation of basic times. If the element is not of sufficient importance to warrant so much detail, the study man would probably try plotting basic times against the product: width of plank × thickness of plank, thus combining two of the main variables. He would also try to establish a factor by which to multiply the relationship discovered to take account of different hardnesses of wood. The statistical technique of multiple regression analysis is highly suitable for the calculation of variable times. However, a full explanation of this technique falls outside the scope of this book.

It will be evident that, in general, many more observations of a variable element than of a constant element will be necessary before reliable representative basic times can be established. It is well to recognise this at the outset, so that the studywork can be planned to span all the different conditions and variables which are likely to be encountered in practice. It is well also to give close attention from the beginning to discovering the best basis against which to plot the times, essaying trial plots against different possibilities until some satisfactory indicator of the cause of the variable times is revealed. When the basis of the relationship has been discovered, further studywork can be directed to filling any gaps in the information so far compiled. If the essential analysis is left until a later stage, many of the studies taken may turn out to be needless duplication.

It is not possible to prescribe any one method of approach which will yield satisfactory results in the analysis of all variable elements. Each must be treated on its merits. It is here, perhaps more than anywhere else in time study, that close attention to the detailed methods of working is amply repaid; otherwise it will rarely be possible to discover just what it is that causes basic times to vary. Even when the causes are known, there is often scope for considerable ingenuity in devising a simple basis which will reflect the major variables and reveal a definite and repeatable relationship.

## 18.6   Completing the study summary sheet

Having completed his calculations, the study man is now ready to enter on the study summary sheet the information which will make it a clear and concise record of all the results obtained from his observations at the workplace. Against each of the constant elements listed on the sheet he will enter the selected basic time for the element and the number of occasions on which the element was observed. The frequencies of occurrence have already been entered. Against the variable elements he will note the relationship between basic time and the controlling variable, if he has discovered this, or will record a reference to the graph sheet or other study analysis sheet on which the basic times derived have been analysed.

To complete the summary, he must enter a record of any occasional elements observed which have not already been included, and also any foreign elements which may have appeared during the study. Contingency elements and any contingency time extracted during the calculations must be shown. It is usual to express the contingency basic minutes as a percentage of the total basic minutes of repetitive work observed during the whole of the study, so that there may be a basis for comparing the contingencies occurring during one study with those in another.

All the entries which have so far been made represent work, in one form or another. All except any foreign elements will figure later in the calculation of a standard time for the operation, and since they are all work they will all attract relaxation allowances (see section 11). Besides the elements of work, however, there may well have been periods when no work was done during the study, either because the operative was resting or because he was engaged on one or other of the activities which were described earlier in this book as "ineffective time". The time so spent must now be totalled and entered on the summary. It is useful to break down such time into a few main categories, such as 'relaxation' and "ineffective time". The entries will all be in terms of observed times, of course—periods when no work is done cannot be rated.

## 18.7   Number of elements and cycles to be studied

We dealt with this problem in Chapter 16, outlining a statistical and a conventional method for determining the number of elements and cycles to be studied. When the working conditions vary, studies must be made in each of the different sets of conditions which will be met with in practice: at different times of day if atmospheric conditions change markedly during the shift, for instance, and on all the types of material which have to be processed if the material is not rigidly standard.

The study man must be prepared to study all the work involved in starting up at the beginning of a shift and in shutting down at the end of it. Start-up and shut-down times are part of the work and may need a separate work value, or they may be taken into account (if appropriate) by making an allowance for them when calculating the standard times for individual jobs. In industries such as printing, presses are not normally left inked up overnight, as the ink would dry before morning. Time may have to be allowed for cleaning machines and the workplace, and for changing clothes in industries where special clothing is required. Activities of this sort are not usually taken into account in the calculation of standard times for individual jobs but are more often dealt with by time allowances. Allowances are discussed later in this chapter; at this point it is sufficient to note that studies will have to be made on all the ancillary and incidental activities which are undertaken during the working day before the matter of allowances can be properly considered.

A simple method of determining when enough cycles of a constant element have been observed—enough, that is, to permit a representative basic time for the element to be selected—is to plot the cumulative average basic time for the element each time a study is made on it and summarised. The plot is started with the basic time derived from the first study. When the second study comes in the figure then plotted is the average, calculated by adding the basic time from the first study times the number of observations during the first study to the product (basic time × observations) from the second study, and then dividing by the total number of observations made during both studies. Further plots are made in the same fashion as successive studies are worked up. When the line on the graph ceases to 'wag' and settles down at a constant level, enough studies have been made on this element. An example is shown in Figure 80.

With variable elements it is convenient to start by making several short studies which together span the full range of variability, so that an early attempt may be made

*Figure 80. Cumulative average basic times for a constant element*

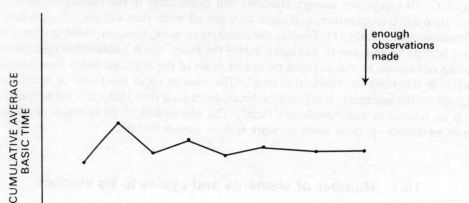

to establish the relationship between basic time and the indicative variable. Subsequent studywork may then be directed to obtaining the information needed to complete, modify or validate the apparent relationship suggested by the first studies.

## 18.8   The analysis of studies sheet

An example of an "analysis of studies sheet" is shown in Figure 74. The results obtained in each study on an operation are entered on this sheet by copying from the study summary sheet, as soon as the study has been worked up. A form of the type illustrated provides for a list of all the elements which makes up a job or operation, and also for full details in respect of repetitive and occasional elements, together with a record of the contingency and ineffective times observed. Graphs are appended to the sheet to record the results obtained from studying variable elements.

When it is considered that enough observations have been made, the next step is to calculate the final representative basic times for each element. This is done on the analysis of studies sheet. The process of selection is essentially similar to that described in section 4 of this chapter, the usual method being to calculate the overall weighted average of all the basic times recorded for each element, disregarding any entries which subsequent studywork has shown to be erroneous. The weighted average is obtained by multiplying the basic time recorded from a study by the number of observations of the element made in that study, adding up the products so derived for all the studies, and dividing the total by the sum of all the observations made in all the studies.

*Figure 81. Allowances*

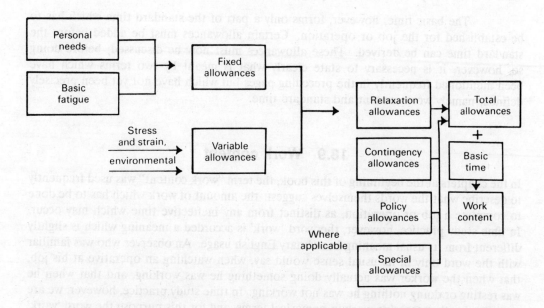

When these final representative basic times have been calculated for each constant element, it is a simple matter to calculate the basic time per cycle, per job or per operation for these elements, by multiplying the time per occasion and by the frequency per cycle with which each element recurs. Variable elements cannot be dealt with in this way, of course. For them, the basic time may have to be read off the appropriate graph, or, if a straight-line relationship has been established, be calculated from the formula which expressed the line in algebraic terms, or be derived by regression analysis.

If it is considered appropriate to make a provision in the job time for contingencies, the allowance necessary is also calculated on the analysis of studies sheet. The first step in doing this is to calculate the percentage which the total observed contingencies represent of the total other work observed. Time spent on contingencies is just as much work as that devoted to repetitive and occasional elements, so contingency time will also be recorded in basic minutes. If the percentage is a very small one, it will probably be convenient to adopt the figure as the percentage allowance to be made; but if it comes out at more than about 4 or 5%, the better course is to inquire into the causes of the contingencies so as to eliminate or reduce them as far as possible. When action of this sort has been taken as a result of the studies, the percentage observed during the earlier studywork will no longer be valid and it will be necessary to make fresh observations.

At the stage now reached, a basic time has been built up for the job or operation, including all repetitive and occasional elements and also any small amount of extra work which may be met with occasionally as a contingency. The compilation has been done element by element, so that, if at any time in the future the job is changed slightly by deleting or changing an element or by adding a fresh one, it will not be necessary to restudy the whole job. The entries on the analysis of studies sheet will still hold good for all the unchanged elements in the new job sequence, and therefore it will be possible to make a fresh compilation after studying the new elements only.

The basic time, however, forms only a part of the standard time which has to be established for the job or operation. Certain allowances must be added before the standard time can be derived. These allowances must now be discussed; before doing so, however, it is necessary to state clearly what is meant by two terms which have been mentioned frequently in the preceding pages but which have not yet been precisely defined; namely work content and standard time.

## 18.9  Work content

In the chapters at the beginning of this book, the term "work content" was used frequently to describe what the words themselves suggest: the amount of work which has to be done to complete a job or operation, as distinct from any ineffective time which may occur. In time study practice, however, the word 'work' is accorded a meaning which is slightly different from its usual meaning in ordinary English usage. An observer who was familiar with the word only in its usual sense would say, when watching an operative at his job, that when the worker was actually doing something he was working, and that when he was resting or doing nothing he was not working. In time study practice, however, we are concerned with measuring work in numerical terms, and for this purpose the word 'work' is extended to include not only the physical labours performed but also the proper amount of relaxation or rest necessary to recover from the fatigue caused by those labours. We shall see later that relaxation allowances are made for other purposes besides recovery from fatigue; but for the moment the important point is that, when in time study we speak of 'work' and set out to measure it, we define work to include the appropriate relaxation allowance, so that the amount of work in a job is taken to be not only the time needed at standard performance to do whatever the job requires but also the additional time which is considered necessary for relaxation.

---

**The work content of a job or operation is defined as: basic time + relaxation allowance + any allowance for additional work—e.g. that part of contingency allowance which represents work**

---

## 18.10  Allowances

We have seen that, during the method study investigation which should be carried out before any job is timed, the energy expanded by the worker in performing the operation should be reduced to a minimum through the development of improved methods and procedures, in accordance with the principles of motion economy and, wherever practicable, by mechanisation. Even when the most practical, economic and effective method has been developed, the job will still require the expenditure of human effort, and some allowance must therefore be made for recovery from fatigue and for relaxation. Allowance must also be made to enable a worker to attend to his personal needs; and other allowances (e.g. contingency allowances) may have to be added to the basic time in order to give the work content.

The determination of allowances is probably the most controversial part of Work Study. For reasons that will be explained later, it is very difficult to determine precisely the allowances needed for a given job. What should therefore be attempted is an objective assessment of the allowances that can be consistently applied to the various elements of work or to various operations.

The fact that the calculation of allowances cannot be altogether accurate under all circumstances is no excuse for using them as a dumping ground for any factors that have been missed or neglected in making the time study. We have seen how the study man can go to great lengths to arrive at fair and accurate time standards. These should not be spoilt by the hasty or ill-considered addition of a few percentage points here and there "just in case". Above all, allowances should not be used as 'loosening' factors.

The difficulty experienced in preparing a universally accepted set of precise allowances that can be applied to every working situation anywhere in the world is due to various reasons. The most important among them are—

1) **Factors related to the individual.** If every worker in a particular working area was to be considered individually, it might well be found that a thin, active, alert worker at the peak of physical condition required a smaller allowance to recover from fatigue than an obese, inept worker. Similarly, every worker has a unique learning curve which can affect the manner in which he conducts his work. There is also some reason to believe that there may be ethnic variations in the response to the degree of fatigue experienced by workers, particularly when engaged on heavy manual work. Undernourished workers take a longer time than others to recover from fatigue.

2) **Factors related to the nature of the work itself.** Many of the tables developed for the calculation of allowances give figures which may be acceptable for light and medium work in industry but which are inadequate when applied to operations involving very heavy and strenuous work, such as work beside furnaces in steel mills. Moreover, every working situation has its own particular attributes which may affect the degree of fatigue experienced by the worker or may lead to unavoidable delay in the execution of a job. Examples of these factors are: whether a worker has to perform his work standing up or sitting down, and his posture during work; whether he has to exert force to move or carry loads from one place to another; whether the work itself results in undue eye or mental strain, and so on. Other factors inherent in the job can also contribute to the need for allowances, although in a different way—for example, when protective clothing or gloves have to be worn, or when there is constant danger, or when there is a risk of spoiling or damaging the product.

3) **Factors related to the environment.** Allowances, in particular relaxation allowances, have to be determined with due regard to various environmental factors such as heat, humidity, noise, dirt, vibration, light intensity, dust, wet conditions, and so on. Each of these will affect the amount of relaxation allowances needed. Environmental factors may also be seasonal in nature. This is particularly so for those who work in the open air, such as workers in the construction industry or in shipyards.

It should now be more clear to the reader why it is so difficult to devise an internationally accepted scheme of allowances to meet every working situation. It should also be stated here, in very clear terms, that the ILO has not adopted, nor is it likely to adopt, any standards relating to the determination of allowances. The following discussion quotes examples of the calculation of allowances under different conditions. They are quoted here as examples for training purposes and not as an ILO stand on the matter.

It should also be mentioned that this particular aspect of Work Study has been the subject of extensive research by various organisations which have put forward their own recommendations for the calculation of allowances. Of the more important research carried out, mention should be made of the work of the Max Planck Institut für Arbeitsphysiologie[1], of REFA Verband für Arbeitsstudien[2] and of G.C. Heyde in Australia[3].

## 18.11   Calculation of allowances

The basic model for the calculation of allowances is shown in Figure 81. It will be seen from this model that relaxation allowances (which are intended to aid recovery from fatigue) are the only essential part of the time added to the basic time. Other allowances, such as contingency, policy and special allowances, are applied under certain conditions only.

## 18.12   Relaxation allowances

> **Relaxation allowance is an addition to the basic time intended to provide the worker with the opportunity to recover from the physiological and psychological effects of carrying out specified work under specified conditions and to allow attention to personal needs. The amount of allowance will depend on the nature of the job.**

Relaxation allowances are calculated so as to allow the worker to recover from fatigue. Fatigue may be defined as a physical and/or mental weariness, real or imagined, existing in a person and adversely affecting his ability to perform work. The effects of fatigue can be lessened by rest pauses, during which the body recovers from its exertion, or by slowing down the rate of working and thus reducing the expenditure of energy.

Allowances for fatigue are normally added element by element to the basic times, so that a work value for each element is built up separately, the element standard times being combined to yield the standard time for the whole job or operation. In this way it is possible to deal with any extra allowance which may be required to compensate for severe climatic conditions, since the element may sometimes be performed in cool weather and sometimes when it is very hot. Allowances for climatic conditions have to be applied to the working shift or working day rather than to the element or job, in such a way that the amount of work which the worker is expected to produce over the day or the shift is reduced. The standard time for the job remains the same, whether the job is performed in summer or winter, since it is intended to be a measure of the work that the job contains.

Relaxation allowances have two major components; fixed allowances and variable allowances.

Fixed allowances are composed of—

1) Allowances for personal needs. This allowance provides for the necessity to leave the workplace to attend to personal needs such as washing, going to the lavatory and getting a drink. Common figures applied by many enterprises range from 5 to 7%.

2) Allowances for basic fatigue. This allowance, always a constant, is given to take account of the energy expended while carrying out work and to alleviate monotony. A common figure is 4% of basic time. This is considered to be adequate for a worker who carried out the job while seated, who is engaged on light work in good working conditions, and who is called upon to make only normal use of hands, legs and senses.

Variable allowances are added to fixed allowances when working conditions differ markedly from those stated above, for instance because of poor environmental conditions that cannot be improved, added stress and strain in performing the job in question, and so on.

As was mentioned above, a number of important studies have been carried out by various research organisations to try to develop a more rational approach to the calculation of variable allowances. Most management consultants in all countries have their own tables. In Appendix 3, we give an example of relaxation allowances tables using a points system. Many of these tables appear to work satisfactorily in practice; however, recent evidence indicates that, although many of the fatigue allowance scales established empirically in a laboratory are satisfactory on physiological grounds for work involving normal or moderately intensive effort, they provides inadequate allowances when applied to very heavy operations such as those connected with furnaces.

For the various reasons mentioned earlier in the chapter, when using one of the standard scales it is always preferable to check the amount of relaxation time they yield by carrying out whole-day studies at the workplace, noting the amount of time which the workers actually spend in relaxation (in one form or another) and comparing this with the calculated allowance. Checks of this sort do at least show whether the scale is, in general, too tight or too loose.

Relaxation allowances are given as percentages of the basic time. As mentioned earlier, they are normally calculated on an element-by-element basis. This is particularly the case when the effort expended on different elements varies widely (for example, where a heavy workpiece has to be lifted on or off a machine at the beginning and end of an operation). If, on the other hand, it is considered that no one element of a job is any more or any less fatiguing than any of the other elements, the simplest course is to add up all the elemental basic time first and then add the allowance as a single percentage to the total.

## REST PAUSES

Relaxation allowances can be taken in the form of rest pauses. While there is no hard and fast rule governing rest pauses, a common practice is to allow a 10 to 15 minutes break at mid-morning and mid-afternoon, often coupled with facilities for tea, coffee or cold drinks and snacks, and to permit the remainder of the relaxation allowance to be taken at the discretion of the worker.

Rest pauses are important for the following reasons:

1) They decrease the variation in the worker's performance throughout the day and tend to maintain the level nearer the optimum.

2) They break up the monotony of the day.

3) They give workers the chance to recover from fatigue and to attend to personal needs.

4) They reduce the amount of time off taken by workers during working hours.

## 18.13    Other allowances

It is sometimes necessary to incorporate allowances other than relaxation allowances in the compilation of standard time. Three such allowances are described below.

### CONTINGENCY ALLOWANCES

> A contingency allowance is a small allowance of time which may be included in a standard time to meet legitimate and expected items of work or delays, the precise measurement of which is uneconomical because of their infrequent or irregular occurrence.

Contingency allowances have already been mentioned when we described the calculations which have to be made to complete the study summary sheet and the analysis of studies sheet. The allowance provides for small unavoidable delays as well as for occasional and minor extra work, and so it would be proper to split the allowance into these components, the contingency allowance for work being allowed to attract fatigue allowance, just as any other items of work does, and the delay part of the allowance being given with only a personal needs increment. In practice this is a distinction which is often ignored. Contingency allowances are always very small, and it is usual to express them as a percentage of the total repetitive basic minutes in the job, adding them to the rest of the work in the job and adding a relaxation percentage to the whole contingency allowance. Contingency allowances should not be greater than 5%, and should only be given in cases where the study man is absolutely satisfied that the contingencies cannot be eliminated and that they are justified. On no account should such allowances be used as 'loosening' factors or to avoid carrying out proper time study practice. The duties for which the contingency allowance is given should be specified. However, in fairness, it may be necessary to give contingency allowances as a matter of course in enterprises where the production work is not well organised. This further stresses the need to make the conditions and organisation of work as good as possible before setting time standards and is an incentive to the management to do so.

## POLICY ALLOWANCES

> **A policy allowance is an increment, other than bonus increment, applied to standard time (or to some constituent part of it, e.g. work content) to provide a satisfactory level of earnings for a specified level of performance under exceptional circumstances.**

Policy allowances are not a genuine part of time study and should be used with the utmost caution and only in clearly defined circumstances. They should always be dealt with quite separately from basic times, and, if used at all, should preferably be arranged as an addition to standard times, so as not to interfere with the time standards set by time study.

The usual reason for making a policy allowance is to line up standard times with the requirements of wage agreements between employers and trade unions. In several enterprises in the United Kingdom, for example, the incentive performance is generally set at such a level that the average qualified worker, as defined, can earn a bonus of $33\frac{1}{3}\%$ of his basic time rate if he achieves standard performance. There is no need to apply a policy allowance to achieve this state of affairs; it is simply necessary to arrange for the rate paid per standard minute of work produced to be $133\frac{1}{3}\%$ of the basic time rate per minute, and in general it is better to accommodate any special wage requirements in this way, by adjusting the rate paid per unit of work rather than the standard time.

There are, however, certain employer-union agreements under which higher bonuses can be earned, and it may not be politic to seek a revision of the terms of these agreements to permit the achievement of their terms by modifying the rates paid rather than the times set. In these circumstances a policy allowance is given to make up the difference. It may be applied as a factor to the work content or to the standard time.

This might be an appropriate course to take when standard times are being introduced to only a small proportion of the total work-force covered by the agreement. Similar policy allowances are sometimes made as temporary additions to cover abnormal circumstances, such as the imperfect functioning of a piece of plant or disruption of normal working caused by rearrangements or alterations.

## SPECIAL ALLOWANCES

Special allowances may be given for any activities which are not normally part of the operation cycle but which are essential to the satisfactory performance of the work. Such allowances may be permanent or temporary; care should be taken to specify which. Wherever possible, these allowances should be determined by time study.

When time standards are used as the basis for a payment-by-results scheme, it may be necessary to make a start-up allowance to compensate for time taken by any work and any enforced waiting time which necessarily occurs at the start of a shift or work period before production can begin. A shut-down allowance may similarly be given for work or waiting time occurring at the end of the day. A cleaning allowance is of much the same character: it is given when the worker has to give attention from time to time to cleaning his machine or workplace. Tool allowance is an allowance of time to cover the adjustment and maintenance of tools.

It would be possible, after the time necessary to perform any or all of these activities has been studied, to express the result as a percentage of the total basic time for the operations expected to be performed during a day and to give the allowance as an increment included in the compilation of standard times. Indeed, this is sometimes thought to be the better course with tool allowance; but, in general, it is preferable to give all these allowances as periods of time *per day* rather than embodying them in the standard times. Usually this is fairer to the operatives, and it has the signal advantage of bringing to the attention of the management the total amount of time which has to be devoted to these activities, thus prompting thoughts about how it could be reduced.

Some allowances are normally given *per occasion* or *per batch*. One such allowance is set-up allowance, given to cover the time required for preparing a machine or process for production, an operation which is necessary at the start of production on a batch of fresh products or components. Set-up time is sometimes called make-ready time: its opposite is tear-down or dismantling time, for which a dismantling allowance may be given, to cover the time needed for making alterations to machine or process settings after completing a run of production. Very similar is change-over allowance, usually given to operatives who are not actually engaged in setting-up or dismantling, to compensate them for time on necessary activities or waiting time at the start and/or the end of a job or batch. These allowances should be denoted as "job change-over allowance" or "batch change-over allowance", as appropriate.

A reject allowance may be included in a standard time when the production of a proportion of defective products is *inherent* in the process, but is perhaps more usually given as a temporary addition to standard times, per job or per batch, if an occasional bad lot of material has to be worked. An excess work allowance, if necessary, would also be given as an addition to the standard time, to compensate for extra work occasioned by a temporary departure from standard conditions.

Learning allowances may be given to trainee operatives engaged on work for which standard times have been issued, as a temporary benefit while they develop their ability. A training allowance is a similar allowance given to an experienced worker to compensate him for the time he is required to spend instructing a trainee, while both are working on jobs for which standard times have been set. These allowances are often given as so many minutes per hour, on a declining scale so that the allowances taper off to zero over the expected learning period. Very similar is an implementation allowance, given to workers asked to adopt a new method or process to encourage them to attempt an enthusiastic implementation of the new ways and prevent their losing earnings by doing so. In fact, it is sometimes arranged that their earnings will actually be increased during the change-over period, so as to give the new method every chance of success. One system of implementation allowances credits the workers with ten minutes per hour on the first day, nine on the second, and so on down to zero.

A small batch allowance is required to enable a worker working on small batches to decide what to do and how to go about it (from instructions, by experience, or by trial and error) and then to work up to a standard performance by practice and repetition. The calculation of this allowance will depend on whether it is a one-of-a-type batch or not, on the length and batch size or run length and on the frequency of similar work and its degree of complexity.

## 18.14   The standard time

It is now possible to obtain a complete picture of the standard time for a straightforward manual job or operation, one which is considered to attract only the two allowances which have so far been discussed in detail: contingency allowance and relaxation allowance. The standard time for the job will be the sum of the standard times for all the elements of which it is made up, due regard being paid to the frequencies with which the elements recur, plus the contingency allowance (with its relaxation allowance increment). In other words—

> **Standard time is the total time in which a job should be completed at standard performance.**

The standard time may be represented graphically as shown in Figure 82.

*Figure 82. How the standard time for a simple manual job is made up*

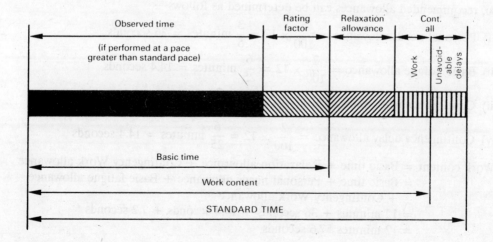

In a case where the observed time is rated at less than standard pace, the rating factor will, of course, be shown inside the observed time. The contingencies and relaxation allowances, however, are still percentages of the basic time. The standard time is expressed in standard minutes or standard hours.

In Chapter 19 we shall discuss the application of time study to operations involving the use of machinery, in which part of the operation time is taken up by work done by the machine while the operative stands by. An example of a fully worked time study is shown in Chapter 20. To explain the calculational steps, the following example is stated.

## 18.15   Example

The observed time is recorded to be 15 minutes for a job done by a worker whose rating is 80. Following allowances are recommended by the management —

i)   Personal needs allowance      — 5% of Basic time
ii)  Basic fatigue allowance       — 2% of Basic time
iii) Contingency work allowance    — 1% of Basic time
iv)  Contingency delay allowance   — 2% of Basic time

Determine basic time, work content and standard time for the job.

From the relationship,

$$\text{Basic Time} = \text{Observed time} \times \frac{\text{Rating}}{\text{Standard Rating}},$$

basic time for the job in the above example is calculated as,

$$\text{Basic time} = 15 \times \frac{80}{100} = 12 \text{ minutes.}$$

So, recommended allowances can be determined as follows—

i) Personal needs allowance $= \frac{5}{100} \times 12 = \frac{3}{5}$ minutes = 36 seconds

ii) Basic fatigue allowance $= \frac{2}{100} \times 12 = \frac{6}{25}$ minutes = 14.4 seconds

iii) Contingency work allowance $= \frac{1}{100} \times 12 = \frac{3}{25}$ minutes = 7.2 seconds

iv) Contingency delay allowance $= \frac{2}{100} \times 12 = \frac{6}{25}$ minutes = 14.4 seconds

Work content = Basic time + Relaxation allowance + Contingency Work allowance
= Basic time + Personal needs allowance + Basic fatigue allowance
+ Contingency Work allowance
= 12 minutes + 36 seconds + 14.4 seconds + 7.2 seconds
= 12 minutes 57.6 seconds

Standard time = Work content + contingency delay allowance
= 12 minutes 57.6 seconds + 14.4 seconds
= 13 minutes 12 seconds

## NOTES

1. G. Lehmann: *Praktische Arbeitsphysialogie* (Stuttgart, George Thieme Verlag, 1953).
2. REFA: *Methodenlehre des Arbeitsstudiums*, Vol. 2: *Datenermittlung* (Munich, Carl Hanser Verlag, 1971), pp. 299–335.
3. Chris Heyde: *The sensible taskmaster* (Sydney, Heyde Dynamics, 1976).

# Chapter 19

# Setting time standards for man-machine systems

## 19.1 Introduction

Advancement of Science and Technology has led to mechanisation and automation of production processes all over the world. It is becoming increasingly common for industrial jobs to be made up partly of elements performed manually and partly of elements carried out automatically by machines. The output for such jobs is not totally under the control of the worker. Consequently, setting time standards in such cases involves techniques that are different from those described in the last few chapters. The objective of this chapter is to present the procedure for finding standard times for jobs which are produced through man-machine interactions. For the purpose, the basic concepts and definitions are first presented in the next two sections, followed by a presentation of the procedure for finding the time required to produce one unit of output. The allowances to be provided and the determination of standard time are then discussed. In the concluding section, we discuss other areas of application of the above procedure, and briefly outline some measures for machine utilisation that provide valuable insights to setting time standards for man-machine system.

## 19.2 The concepts of restricted work and cycle time

The concepts of restricted work and cycle time are central to the understanding of the procedure of obtaining standard times for man-machine systems.

> **Restricted Work is defined as those work in which the output of the worker is limited by factors outside the control of the worker.**

> **Cycle time is the total time taken to complete all the elements in a work cycle.**

Any job produced through a man-machine system is essentially restricted work. In order to understand the dynamics behind setting time standards for such jobs, it is necessary to have a clear idea of the following:

a) What exactly constitutes a man-machine system.

b) How the cycle time for such systems is affected by the presence of machines.

We elaborate these issues in the following paragraphs:

A man-machine system is essentially a semi-automatic production process where men and machines are engaged together in continuous production of a job. Such jobs are composed of two components; one to be performed manually, and the other to be carried out by machines independently. A common example of such jobs is that of a worker running a single machine, with the machine working automatically for a part of the work cycle. The worker may perform the manual elements of his task at standard pace, or faster or slower; while this will influence the rate at which the operation is completed, it will not govern it, because the time during which the machine is working automatically will remain the same, irrespective of what the worker does. Cycle time and hence the output rate are thus determined by the nature of relationship between the manual and the machine component of the job. By our definition, these jobs are restricted work, compared to those jobs which are totally manual, or where the worker uses hand tools or power tools. The latter type of jobs can be called unrestricted work, in the sense that the output of the worker is limited by factors within his control. A man grinding a cutting tool on an electrically operated grindstone is engaged on unrestricted work, and so is a worker polishing a metal component by holding it against a power-driven polishing mop. In neither of these cases does the worker clamp the workpiece securely in position and leave the machine to get on with the work. The techniques for setting time standards discussed in the last few chapters pertain to unrestricted work. Our concern here is to arrive at a time standard for restricted work.

As noted earlier, the output of a man-machine system is dependant on the nature of relationship between the manual and the automatic component of the job in question. We demonstrate this through a simple example. Let us consider a job comprising the following elements:

| Elements | Time* (in mins.) | Remarks |
|---|---|---|
| 1. Load the machine | 0.6 | Manual |
| 2 Machining | 2.0 | Automatic |
| 3. Unload | 0.4 | Manual |
| 4. Inspect the job | 1.0 | Manual |

(*The time for the manual elements are as calculated at standard pace, while the time for the automatic element represents the actual time.)

We thus have a situation, where a worker is working on a single machine, with part of the job as manual and part as automatic. Once the machine is loaded, it starts machining on its own, apparently leaving the worker idle for this time. At the end of the machining process, the worker unloads the job and inspects it. This gives us a cycle time of 4 minutes. However, a closer look at the relationship between the manual and the automatic component of the job reveals a possibility of reducing the cycle time.

Assuming that the worker can carry out the inspection while the machine is running, we find that the cycle time reduces to 3 minutes. This also assumes a continuous process of production. Thus, it is not necessary that the worker carries out all his activities with the machine stopped. Examination of the elements of such jobs will reveal that the manual elements can be divided into two parts:

a) that which must be done with the machine stopped, such a loading and unloading in this case; and

b) that which can be done while the machine is running, such as inspection.

It is obviously an advantage to do as much as possible while the machine is running, as this reduces the cycle time. The multiple activity chart explained in Chapter 10, can be used here to depict the repetitive cycle. For the given example, we thus have:

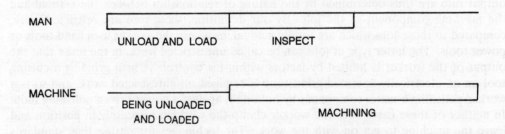

In the above example, we have not taken into account any allowances. Manual work has been calculated at standard pace and is thus shown in basic minutes. Machining time is shown in actual minutes, and so, using the 0–100 rating scale advocated in this book, basic minutes for manual work and actual minutes of machine operation are comparable and can be drawn to the same scale. The allowances that need to be considered are discussed separately in section 5.

Through a one-man one-machine example, we have illustrated how the presence of machine affects the output rate. Essentially the same concepts would apply even when the number of men or machines engaged in the job are more than one. As the number of machines per worker goes up, we are faced with one more complication, that of a machine waiting for the worker to attend to it. Proper synchronisation in such cases is extremely necessary to reduce the cycle time. Before we take this up in section 4, we present in the next section certain basic definitions pertaining to a man-machine system, essentially formalising certain concepts that have been discussed here.

## 19.3   Basic definitions pertaining to a man-machine system

It was apparent from the last section that a man-machine system may give rise to the following:

a) a period during which the machine is working independently;

b) a period in 'a' above, where the worker is also engaged;

c) a period in which the worker works, while the machining is not on;

d) a period during which the worker is idle; and

e) a period for which a machine waits for a worker to come and attend to it, (if the number of machines per worker is more than one).

The above are formalised through the definitions presented in the Table below:

Table 18: Basic definitions pertaining to a man-machine system

---

**Machine-controlled time** is the time taken to complete that part of the work cycle which is determined only by technical factors peculiar to the machine.

**Outside work** comprises elements which must necessarily be performed by a worker outside the machine-controlled time.

**Inside work** comprise those elements which can be performed by a worker within the machine-controlled time.

**Unoccupied time** comprises the periods during machine controlled time when a worker is neither engaged on inside work nor taking authorised rest.

**Load factor** is the proportion of the cycle time required by the worker to carry out the necessary work at standard performance, during a machine-controlled cycle.

**Machine Interference** is the queuing of machines for attention—e.g. when one worker is responsible for attending to more than one machine.

---

Following the above definitions, we find that in the example given in the last section, the machine controlled time, outside work, inside work and the load factor are 2 minutes, 1 minute, 1 minute and 67% respectively. While calculating the unoccupied time, the working time must first be calculated at standard performance and proper allowances are to be given.  In our case, we have assumed that the times are as per standard pace only; as such, once provisions for allowances are made, the idle time of the worker will be less than that depicted in the diagram.

Another convenient way of presenting these times is through a pump diagram. The pump diagram for the example is shown in Figure 83. The name is derived from the schematic representation which resembles a bicycle pump.

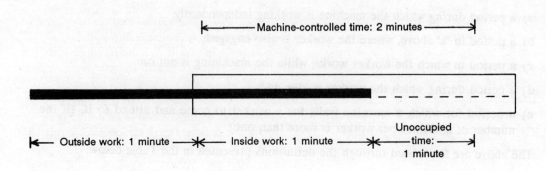

Figure 83. Pump diagram

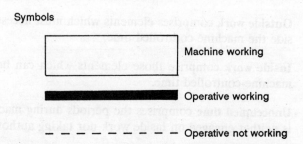

When seeking to improve the method, the work study man follows two main approaches. First, he tries to push the handle down into the pump—i.e. to arrange for some of the manual elements which are being performed outside the machine-controlled time to be carried out as inside work, thus shortening the work cycle (this has been done in the present example). Second, he gives close attention to shrinking the pump—making the machine-controlled time as short as possible by ensuring that the machine is being used to the best advantage, at the correct speeds and feeds, and using cutting tools which are correctly ground and made of the best type of cutting steel for the sort of work in hand, so that the machine running time is machine running time at standard.

Load factor and machine interference assume importance when the number of machines allotted per worker exceeds one. We elaborate on these in the following paragraphs.

The load factor is sometimes known by the alternative term extent occupied or work load. In the simplest case of one man operating one machine, if the overall cycle time is ten minutes and the amount of manual work contained within the cycle totals only 1 standard minute, the load factor would be one-tenth, or 10% per cent.

The reciprocal of the load factor therefore indicates the number of machines which the worker could theoretically tend. In practice, other factors have to be taken into account, so that the load factor can be taken only as a very rough first indication of the number of machines which can usefully be allocated to a worker. It does sometimes occur that the work elements consist solely of unloading finished pieces from machines which have stopped automatically, and by loading fresh pieces and restarting the machines; and if all the machines are alike and are working on exactly similar pieces, it may be possible to achieve the ideal sequence of operation, with the worker able to operate the number of machines indicated by the reciprocal of the load factor. Much more commonly, however, differences occur in the machines or in the work, and frequently attention has to be given to the machines while they are running, with the result that the worker cannot always get to a machine at the exact moment when attention is needed. The delays which then occur are known as machine interference.

When studying multiple machine working, the work study man has first to examine the methods of working with the object of devising a sequence of operations which will result in the least interference, and then to use time study techniques to measure the amount of interference which will occur even when the best sequence has been determined. These tasks may sometimes be extremely complicated. If one or two workers are operating only a few machines between them, simpler methods will suffice. Operation sequences can be plotted and examined on multiple activity charts (described in Chapter 10), supplemented by cycle diagrams similar to those shown in Figure 83. The diagrams for each machine are drawn one below the other, to the same time scale. A simple example, that of an operative working three machines, is shown in Figure 84.

*Figure 84. Machine interference*

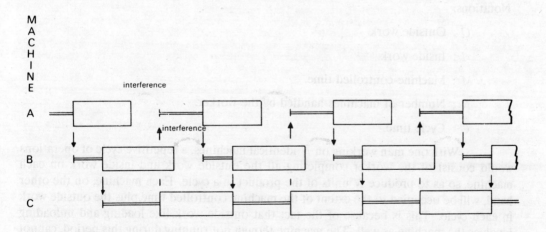

In this example there is no inside work, so that when a machine has been started the operative can turn his attention to another. The sequence in which he does so is indicated by the small vertical arrows. It will be seen that, with this particular routine, machine C is operated without any delays occurring; but the result of doing this is that both machine A and machine B switch themselves off at the end of their respective operations and then have to wait awhile before the operative can get to them. The interference is indicated on the cycle diagrams for machines A and B by arcs.

Having clarified the basic concepts and definitions, we are now in a position to discuss the theory behind determination of cycle time. We present this in the next section.

## 19.4 Determination of cycle time

The objective of this section is to formalise the procedure for determining the cycle time for man-machine systems. The assumptions and notations used in the process of derivation are given below:

Assumptions:

a) The production process is continuous and all the work elements are repeated in each cycle.

b) A worker is working on one or more number of machines of similar type.

c) A machine starts operating as soon as it is loaded and stops on its own at the end of the machining process.

d) The timings for each of the manual elements are determined at standard performance, and are in basic minutes.

e) Machine running times are the actual times, and the machines are working under optimum conditions.

Notations:

$O$ : Outside work.

$I$ : Inside work.

$M$ : Machine-controlled time.

$N$ : Number of machines handled by the worker.

$C$ : Cycle time.

With one man working on $N$ identical machines, a repetitive cycle of operations would consist of the worker completing all the outside work and inside work on each machine, so as to produce $N$ units of the product in a cycle. Each machine, on the other hand, will be occupied to the extent of the machine controlled time plus the outside work in each cycle. This is because of the fact that outside work like loading and unloading involves the machine as well. The machine though not running during this period, cannot be utilised for any other work.

Thus, we have:

Time for which the worker is busy in a cycle = $t_1$ (say) = $N(O + I)$
Time for which each machine is busy in a cycle = $t_2$ (say) = $O + M$
The cycle time $C$ is given by: $C = \text{Max} \{t_1, t_2\}$

This is illustrated now through an example. In the example of the preceding section, we find:

$$O = 1 \text{ minute}$$
$$I = 1 \text{ minute}$$
$$N = 1$$
$$M = 2 \text{ minutes}$$

Thus, $t_1 = 1(1+1) = 2$ minutes and $t_2 = 1 + 2 = 3$ minutes

Therefore, $C = \text{Max. } \{1, 3\} = 3$ minutes.

Suppose, in the above example, $M$ is charged to 3 minutes, $O$, $I$ values remain unchanged and the number of machines operated by the worker is two instead of one.

Then $t_1 = 2(1 + 1) = 4$ minutes, and $t_2 = (1 + 3) = 4$ minutes.

Hence $C = 4$ minutes.

Thus, in the first case one unit will be produced every three minutes, while in the second case, two units will be produced every four minutes. The above also gives us an idea about the operator and machine idle time per cycle. If $t_1$ dominates (i.e. $t_{-1} > t_{-2}$), then each machine remains idle for $(t_1 - t_2)$ time in a cycle; whereas $t_2 > t_1$ implies that the worker will be idle for $(t_2 - t_1)$ time. Thus in the first case shown here, the worker is idle for 1 minute per cycle. Finally, $t_1 = t_2$ implies that both the worker and the machines are fully utilised (as in the second case). It is instructive to draw the multiple activity chart in such cases.

In the above procedure, allowances have not been taken into account. To arrive at the standard time for a job, we need to incorporate allowances in the cycle time. The procedure for providing proper allowances for restricted work are discussed in the next section.

## 19.5  Determining the standard time

Standard times for jobs or operations are calculated on the basis of the work done by operatives. For a job made up totally of manual elements, the standard time is essentially a measure of the work which the job contains. With restricted work, however, the standard time expresses something more than this. It will be recalled that the definition of standard time is as follows:

> **Standard time is the total time in which a job should be completed at standard performance.**

In order to compile the standard time for a restricted operation, it is not sufficient simply to calculate the manual work content, as the machine influences the time as well. The normal procedure is to add the following allowances depending on the situation:

a) Relaxation allowance: As discussed in a previous chapter, this consists mainly of the personal need and fatigue allowances. The procedure for providing for such allowances for restricted work are different from that for unrestricted work and will be discussed here.

b) Unoccupied time allowance is provided to take care of idle time during machine-controlled time. As we have already seen, the worker may remain idle for a part of the cycle due to no fault of his own. Similarly, machine interference may actually reduce the output rate. These make it necessary to provide for allowances to account for the unavoidable unoccupied time.

c) Contingency Allowance: is the same as discussed under unrestricted work.

We now elaborate on the procedure for providing these allowances, to arrive at the standard time. Specifically, relaxation, unoccupied time and interference allowances are discussed here.

## RELAXATION ALLOWANCE

In restricted work, it is essential that the personal needs allowance and the fatigue allowance be calculated quite separately. The reason for this is that the personal needs allowance has to be calculated not simply on the elements of manual work contained in the work cycle but on the whole of the cycle time, including the machine-controlled time. This is because the percentage figures for the allowance are based on time spent at the workplace rather than on the time actually devoted to work. Fatigue allowance, on the other hand, is necessitated by work and is calculated on the basic minutes of work actually performed.

Apart from this difference, relaxation allowance is calculated in exactly the same way as was described in Chapter 18. When the allowance has been calculated, it is next necessary to consider whether the operative can be expected to take any or all of it within the work cycle or whether it must be added to the sum of outside work plus machine-controlled time to derive the true cycle time. If the work cycle is a very long one, and there are lengthy periods of unoccupied time within it, it may be possible in certain circumstances for the whole of the personal needs allowance and the fatigue allowance to be taken within the cycle, during the time when the operative is not working. Such periods can only be considered adequate for personal needs allowance if they are long enough (say, 10 or 15 minutes), if they occur in an unbroken stretch, and if it is possible for the operative to leave his machine unattended meanwhile. This may be done safely if the machine has an autostop mechanism and needs no attention whatever while it is running; alternatively, when groups of operatives work together, it is sometimes possible to arrange for a neighbour to use some of his own unoccupied time in giving attention to the absent worker's machine. In textile factories and in other industries in which the processing machinery is run continuously, perhaps 24 hours a day, it is common to provide floating workers who can fill in at work stations for odd moments and can help to keep the machines running during short meal breaks if these are taken at staggered times.

It is much more usual, however, especially with cycles of short duration, for the whole of the personal needs allowance to be taken outside the working cycle. In the example which has been illustrated earlier, which has a cycle time of 3 minutes, it would obviously be impossible for the operative to take any of his personal needs allowance within the cycle.

Fatigue allowance is a rather different matter. Quite short periods of unoccupied time can be used for recovery from fatigue, provided that the operative can truly relax during them and is not required to be constantly on the alert or to give attention to the machine during them, and that he has a seat nearby. It is generally considered that any period of 0.50 minutes or less is too short to be counted as available for relaxation, and that any unbroken period of 1.5 minutes or longer can be reckoned as fully available for recovery from fatigue. Periods of 0.50 minutes or less would thus be disregarded. For periods between 0.50 and 1.50 minutes, it is common to calculate the time which may be considered as effectively available for relaxation by deducting 0.50 minutes from the actual length of the period and multiplying the result by 1.5. The effect of applying this calculation to four periods between 0.50 and 1.50 minutes is shown below—

| Actual unbroken period of unoccupied time | Time calculated as effectively available for recovery from fatigue |
|---|---|
| 0.50 minutes | nil |
| 1.00 | 0.75 minutes |
| 1.25 | 1.12 |
| 1.50 | 1.50 |

It is quite common in machine operations for the workers to have to make adjustments or attend to the machine at intervals, or perhaps carry out manual elements on other workpieces from time to time while the machine is working, so that within the machine-controlled time there will be separated periods of inside work and unoccupied time. The length of the cycle and the manner in which any inside work occurs both affect the way in which relaxation allowance must be treated. Four cases can be distinguished:

1) All the personal needs allowance and all the fatigue allowance must both be taken outside the working cycle.

2) The personal needs allowance must be taken outside the cycle, but all the fatigue allowance can be taken within it.

3) The personal needs allowance and some of the fatigue allowance must be taken outside the cycle, but the rest of the fatigue allowance can be taken within it.

4) All the personal needs allowance and all the fatigue allowance can be taken within the working cycle.

The effect of these four cases for four different operation sequences is illustrated in Figure 85. All the four operations have the following characteristics in common:

| | |
|---|---|
| Machine-controlled time | 15 minutes |
| Outside work | 10 basic minutes |
| Inside work | 5 basic minutes |
| Personal needs allowance: 5% of outside work plus machine-controlled time | 1.25 minutes |
| Fatigue allowance: 10% of total basic minutes | 1.50 minutes |

*Figure 85. Four operations with machine elements*

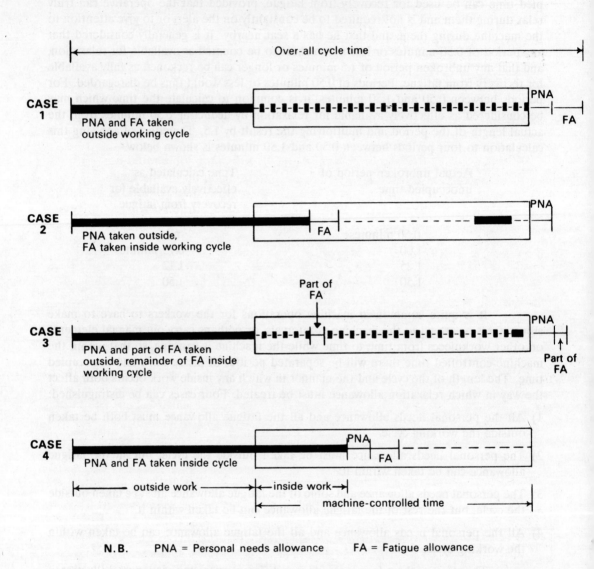

Over-all cycle time

**CASE 1**
PNA and FA taken outside working cycle
PNA
FA

**CASE 2**
PNA taken outside, FA taken inside working cycle
FA
PNA

**CASE 3**
Part of FA
PNA and part of FA taken outside, remainder of FA inside working cycle
PNA
Part of FA

**CASE 4**
PNA and FA taken inside cycle
PNA
FA
outside work — inside work
machine-controlled time

**N.B.**   PNA = Personal needs allowance   FA = Fatigue allowance

In Case 3 above, there is a period of 1 minute within the machine-controlled time when the operative is not working. By using the method of calculation described above, 0.75 minutes of this is considered to be available for recovery from fatigue, so that the remaining 0.75 minutes of the fatigue allowance has to be taken outside the working cycle. In Case 4, the assumption has been made that a neighbouring worker could attend to the operation if it should be necessary for the operative to leave his work station for longer than the ten minutes of non-working time available during the machine element. It will be seen that the over-all cycle time differs in each of the four cases, so that the number of units of output which could be expected over eight-hour days also differs:

| | Overall cycle time (minutes) | Anticipated daily output (units) |
|---|---|---|
| Case 1 | 27.75 | 17.3 say, 17 |
| Case 2 | 26.25 | 18.3 say, 18 |
| Case 3 | 27.00 | 17.7 with overtime, 18 |
| Case 4 | 25.00 | 19.2 say, 19 |

> **The overall cycle time is the total time in which the job should be completed at standard performance, and is made up (in the case of operations of the types so far discussed) of outside work at standard pace, machine-controlled time, and any portion of the relaxation allowance which has to be allowed outside the machine-controlled time.**

If there are no other allowances to be taken into account (e.g. contingency allowance), and an allowance is made for unoccupied time in actual minutes, the overall cycle time will be numerically equal to the standard time for the operation.

## UNOCCUPIED TIME ALLOWANCE

In the example of one man working on a machine (section 19.2), we have seen that the worker was idle for 1 minute in each cycle. Thus, if standard time is calculated based on the manual work only (i.e. outside plus inside work), it will be unfair, as this idle time is not due to the fault of the worker. It is thus customary to incorporate this time in the standard time. This is called unoccupied time allowance. However, in the earlier example, we have noted that the unoccupied time is to be calculated after the work has been measured at a standard performance and the relaxation allowances are given.

The unoccupied time is calculated by subtracting from the machine-controlled time the sum of all periods of inside work (in basic minutes), plus any part of the relaxation allowance which may be taken within the machine-controlled time. Before providing this as an allowance, the work study man has to verify that this time is truly unavoidable and it cannot be reduced further by method improvement or by a reallocation of work or machines. It may be noted that it is worthwhile to have some in-built unoccupied time, if by so doing, costly machines can be kept more fully employed.

When standard times are used as a basis for payment-by-results schemes, the inclusion of unoccupied time allowances in standard times for restricted work may give rise to payment anomalies, unless special measures are taken to deal with the problems which arise. The sort of difficulty which can occur is most easily seen by considering an example. Let us assume that in a given enterprise there are three jobs, for each of which the standard time has been calculated as 100 minutes. The first job is made up wholly of manual elements. The other two are both restricted operations and for both the standard times include allowances for unoccupied time—say, 15 minutes in one case, and

45 minutes in the other. If all three workers perform the manual elements of their tasks at standard pace and all take exactly the allotted relaxation periods, all three jobs will be completed in the same time (100 minutes). But the operative on unrestricted work will have been working all the time (except, of course, for the relaxation period) while the other two will have been idle for 15 and 45 minutes respectively. If payment is made for unoccupied time at the same rate as that for working time, the more heavily loaded workers will soon become discontented; jobs will become known as good jobs or bad jobs according to the amount of unoccupied time they contain; and there will be reluctance to undertake tasks with the higher work contents. Usually this difficulty is dealt with not by modifying the standard times but by establishing different rates of payment for work and for idle time. To enable this to be done, it is usual to express standard times not only as totals but also as work credits plus idle time credits (or in similar terms). Thus, in the example cited above, the standard time (100 minutes in each instance) would be shown as being made up of 100, 85 and 55 work credits plus 0, 15 and 45 idle time credits respectively. It may be noted in passing that idle time credits included in a standard time may be allocated for reasons other than unoccupied time as discussed above. Idle time credits may sometimes be necessary to compensate for delays caused by waiting for work or for instructions, or by machine breakdowns.

The scheme to be adopted to make differential payments for work and for idle time in a particular enterprise is a matter of wages administration, rather than of time study practice, and is thus outside the scope of this introductory book. It may be noted, however, that any such scheme should be simple to understand, so that the workers may readily comprehend why jobs taking the same time to complete attract different payments. The scheme should be negotiated and agreed with the workers' representatives before it is applied. In a typical scheme, idle time credits amounting in total to less than 5% of the work credits may be paid for at the same rate as work credits; idle time amounting to 40% or more of the work credits at three-quarters of the rate of working; and idle times between 5% and 40% at varying rates in between. The scheme which will be most appropriate for a particular organisation will depend on local circumstances, and especially on whether jobs with large amounts of unoccupied time are exceptional or common. Sometimes variable rates which have to be read off a curve are adopted, but in general a linear relationship is to be preferred, and always one which is simple.

The time study man is concerned primarily with measuring the amount of time needed to complete a job or operation, rather than with whatever arrangements are agreed for making payment for that time. It is common in industrial wage agreements to take account of different levels of skill required for different operations, by paying differing rates per minute or per hour of work. Other factors may also be taken into account in setting payment rates. None of these matters will affect the calculation of any unoccupied time allowance which may be necessary to compile the standard time for a job. The time allowance will be in minutes or hours: payment for those minutes or hours will be negotiable quite separately. In the scheme mentioned above, relatively long periods of unoccupied time are paid for at lower rates than those paid for working. In some circumstances, however, it may be appropriate to pay for both working time and unoccupied time at very high rates indeed, in which case the payment actually made to a particular operative for a minute of unoccupied time may be greater than that paid to another for a minute spent working. An example is the final machining of a shaft for a turbine-driven electricity generating set. Such a shaft may be several metres in length, and by the time that the last stages of machining are undertaken the component will represent a large investment, in terms of both labour and the costly materials of which it is made. A faulty cut may result in a diameter becoming undersize, with the result that the whole shaft would have to be scrapped. The operative is thus burdened with a very heavy responsibility, although the actual operation itself is not particularly complex. Because of this responsibility the rates paid to the operative, both for working and for any necessarily unoccupied time, may be higher than those for the general run of turning operations. Similar, key operations or tasks occur in many industries.

## INTERFERENCE ALLOWANCE

An interference allowance is an allowance of time for production unavoidably lost through synchronisation of stoppages on two or more machines attended by one worker.

The relationship between interference and unoccupied time can be understood better with the help of an example. Let us assume that the job consists of the following outside and inside work:

$$\text{Outside work: loading and unloading } = O = 1 \text{ minute}$$
$$\text{Inside work: attending the machine } = I = 1 \text{ minute}$$

Let the machine controlled time be 3 minutes. With one man working on one machine, the question of interference does not arise. The unoccupied time can be calculated as before, with

$$t_1 = N(O + I) = 1(1 + 1) = 2 \text{ minutes}$$
$$t_2 = O + M = 1 + 3 = 4 \text{ minutes}$$

Cycle time = Max $\{t_1 + t_2\}$ = 4 minutes

and unoccupied time = $t_2 - t_1$ = 2 minutes (assuming that the relaxation allowances have been provided for as discussed).

From the reciprocal of the load factor (50% in this case), it is clear that the worker should be able to handle two machines. With two machines, the cycle time works out to 4 minutes and the unoccupied time becomes zero. However, this assumes that during the machine-controlled time of the first machine, the worker performs the inside work of both the machines and also the outside work of the second machine. This assumption may not always be valid. In this case especially, it may be logical to assume that the inside work of attending the machine when it is running is not in a single stretch of 1 minute, instead, it is to be done in two stretches of 0.5 minutes each. It may so happen that the second period of inside work of 0.5 minutes on the second machine occurs simultaneously with the outside work on the first machine, both demanding the attention of the worker. The worker thus faces machine interference. This results in the lengthening of the cycle time by a period equal to the interference (here 0.5 minutes). The interference allowance is given to take care of the output lost due to the interference. It may be noted that the worker is not unoccupied during the period of interference.

Thus, for restricted work with a worker working on a single machine, the standard time is given by:

> **Standard time = Outside work + Inside work + Relaxation allowance + Unoccupied time Allowance**

The same formula would apply for multi-machine case also, provided the following conditions are met:

a) O.I.M. are the same for all the machines and they remain constant in each cycle of operation.

b) I is done in a single stretch, just after or just before the outside work.

In case, the conditions are violated, interference occurs and interference allowances are to be calculated and used in the standard time calculation. Detailed discussion on interference allowance is beyond the scope of this introductory book and the reader may refer to the readings mentioned at the end of this part.

## 19.6   Conclusion

The objective of this chapter has been to introduce the reader to the procedure for determining standard time for man-machine system. Man-machine system has been defined in the context of restricted work situation, as a system where the output is not totally under the control of the worker. There are other types of system which can be categorised as restricted work, and the procedure outlined in this chapter applies for those systems as well. Other examples of restricted work occur when—

a) Operatives are in control of processes, their principal duties being to observe the behaviour of the processes or instruments recording their behaviour and to take action only in response to changes in behaviour, state or reading.

b) Two or more operatives are working as a team, dependent on one another, and it proves impossible completely to balance the work-load of each, with the result that some workers are left with periods of idleness within the work cycle.

Team working can give rise to restricted work even when no machines are used. Assembly work carried out in conjunction with moving conveyors usually does. Even if the conveyor is used simply to transport pieces from one work station to the next, with each operative taking a component off the belt to work on it and returning it when he has finished, a restriction may be imposed by having to wait for the next piece. Again, when assembly operations are carried out directly on the moving conveyor, as is done in motor vehicle manufacture, the conveyor produces conditions equivalent to those imposed by a static production machine.

Finally, before embarking on any study on man-machine system, it is important for the management and the work study man to be clear about the status of the system to be studied. One should be clear about the degree of importance laid on labour and machine utilisation. Machine utilisation studies thus can be used as an input before attempts are made for the setting of time standards for a man-machine system. The terms and concepts used in the study of machine utilisation (or plant or process utilisation) are described below. They are largely self-explanatory. The relationship between them is shown graphically in Figure 86.

---

**Machine maximum time** is the maximum possible time during which a machine or group of machines could work within a given period, e.g. 168 hours in one week or 24 hours in one day.

**Machine available time** is the time during which a machine could work based on attendance time—i.e. working day per week plus overtime.

**Machine idle time** is the time during which a machine is available for production or ancillary work but is not used owing to shortage of work, materials or workers, including the time that the plant is out of balance.

**Machine ancillary time** is the time when a machine is temporarily out of productive use owing to change-overs, setting, cleaning etc.

**Machine down time** is the time during which a machine cannot be operated on production or ancillary work owing to breakdown, maintenance requirements, or for other similar reasons.

**Machine running time** is the time during which a machine is actually operating. i.e., the machine available time less any machine down time, machine idle time, or machine ancillary time.

---

The machine running time is observable by direct study at the workplace. It does not follow, however, that the machine, though running, is actually operating in the manner in which it should, or has been set so as to perform in the very best manner of which it is capable. It is useful therefore to introduce another concept—

---

**Machine running time at standard**. This is the running time that should be incurred in producing the output if the machine is working under optimum conditions.

---

The most useful work measurement method for studying machine utilisation is work sampling, as described in Chapter 14. This technique gives the information required with much less effort than would be needed with time study, especially when many machines are involved.

It is convenient to express the results obtained from studies on machine utilisation in the form of ratios or indices. For this purpose three indices are commonly used:

Figure 86. Explanatory diagram of machine time

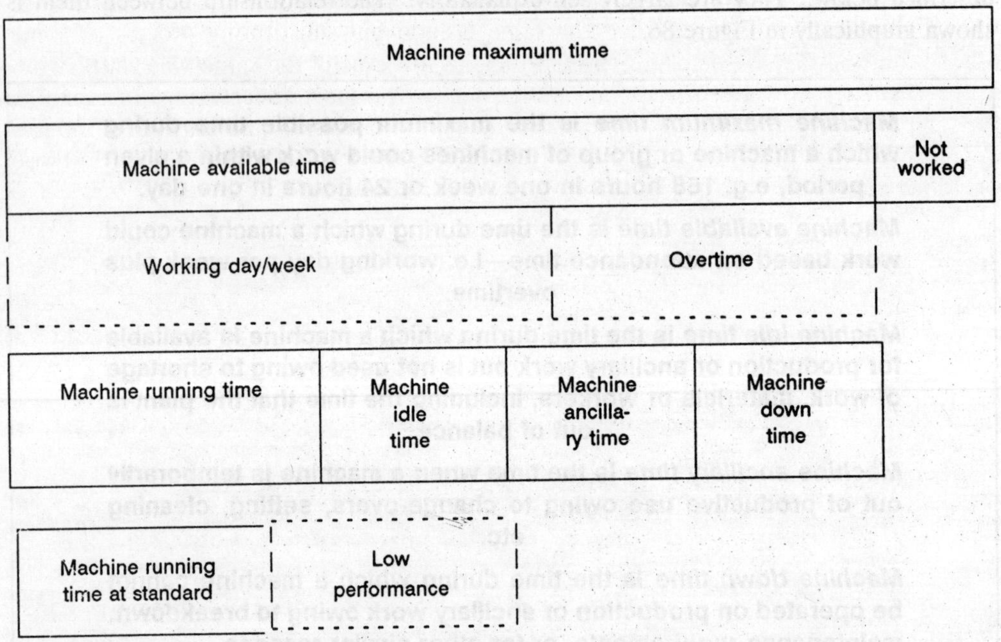

| Machine maximum time | | |
|---|---|---|

| Machine available time | | Not worked |
|---|---|---|

| Working day/week | Overtime |
|---|---|

| Machine running time | Machine idle time | Machine ancilla- ry time | Machine down time |
|---|---|---|---|

| Machine running time at standard | Low performance |
|---|---|

Source: Based on a diagram contained in the B.S. Glossary, op. cit.

1) Machine utilisation index, which is the ratio of

        machine running time to

        machine available time

and thus shows the proportion of the total working hours during which the machine has been kept running.

2) Machine efficiency index, the ratio of

        machine running time at standard to

        machine running time.

A ratio of 1.0 (or 100%, as it would usually be expressed) would indicate the ideal state, with the machine always performing to the best of its capability whenever it is running.

3) Machine effective utilisation index, the ratio of

        machine running time at standard to

        machine available time.

This ratio can be used to provide an indication of the scope for cost reduction that would be available if the machine were operated at full efficiency for the whole of the working time.

      When work measurement has been applied throughout an organisation, it is an easy matter to arrange for these indices and others like them to be reported to top management as routine at regular intervals, for they can be calculated quite simply from the records instituted to maintain labour, output and machine controls. The incidence of idle time, down time and ancillary time can be highlighted by expressing these figures as ratios in a similar way, using either machine available time or machine running time as the base.

In process industries, utilisation studies are carried out in much the same way, the terms and concepts applied in the same fashion but substituting process or some other suitable word for machine. The principles are exactly the same when utilisation in service undertakings is considered: in a passenger transport undertaking, for example, the same useful results could be expected to accrue from studying the utilisation of buses or trains and expressing the results being achieved in the form of indices similar to those described above.

# Chapter 20

# Example of a time study

A complete example of a time study is presented in this Chapter to enable the reader to develop an understanding of the detailed process of compilation of standard time.

This particular example has been chosen because—

a) it is simple;

b) it has already been the subject of a method study;

c) it includes both manual and machine elements; and,

d) it is typical of the sort of operation met everywhere in the engineering industry and in other industries using machines and semi-automatic processes.

The forms used are simple general-purpose forms such as those illustrated in Chapter 15. Although all the entries made on the forms will be handwritten, it is usual 'o space the lines for use with a typewriter because occasions may arise on which it is required to produce fair copies of original studies for discussion or circulation.

The study illustrated in this chapter was not the first one on this operation. The elements and break points were defined at the time the method study was undertaken, and were then set out on a card prepared and filed by the work study department. This is a useful practice when it is expected that an operation will be studied several times, perhaps by different study men. It ensures that the recordings made on all the studies are comparable.

Although the example which has been studied in detail is a simple one for a manufacturing industry, exactly the same procedure is carried out for non-manufacturing operations or for any other work which is time-studied for the purpose of setting time standards. Entirely manual operations, such as assembly, would be treated in exactly the same way.

The study is illustrated through Figures 87 to 98. Explanatory notes are provided with the figures wherever necessary.

*Figure 87. Card giving details of elements and break points*

---

Card No. 1264

| | |
|---|---|
| *Part:* | B. 239 Gear case |
| Material: | ISS 2 Cast iron |
| Operation | Finish-mill second face |
| Machine | No. 4 Cincinnati vertical miller |
| Fixture | F.239 |
| Cutter | 25 cm. TLF |
| Gauge | 239.7.  Surface plate. |

Drawing: 239/1

---

### Elements and Break Points

A. Pick up casting, locate in fixture, lock two nuts, set guard, start machine and auto feed. Depth of cut 2.5 mm. Speed 80 r.p.m. Feed 40 cm/min.

   *Break point:* Machine commences cut.

B. Hold casting, break milled edge with file, clean with compressed air.

   *Break point:* Air gun dropped on to hook.

C. Move depth gauge to casting, check machined surface, move gauge away.

   *Break point:* Left hand releases gauge.

D. Pick up machined casting, carry to finished parts box and place aside, pick up next part and position on machine table.

   *Break point:* Casting hits table.

E. Wait for machine to complete cut.

   *Break point:* Machine ceases to cut.

F. Stop machine, return table, open guard, unlock fixture, remove machined casting and place on surface plate.

   *Break point:* Casting hits surface plate.

G. Clear swarf from machine table with compressed air.

   *Break point:* Air gun dropped on to hook.

---

*Note:* Elements B. C and D are inside work, and are performed on a casting which has already been machined while the milling machine is cutting the next casting. Element D includes bringing up into a handy position a fresh casting which will be machined after the one now in the machine.

*Figure 88. Sketch of part and of workplace layout*
*(on reverse of time study top sheet)*

A sketch of the workplace layout is generally more necessary in assembly or material-handling studies than in studies of machine shop operations where workplaces are likely to be the same for all jobs on the machines. The part should be sketched showing the surfaces machined; in the case of capstan lathes, tool set-ups should be included. This is best done on squared paper and may be on the back of the time study top sheet, if desired, in order to keep all the information relevant to the study on one sheet. To facilitate sketching, the reverse of the top sheet is often printed as squared paper.

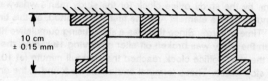

*(b)* Layout of **workplace**

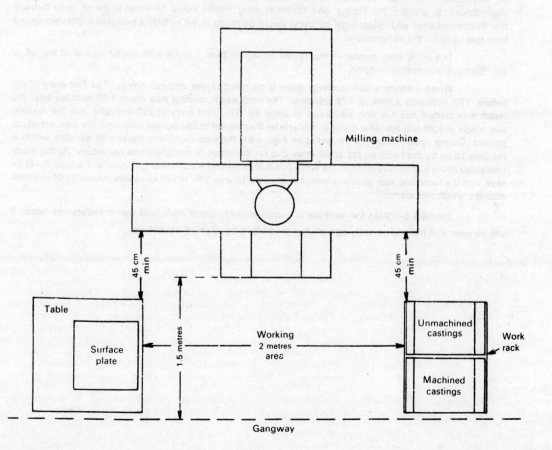

*Figure 89. Time study top sheet*

All the information in the heading block at the top of the form (except time off and elapsed time) was entered before the stop-watch was started and study commenced.

If the study had been the first one on this operation, the study man would have entered in full the element descriptions and break points in the column headed "Element description" on the left-hand side of the page. In the present instance this was not necessary, as the card shown in Figure 87 listed all the details. The study man should watch a few cycles of the operation to make sure that the listed method is being used, and to familiarise himself with the break points, before starting to record. The elements were identified simply by the letters A to G.

At exactly 9.47 a.m. by the study office clock (or the study man's wrist-watch) the stop-watch was started. It ran for 1.72 minutes before element A of the first cycle started, so this time is entered at the beginning of the study as the "Time before". Since this was a study using cumulative timing, the watch ran continuously throughout. When the study was broken off after observing 18 cycles, the study man allowed his stop-watch to run on until the study office clock reached the next full minute (at 10.25 a.m.), noted the "Time after", and stopped his stop-watch. These terminal entries will be found at the end of the recordings in Figure 91.

The four columns used in cumulative timing are respectively 'Rating' (R), 'Watch reading' (WR), 'Subtracted time' (ST) and 'Basic Time' (BT). The placing of the rating column first is logical and encourages the observer to rate while the element is in progress and not to wait for the watch reading. If flyback timing had been used, the WR column on the form would not be necessary.

Only the entries in the two columns headed R and WR were made during observations at the workplace. The other two columns were completed in the study office after observations had been discontinued. In practice, the 'Rating' and 'Watch reading' entries would be made in pencil while those in the 'Subtracted time' and 'Basic time' columns would be made in ink or with a pencil of a different colour from that used for the observations.

The study man numbered the cycles observed, from 1 to 18, with ringed figures at the left of the "Element description" column.

When entering watch readings there is no need to use decimal points. The first entry (Time before, 172) indicates a time of 1.72 minutes. The next watch reading was made 1.95 minutes after the watch was started, but it is only necessary to enter 95. The third entry of 220 indicates that the reading was made at 2.20 minutes after starting; the entries then revert to two figures only until the next minute is passed. During cycle number 15 (recorded on Figure 91) the total study time passed 30 minutes, which is the time taken by the hand on the small inner dial on the watch to complete one revolution. As the study continued into a further revolution of the small hand, subsequent watch readings revert to 1 again. It will be seen that the recording was against element F of cycle 15 was 106, which of course means 31.06 minutes after the watch was started.

Element E—"Wait for machine to complete cut"—is not work, and was therefore not rated. It will be seen that there is no entry against this element in the 'Basic time' column.

# TIME STUDY TOP SHEET

| | | |
|---|---|---|
| DEPARTMENT: *Machine Shop – Milling Section* | | STUDY No. *17* |
| OPERATION: *Finish-mill second face*     M.S. No. *9* | | SHEET  No. *1*  OF *5* |
| | | TIME OFF: *10.25 a.m.* |
| PLANT/MACHINE: *Cincinnati No. 4 vertical miller* No. *26* | | TIME ON: *9.47 a.m.* |
| | | ELAPSED TIME: *38.00* |
| TOOLS AND GAUGES: *Fixture F 239 : Cutter 25 cm TLF* | | OPERATIVE: |
| *Gauge 239/7 : Surface plate* | | CLOCK No. *1234* |
| PRODUCT/PART: *B. 239 Gear Case*        No. *239/1* | | STUDIED BY: |
| DWG. No. *B. 239/1*   ISS. *2*   MATERIAL: *Cast iron* | | DATE: |
| QUALITY: *As drawing* | | CHECKED: |

NOTE: Sketch the WORKPLACE LAYOUT/SET-UP/PART on the reverse or on a separate
sheet and attach

| ELEMENT DESCRIPTION | R | WR | ST | BT | ELEMENT DESCRIPTION | | R | WR | ST | BT |
|---|---|---|---|---|---|---|---|---|---|---|
| *Time before* | – | 172 | – | – | ④ | A | 80 | 622 | 32 | 26 |
| ①            A | 110 | 95 | 23 | 25 | | B | 85 | 50 | 28 | 24 |
| B | 100 | 220 | 25 | 25 | | C | 85 | 63 | 13 | 11 |
| *Elements & B.P.*  C | 100 | 32 | 12 | 12 | | D | 85 | 83 | 20 | 17 |
| *as Card No. 1264* D | 95 | 52 | 20 | 19 | | E | – | 703 | 20 | – |
| E | – | 77 | 25 | – | | F | 105 | 26 | 23 | 24 |
| F | 110 | 300 | 23 | 25 | | G | 85 | 38 | 12 | 10 |
| G | 110 | 08 | 08 | 09 | | | | | | |
| | | | | | ⑤ | A | 80 | 70 | 32 | 26 |
| ②            A | 110 | 31 | 23 | 25 | | B | 85 | 97 | 27 | 23 |
| B | 95 | 58 | 27 | 26 | | C | 85 | 810 | 13 | 11 |
| C | 95 | 71 | 13 | 12 | | D | 85 | 30 | 20 | 17 |
| D | 100 | 89 | 18 | 18 | | E | – | 53 | 23 | – |
| E | – | 412 | 23 | – | | F | 105 | 76 | 23 | 24 |
| F | 105 | 37 | 25 | 26 | | G | 85 | 88 | 12 | 10 |
| G | 100 | 47 | 10 | 10 | | | | | | |
| | | | | | ⑥ | A | 95 | 915 | 27 | 26 |
| ③            A | 105 | 72 | 25 | 26 | | B | 95 | 42 | 27 | 26 |
| B | 105 | 97 | 25 | 26 | | C | 105 | 54 | 12 | 13 |
| C | 95 | 510 | 13 | 12 | | D | 80 | 77 | 23 | 18 |
| D | 110 | 28 | 18 | 20 | | E | – | 97 | 20 | – |
| E | – | 53 | 25 | – | | F | 95 | 1020 | 23 | 22 |
| F | 100 | 78 | 25 | 25 | | G | 100 | 30 | 10 | 10 |
| G | 95 | 90 | 12 | 11 | | | | | | |
| | | | | | | | | | | |
| | | | 418 | | | | | | 440 | |

*Figure 90. Time study continuation sheet*

The recordings covered three sheets in all. Figure 90 shows the first of the two continuation sheets, and it will be seen that it is numbered in the top right-hand corner: Sheet No. 2 of 5. The analysis sheet and study summary sheet eventually completed the set of five sheets, all of which were stapled together after the study was worked up.

Besides the element ratings and timings, continuing as on the top sheet, two interruptions were recorded on this sheet: "Talk to foreman", and "Break for tea". Neither of these was rated, of course. The first was taken account of when considering contingencies, while the second was covered by the relaxation allowance made when the standard time for the operation was compiled.

| STUDY No.: 17 | TIME STUDY CONTINUATION SHEET | | | | | SHEET No. 2 OF 5 | | | |
|---|---|---|---|---|---|---|---|---|---|
| ELEMENT DESCRIPTION | | R | WR | ST | BT | ELEMENT DESCRIPTION | R | WR | ST | BT |
| ⑦ | A | 105 | 55 | 25 | 26 | ⑪ A | 115 | 86 | 25 | 29 |
| | B | 115 | 78 | 23 | 26 | B | 95 | 1713 | 27 | 26 |
| | C | 95 | 91 | 13 | 12 | C | 75 | 28 | 15 | 11 |
| | D | 85 | 1113 | 22 | 19 | D | 85 | 50 | 22 | 19 |
| | E | – | 36 | 23 | – | E | – | 68 | 18 | – |
| | F | 80 | 68 | 32 | 26 | F | 115 | 90 | 22 | 25 |
| | G | 95 | 80 | 12 | 11 | G | 80 | 1803 | 13 | 10 |
| | | | | | | | | | | |
| ⑧ | A | 75 | 1218 | 38 | 28 | ⑫ A | 95 | 30 | 27 | 26 |
| | B | 110 | 40 | 22 | 24 | B | 95 | 55 | 25 | 24 |
| | C | 105 | 52 | 12 | 13 | C | 100 | 67 | 12 | 12 |
| | D | 100 | 70 | 18 | 18 | D | 95 | 87 | 20 | 19 |
| | E | – | 1300 | 30 | – | E | – | 1902 | 15 | -- |
| | F | 115 | 25 | 25 | 29 | F | 95 | 30 | 28 | 27 |
| | G | 105 | 35 | 10 | 10 | G | 75 | 42 | 12 | 09 |
| | | | | | | | | | | |
| Talk to foreman | | – | 75 | 40 | – | Break for tea | – | 2554 | 612 | – |
| | | | | | | | | | | |
| ⑨ | A | 105 | 1400 | 25 | 26 | ⑬ A | 85 | 86 | 32 | 27 |
| | B | 100 | 25 | 25 | 25 | B | 80 | 2618 | 32 | 26 |
| | C | 95 | 38 | 13 | 12 | C | 85 | 33 | 15 | 13 |
| | D | 95 | 56 | 18 | 17 | D | 100 | 53 | 20 | 20 |
| | E | – | 81 | 25 | – | E | – | 68 | 15 | – |
| | F | 100 | 1509 | 28 | 28 | F | 85 | 96 | 28 | 24 |
| | G | 85 | 21 | 12 | 10 | G | 95 | 2708 | 12 | 11 |
| | | | | | | | | | | |
| ⑩ | A | 95 | 43 | 22 | 21 | ⑭ A | 80 | 40 | 32 | 26 |
| | B | 80 | 75 | 32 | 26 | B | 100 | 65 | 25 | 25 |
| | C | 95 | 88 | 13 | 12 | C | 85 | 80 | 15 | 13 |
| | D | 95 | 1608 | 20 | 19 | D | 95 | 2800 | 20 | 19 |
| | E | – | 25 | 17 | – | E | – | 22 | 22 | – |
| | F | 105 | 48 | 23 | 24 | F | 80 | 54 | 32 | 26 |
| | G | 85 | 61 | 13 | 11 | G | 105 | 64 | 10 | 10 |
| | | | | | | | | | | |
| | | | | 631 | | | | 1203 | | |

*Figure 91. Second continuation sheet*

The first entry on this sheet recorded another interruption—the patrol inspector, having checked three workpieces, drew the operative's attention to some features of them and discussed them with him. The time taken to do this, like that recorded on the previous sheet against "Talk to foreman", was later entered as a contingency.

After cycle number 16, a fresh element of work occurred—helping the labourer to move boxes of work off and on to the truck. This was an occasional element, in contrast with elements A to G which were repetitive. The study man rated and timed the element, and it will be noted that, since the element ran on for over a minute in all, the study man made a rating and a watch reading at the end of each of the first two half-minutes, as well as during the last part of the element. This practice, which makes for greater accuracy, was referred to in section 9 of Chapter 17.

Back in the study office after breaking off observations, the study man first completed the "Time off" and "Elapsed time" entries in the heading block on the top sheet, and then set about calculating the subtracted times, by deducting each watch reading from the one which follows it and entering the result in the third column, headed ST. It will be seen that he totalled these subtracted times at the foot of each page, and carried forward the subtotals to the sheet shown opposite, where they were added up to yield 35.20 minutes. When the time before and the time after were added to this figure, the result was 38 minutes, which agreed with the elapsed time and thus afforded a check that the work of subtraction had been done correctly.

The next step was 'extension': multiplying each subtracted time by the percentage rating recorded against it to yield the basic time, entered in the fourth column. Extension is easily and quickly done with the aid of a pocket calculator. The calculation is made to the nearest second decimal place: that is, to the nearest one-hundredth of a minute. Thus 0.204 would be shown as 20, and 0.206 minutes as 21—which leaves the problem of what to do with 0.205. Evidently, in this study office the standing rule was to take half-hundredths of a minute down rather than up, as can be seen by the entry against element G of cycle 15. Here, the rating was 105 and the subtracted time 10, so that the extension yields 0.105 minutes to three places. This has been shown as 10, the half-hundredth having been taken down. Other instances will be found in the study. Most study offices apply the reverse rule: that is, taking middle times up.

| ELEMENT DESCRIPTION | R | WR | ST | BT | ELEMENT DESCRIPTION | R | WR | ST | BT |
|---|---|---|---|---|---|---|---|---|---|
| STUDY No.: *17* | | | | | **TIME STUDY CONTINUATION SHEET** | | | SHEET No. *3* OF *5* | |
| *Patrol inspector checks* | | | | | ⑱                A | 100 | 71 | 27 | 27 |
| *3 pieces: discuss* | – | 2966 | 102 | – | B | 100 | 96 | 25 | 25 |
| | | | | | C | 95 | 609 | 13 | 12 |
| ⑮                A | 95 | 93 | 27 | 26 | D | 75 | 34 | 25 | 19 |
| B | 80 | 3023 | 30 | 24 | E | – | 52 | 18 | – |
| C | 100 | 36 | 13 | 13 | F | 100 | 77 | 25 | 25 |
| D | 100 | 56 | 20 | 20 | G | 75 | 92 | 15 | 11 |
| E | – | 74 | 18 | – | | | | | |
| F | 80 | 106 | 32 | 26 | | | | 148 | |
| G | 105 | 16 | 10 | 10 | | | | | |
| ⑯                A | 80 | 49 | 33 | 26 | *Watch stopped 10.25* | | 800 | | |
| B | 85 | 77 | 28 | 24 | *a.m. (elapsed time* | | | | |
| C | 105 | 89 | 12 | 13 | *38.00)* | | | | |
| D | 100 | 207 | 18 | 18 | *Time after* | | | 108 | |
| E | – | 30 | 23 | – | | | | | |
| F | 95 | 57 | 27 | 26 | | | | | |
| G | 85 | 70 | 13 | 11 | | | | | |
| *Help labourer unload* | 85 | 320 | 50 | 43 | | | | | |
| *boxes of new castings* | 95 | 70 | 50 | 48 | | | | | |
| *and load finished work* | 95 | 90 | 20 | 19 | *Check on* subtracted | | | 418 | |
| *on truck (30 new + 30* | | | | | *times* | | | 440 | |
| *fin. in boxes of 10)* | | | | | | | | 631 | |
| ⑰                A | 100 | 417 | 27 | 27 | | | | 1203 | |
| B | 85 | 49 | 32 | 27 | | | | 680 | |
| C | 85 | 64 | 15 | 13 | | | | 148 | |
| D | 85 | 86 | 22 | 19 | | | | 3520 | |
| E | – | 509 | 23 | – | | | | | |
| F | 100 | 34 | 25 | 25 | *Time before* | | | 172 | |
| G | 105 | 44 | 10 | 10 | *Time after* | | | 108 | |
| | | | | | | | | | |
| | | | | | *Elapsed* | | | 3800 | |
| | | | | | | | | | |
| | | | 680 | | | | | | |

*Figure 92. Working sheet*

The repetitive elements A, B, C, D, F and G were all constant elements, and selected basic times for them were obtained by averaging. As was noted in Chapter 15, study analyses take several forms and for this reason it is not usual to have specially printed sheets for them. Ordinary lined or squared paper serves very well, and when the time study top sheet has been printed on the reverse as squared paper (to facilitate sketching), it will do well enough to use the back side of a top sheet, entering at the top the study and sheet numbers. For a simple study the analysis is often made straight on to the study summary sheet, a few extra columns being ruled in the space headed "Element description."

Methods of obtaining the selected basic times are discussed in Chapter 18. In this instance, inspection of the basic times tabulated under elements A, B, C, D, F and G showed no anomalies, and therefore no need to ring out 'rogue' times. For each of these elements the basic times have been totalled, and the selected basic time was calculated by dividing the total by the number of observations (18).

No figures were listed under element E, "Wait for machine to complete cut". This was unoccupied time, which was not rated in the study. The actual length of unoccupied time experienced in the various cycles observed depended on the speed with which the operative carried out the inside work which he performed on another casting while the machine was cutting automatically.

The time the machine took to make the cut, while on automatic feed, did not vary from cycle to cycle because it was determined by the rate of feed at which the machine was set and the length of cut to be made. It could thus be calculated quite easily. In this study the machine-controlled time started at the end of element A and ended with the conclusion of element E. The machine-controlled time can therefore be obtained from the study sheets by subtracting the watch reading against element A from that against E. This has been done, the results being tabulated under 'MCT' at the right-hand side of the working sheet. These times are of course actual minutes, not basic times.

It will be seen that two of the MCT entries have been ringed out. The study man did not enter any explanation of unusual events on his record, and inspection of the observations for the cycles in which these rogue times occurred does not provide any conclusive explanation. Possibly the explanation for the shorter time is to be found in the fact that the operative can start the cut on hand-feed before locking on the auto-feed, and on this occasion, unnoticed by the study man, he spent longer on the switch the machine off quite as quickly as usual on this occasion, and again this escaped notice. The two ringed times were excluded from the total of 13.05 actual minutes for the machine-controlled times, so that this total was divided by 16 instead of 18 to derive the average MCT of 0.816.

Element E, the unoccupied time, was dealt with by subtracting the total of the selected basic times for elements B, C and D, the inside work elements, from the average MCT. The resulting figure for the average unoccupied was 0.257 minutes.

At this stage in the calculations, it is usual to make use of three decimal places for the selected basic times, and to retain the third place on the study summary sheet and the analysis of studies sheet.

| Study No. 17 | | WORKING SHEET | | | | | Sheet 4 of 5 | |
|---|---|---|---|---|---|---|---|---|
| **Element:** | **A** | **B** | **C** | **D** | **E** | **F** | **G** | **MCT** |
| | | (Basic times) | | | | | | (Actual minutes) |
| **Cycle No.** | | | | | | | | |
| 1 | 25 | 25 | 12 | 19 | | 25 | 09 | 82 |
| 2 | 25 | 26 | 12 | 18 | | 26 | 10 | 81 |
| 3 | 26 | 26 | 12 | 20 | | 25 | 11 | 81 |
| 4 | 26 | 24 | 11 | 17 | | 24 | 10 | 81 |
| 5 | 26 | 23 | 11 | 17 | | 24 | 10 | 83 |
| 6 | 26 | 26 | 13 | 18 | | 22 | 10 | 82 |
| 7 | 26 | 26 | 12 | 19 | | 26 | 11 | 81 |
| 8 | 28 | 24 | 13 | 18 | | 29 | 10 | 82 |
| 9 | 26 | 25 | 12 | 17 | | 28 | 10 | 81 |
| 10 | 21 | 26 | 12 | 19 | | 24 | 11 | 32 |
| 11 | 29 | 26 | 11 | 19 | | 25 | 10 | 82 |
| 12 | 26 | 24 | 12 | 19 | | 27 | 09 | (72) |
| 13 | 27 | 26 | 12 | 20 | | 24 | 11 | 82 |
| 14 | 26 | 25 | 13 | 19 | | 26 | 10 | 82 |
| 15 | 26 | 24 | 13 | 20 | | 26 | 10 | 81 |
| 16 | 26 | 24 | 13 | 18 | | 26 | 11 | 81 |
| 17 | 27 | 27 | 13 | 19 | | 25 | 10 | (92) |
| 18 | 27 | 25 | 12 | 19 | | 25 | 11 | 81 |
| **Totals** | 4.69 | 4.52 | 2.20 | 3.35 | | 4.57 | 1.84 | 13.05 |
| **Occasions** | 18 | 18 | 18 | 18 | | 18 | 18 | 16 |
| **Average** | 0.261 | 0.251 | 0.122 | 0.186 | | 0.254 | 0.102 | 0.816 |

$$MCT = 0.816 \quad \text{Actual minutes}$$
$$B + C + D = 0.559 \quad \text{Basic minutes}$$

$$\text{Element E (unoccupied)} = 0.257$$

*Figure 93. Study summary sheet*

The study summary sheet, when completed, was stapled on top of the other four study sheets and was eventually filed with them. The sheets which have been used for recording observations at the workplace often become somewhat dirty as a result of the conditions in which they have to be used. Moreover, because of the speed with which the observations have to be written down, the study man may have used many abbreviations, and perhaps his hurried writing may be difficult for anyone except the study man himself to read. The study summary sheet therefore not only presents concisely all the results obtained from the study but also records in the heading block, in ink and neatly written, all the information about the operation which was originally entered on the time study top sheet.

The repetitive elements A to G, excluding E, were entered first, and it has been noted that three of these were inside work and the other three outside work. The entries in the column headed BT are the basic times per occasion, and were taken from the working sheet shown in Figure 92. For each of these elements the frequency of occurrence is shown as 1/1, indicating that each occurred once in every cycle of the operation. The time calculated for the machine element, and hence the unoccupied time (element E), is shown below. The column headed Obs, shows the number of observations of the element which have been taken into account in deriving selected basic times. This information will be carried to the analysis of studies sheet where it will be of use when the final selected basic times are derived for the compilation of the standard time.

Under the heading "Occasional elements and contingencies" is shown the basic time for the element of helping the labourer to load and unload boxes of castings. It is noted that this element was observed once only, and that its frequency ought to be 1/30 since three boxes of ten fresh castings were brought, and three boxes of finished castings loaded. The other two non-repetitive occurrences observed were "Talk to foreman", and "Inspector checks three pieces and discusses." Neither of these periods was rated, so the times are shown in actual minutes (a.m.).

Finally, the study man recorded, in actual minutes, the amount of relaxation taken during the period of the study.

Basic times were entered to the third decimal place, and have been carried forward in this form to the analysis of studies sheet. It may be thought that this is a degree of refinement which is not warranted in view of the accuracy of the data on which the entries are based. There is a good reason for the practice, however. If it is eventually decided to make the final selection of basic times, on the analysis of studies sheet, by the process of averaging, each of the entries from this study will be multiplied by the corresponding number of observations to yield the total basic minutes observed for the element. The totals from all the studies taken on this operation will be added, and an average obtained by dividing by the aggregate number of observations. At this stage, when the whole chain of arithmetical calculations has been completed, the final selections will be expressed to the nearest second decimal place only, that is to the nearest one-hundredth of a minute.

## STUDY SUMMARY SHEET

| DEPARTMENT: *Machine Shop* | SECTION: *Milling* | STUDY No.: *17* | |
|---|---|---|---|
| OPERATION: *Finish mill second face* | M.S. No.: *9* | SHEET No.: *5 OF 5* | |
| | | DATE: | |
| PLANT/MACHINE: *Cincinnati No. 4* | No.: *26*    *25 cm TLF* | TIME OFF: | *10.25 a.m* |
| *Vertical Miller* | *Cutter* | TIME ON: | *9.47* |
| TOOLS AND GAUGES: *Fixture F. 239* | *Gauge 239/7 Surface plate* | ELAPSED TIME: | *38.00* |
| PRODUCT/PART: *B. 239 Gear Case* | No.: | CHECK TIME: | *2.80* |
| | | NET TIME: | *35.20* |
| DWG. No.: *B. 239/1* | MATERIAL: *Cast Iron* | OBS. TIME: | *35.20* |
| | *to I.S.S. 2* | UNACC. TIME: | *–* |
| QUALITY: *as dwg.* | WORKING CONDITIONS: | U.T. AS %: | *–* |
| | *m/c 9 cutter OK: light good* | STUDIED BY: | |
| OPERATIVE:    M/F | CLOCK No.: *1234* | CHECKED BY: | |

Sketch and notes on back of sheet 1

| El. No. | ELEMENT DESCRIPTION | | BT | F | Obs. | |
|---|---|---|---|---|---|---|
| | *Repetitive* | | | | | |
| *A* | | *Outside work* | 0.261 | 1/1 | 18 | |
| *B* | | *Inside work* | 0.251 | 1/1 | 18 | |
| *C* | *As card No. 1264* | " " | 0.122 | 1/1 | 18 | |
| *D* | | " | 0.186 | 1/1 | 18 | |
| *F* | | *Outside work* | 0.254 | 1/1 | 18 | |
| *G* | | " " | 0.104 | 1/1 | 18 | |
| | | | | | | |
| | *Machine element* | | 0.816 | 1/1 | 16 | |
| *E* | *Unoccupied time within MCT* | | 0.257 | 1/1 | 18 | |
| | | | | | | |
| | *Occasional elements and contingencies* | | | | | |
| | *Help unload boxes of new castings* | | | | | |
| | *and load boxes of finished castings* | | | | | |
| | *to truck* | | 1.100 | | 1 | *Freq. 1/30 castings* |
| | *(outside work)* | | | | | *(Boxes hold 10 castings)* |
| | | | | | | |
| | *Talk to foreman    (OW)    (a.m.)* | | 0.400 | 1/18 | obs. | |
| | *Inspector checks 3 pieces and* | | | | | |
| | *discusses  (a.m.)* | | 1.020 | 1/18 | obs. | |
| | *(OW)* | | | | | |
| | | | | | | |
| | | | | | | |
| | | | | | | |
| | *Relaxation time  (a.m.)* | | 6.120 | | | |
| | | | | | | |
| | | | | | | |
| | | | | | | |
| | | | | | | |
| | | | | | | |
| | | | | | | |
| | | | | | | |
| | | | | | | |

*Figure 94. Extract from the analysis of studies sheet*

As each time study on the operation was worked up and summarised, the entries from the study summary sheet were transferred to an analysis of studies sheet of the type illustrated in Figure 74. These sheets are often printed on paper of A3 or double foolscap size or larger, and so only a portion of the whole sheet is reproduced opposite.

It will be seen that five studies were made in all on this operation, a total of 92 cycles being observed. The work of three different operatives was studied, by four different study men. Standard times for regular machine shop operations are usually compiled from predetermined time standards (see Chapter 21), and when a considerable body of data has been built up it is often possible to derive accurate time standards with fewer studies, or by observing a smaller number of cycles of the operation.

Inspection of the study results for the elements A, B, C, D, F and G indicated normal consistency, with no reading suggesting a need for further investigation. The work of proceeding to the final selected basic times for the elements was therefore undertaken next. The selection was made by taking the weighted average for each element. All the repetitive elements were constant elements, so that there was no need for graphical presentation. In the first of the four columns in the block at the right-hand side of the sheet, the total basic time was entered against each element. Dividing these totals by 92, the aggregate number of cycles, yielded the figures for basic minutes per occasion, entered in the next column. These are now shown to the second decimal place only; that is, to the nearest one-hundredth of a minute.

The third column records the frequency of occurrence per cycle—for all the repetitive elements 1/1—and thus the entries in the last column, which show the basic minutes per cycle, are for this operation the same as those in the second column of the right-hand block. The unoccupied time, element E, has been arrived at in the same manner as on the study summary, by deducting the sum of the inside work basic minutes from the machine-controlled time. Usually the unoccupied time would not be evaluated until after relaxation allowance had been added to the work elements, but in this instance, as is indicated when discussing these allowances on the next page, there was no need for such a refinement.

The occasional element "Help labourer" was observed on three occasions only, in three different studies. Since it is known that the truck carries three boxes each containing ten castings, it is clear that the frequency with which this element will occur is once every 30 castings, or cycles. The average basic time per occasion was therefore divided by 30 to yield the basic time per cycle of 0.04 minutes.

"Talk to foreman" was dealt with by dividing the total time observed by the 92 cycles observed, giving a time of 0.01 minutes per cycle. The "Inspector checks" element was treated similarly, though in this instance as it was learned from the foreman that the inspector's duty was to check three castings in every 100 the frequency has been taken as 1/100. These two very small periods of time, both entered in actual minutes, were eventually considered to be best dealt with as contingencies and were covered by the contingency allowance given.

| | | Study No.: | 3 | 9 | 17 | 25 | 28 | | TOTALS | SELECTED BASIC TIME PER OCCASION | FREQUENCY OF OCCURRENCE PER CYCLE | BASIC MINUTES PER CYCLE |
|---|---|---|---|---|---|---|---|---|---|---|---|---|
| | | Date: | 27/4 | 1/5 | 4/5 | 7/5 | 11/5 | | | | | |
| | | Operative: | CAA | TBN | CAA | TBN | CRW | | | | | |
| | | Clock No.: | 1234 | 1547 | 1234 | 1547 | 1846 | | | | | |
| | | Machine No.: | 26 | 34 | 26 | 127 | 71 | | | | | |
| | | | | | | | | | | | | |
| | | | | | | | | | | | | |
| | | | | | | | | | | | | |
| | | Study taken by: | BDM | CEP | MN | DFS | BDM | | Cycles | | | |
| | | No. of cycles studied: | 15 | 26 | 18 | 13 | 20 | | 92 | | | |
| El. No. | ELEMENTS | | BASIC TIME PER OCCASION | | | | | | B.T. | B.M. | | B.M. |
| A | P/U casting, locate, lock, set on | | 0.276 | 0.257 | 0.261 | 0.270 | 0.281 | | 24.645 | 0.27 | 1/1 | 0.27 |
| B | Hold, break milled edge, clean | | 0.240 | 0.266 | 0.251 | 0.252 | 0.244 | | 23.305 | 0.25 | 1/1 | 0.25 |
| C | Gauge | | 0.114 | 0.127 | 0.122 | 0.128 | 0.111 | | 11.089 | 0.12 | 1/1 | 0.12 |
| D | Aside finished part, position new | | 0.197 | 0.196 | 0.186 | 0.191 | 0.180 | | 17.485 | 0.19 | 1/1 | 0.19 |
| E | Wait m/c (actual minutes) | | 0.264 | 0.222 | 0.257 | 0.253 | 0.275 | | | | 1/1 | 0.26 |
| F | Stop m/c, unlock, aside part | | 0.271 | 0.270 | 0.254 | 0.250 | 0.245 | | 23.820 | 0.26 | 1/1 | 0.26 |
| G | Clear swarf | | 0.096 | 0.112 | 0.104 | 0.090 | 0.092 | | 9.240 | 0.10 | 1/1 | 0.10 |
| | Machine-controlled time (actual minutes) | | 0.821 | 0.811 | 0.816 | 0.824 | 0.810 | | 75.000 | 0.82 | 1/1 | 0.82 |
| | Help labourer U/L and load boxes of castings | | — | — | 1.100 (1 occ.) | 1.420 (1 occ.) | 1.310 (1 occ.) | | 3.830 | 1.28 | 1/30 | 0.04 |
| | Talk to foreman (actual minutes) | | 1.140 | — | 0.400 | 0.870 | — | | 2.410 | 0.80 | 1/92 | 0.01 |
| | Inspector checks, discuss (a.m.) | | — | 1.470 (1 occ.) | 1.020 (1 occ.) | — | 1.770 (1 occ.) | | 4.260 | 1.42 | 1/100 | 0.01 |

**Figure 95. Calculation of relaxation allowance**

A form such as that shown in the figure reproduced below is often used for the compilation of relaxation allowances. It provides a convenient way of ensuring that no item of relaxation allowance is omitted. The derivation of the allowances is based on the data given in the tables reproduced in Appendix 3. In this example the weight in kgs has been converted into lbs so that the points can be derived from these tables. The total figure for relaxation allowances (which represents both fixed and variable allowances) has also an added 5 per cent personal needs allowance. By deducting this figure for each element from the total allowances figure, one can arrive at fatigue allowances alone.

Since this is an example of restricted work the fatigue allowance has been calculated separately.

**RELAXATION**

PRODUCT: *B. 239 Gear Case*
WEIGHT : *6.8 kg each (15 lbs)*

OPERATION: *Finish-mill second face*

WORKING CONDITIONS: *Good*

| El. No. | ELEMENT DESCRIPTION | AVERAGE FORCE Strain[2] | Pts. | POSTURE Strain | Pts. | VIBRATION Strain | Pts. | SHORT CYCLE Strain | Pts. | RESTRICTIVE CLOTHING Strain | Pts. | CONCENTRATION/ANXIETY Strain | Pts. |
|---|---|---|---|---|---|---|---|---|---|---|---|---|---|
| A | Pick up casting, locate in fixture, lock 2 nuts, set guard, start machine | M | 20 | L | 1 | — | | — | | — | | L | |
| B | Break edges with file, and clean | L | — | L | 1 | — | — | — | — | — | — | L | |
| C | Gauge | L | — | L | 1 | — | — | — | — | — | — | L | |
| D | Pick up casting, place in box, pick up new casting and place near machine | M | 20 | L | 1 | — | | — | | — | | L | |
| E | Wait for machine (unoccupied time) | — | — | — | — | — | — | — | — | — | — | — | — |
| F | Stop machine, open guard, unlock nuts, remove casting, place on surface plate | M | 20 | L | 1 | — | — | — | — | — | — | L | |
| G | Clean fixture with compressed air | — | — | L | 3 | — | — | — | — | — | — | — | |
| Occa-sional Element | Help labourer load and unload boxes of castings (10 per box = 68 kg/2 men, 1/30 cycles) | H | 89 | H | 12 | — | — | — | — | — | — | — | — |

PHYSICAL STRAINS

[1] The percentages of total allowances, as derived from the points conversion table in Appendix 3, cover both basic and variable allowances and a built-in personal needs allowance of 5 per cent. [2] Severity of strain: L = low; M = medium; H= high.

The only period of unoccupied time during the machine-controlled time totalled 0.26 actual minutes. This was considered to be too short a period for recovery from fatigue (see Chapter 19, section 4), so the whole of the relaxation allowance, both the personal needs part and the fatigue allowance, was considered as an addition to outside work and was added to the cycle time.

The personal needs allowance of 5 per cent was calculated on the sum of the outside work **plus** the machine-controlled time. Fatigue allowance was calculated on the work elements only.

It will be seen from figure 96 that the total relaxation allowance amounted to 0.17 minutes. This is less than the period of unoccupied time (0.26 minutes), but is nevertheless to be added outside the machine-controlled time as periods of 0.50 minutes or less of unoccupied time are ignored for fatigue allowance purposes.

## ALLOWANCE

| MONOTONY | Pts | EYE STRAIN | Pts | NOISE | Pts | TEMPERATURE/HUMIDITY | Pts | VENTILATION | Pts | FUMES | Pts | DUST | Pts | DIRT | Pts | WET | Pts | TOTAL POINTS | TOTAL RELAXATION ALLOWANCE¹ (percentage) | FATIGUE ALLOWANCE (Relaxation Allowance less 5 per cent) |
|---|---|---|---|---|---|---|---|---|---|---|---|---|---|---|---|---|---|---|---|---|
| M | 1 | L | 2 | L | 1 | M | 6 | L | 1 | — | — | — | — | — | — | — | — | 33 | 16 | 11 |
| M | 1 | L | 2 | L | 1 | M | 6 | L | 1 | — | — | — | — | — | — | — | — | 13 | 11 | 6 |
| M | 1 | L | 2 | L | 1 | M | 6 | L | 1 | — | — | — | — | — | — | — | — | 13 | 11 | 6 |
| M | 1 | L | 2 | L | 1 | M | 6 | L | 1 | — | — | — | — | — | — | — | — | 33 | 16 | 11 |
| — | — | — | — | — | — | — | — | — | — | — | — | — | — | — | — | — | — | — | — | — |
| M | 1 | L | 2 | L | 1 | M | 6 | L | 1 | — | — | — | — | — | — | — | — | 33 | 16 | 11 |
| — | — | — | — | L | 1 | M | 6 | L | 1 | — | — | — | — | — | — | — | — | 11 | 11 | 6 |
| — | — | — | — | L | 1 | M | 6 | L | 1 | — | — | — | — | — | — | — | — | 109 | 74 | 69 |

*Figure 96. Final calculation of relaxation allowance*

The allowance which resulted from applying the percentage figures built up in Figure 95 is shown opposite. It will be seen that a contingency allowance of 2.5%, inclusive of relaxation, was included under the heading of outside work, to cover the periods spent in discussions with the foreman and the inspector.

| **Fatigue allowance** | | Basic time | Fatigue per cent | Allowance min |
|---|---|---|---|---|
| Inside work elements: | B | 0.25 | 6 | 0.015 |
| | C | 0.12 | 6 | 0.007 |
| | D | 0.19 | 11 | 0.0209 |
| | | 0.56 | | 0.0429 |
| Outside work elements: | A | 0.27 | 11 | 0.0297 |
| | F | 0.26 | 11 | 0.0286 |
| | G | 0.10 | 6 | 0.006 |
| Occasional element help labourer | | 0.04 | 69 | 0.0276 |
| Contingency allowance – 2.5 per cent of total basic time, inclusive of relaxation allowance | | 0.03 | | – |
| | | 0.70 | | 0.0919 |
| Total fatigue allowance . . . . | | | | 0.1348 |

**Personal needs allowance**

5 per cent of Outside work plus
    machine-controlled time:
    5 per cent of (0.70 + 0.82) . . . . . . .    0.076

**Total relaxation allowance**

Fatigue allowance plus personal needs allowance . .    0.2108
                                                    i.e.    0.21 min

*Figure 97. Calculation and issue of the standard time*

The method of calculation shown below is that appropriate to restricted work. When standard times for jobs made up wholly of manual elements are compiled, it is common to add the appropriate relaxation allowances element by element, thus building up standard times for each element, the sum of which of course represents the standard time for the whole job. In such instances it is usual to show the final calculations on a job summary sheet which lists the elements in full, with their descriptions, and all relevant details of the job for which the standard time has been built up. This would be done also for restricted work such as that in the present example, though inside and outside work would be shown separately. It is good practice to add a cycle diagram to the job summary sheet.

The methods adopted to issue—or publish—standard times vary according to the circumstances of the work situation. In jobbing shops, and for non-repetitive work (such as much maintenance work) jobs may be studied while they are in progress and the time standards be issued directly to the workers concerned, by annotation on the job sheet or other work instruction, after approval by the shop foreman. When the work is mainly repetitive, with the same operations being performed many times over, for perhaps weeks or months on end, tables of values, derived after extensive studywork, may be issued by the work study department.

---

**Computation of Standard Time**

| | |
|---|---|
| Outside work. . . . . . . . . | 0.70 basic min |
| Inside work . . . . . . . . | 0.56 basic min |
| Relaxation allowance . . . . . | 0.21 min |
| Unoccupied time allowance . . | 0.26 min |
| **Standard time.** . . . . . . . | **1.73 standard min** |

Alternatively:

| | |
|---|---|
| Outside work. . . . . . . . | 0.70 basic min |
| Machine-controlled time . . . | 0.82 min |
| Relaxation allowance . . . . . | 0.21 min |
| | **1.73 standard min** |

Figure 98. Overall cycle time

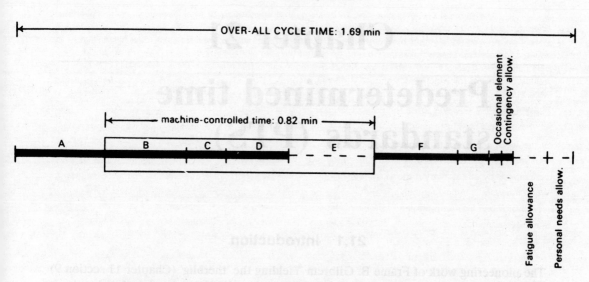

The use to which time standards may be put is discussed in Chapter 23. It should be noted that, although the example which has been studied in detail is a simple one for a manufacturing industry, a very similar procedure is carried out for non-manufacturing operations or for any other work which is time-studied for the purpose of setting time standards.

# Chapter 21

# Predetermined time standards (PTS)

## 21.1  Introduction

The pioneering work of Frame B. Gilbrem Yielding the 'therblig' (Chapter 11 section 9) based on the micromotions involved in gross motions involving hand or hand and eye formed the central theme of motion study. The approach of Gilbreth underlined the idea that a detailed critical analysis of work methods stimulates the ideas for method improvement, even for already improved upon work methods, and also stimulates the idea that a number of motions required to perform a work by any method can serve as a good and reliable evaluation index of the method—the criteria being that the method requiring fewer motions is a better one.

The addition of time dimension to motion study goes to the credit of A.B. Segur, who in 1927 stated that "within practical limits the time required for all experts to perform the fundamental motions is a constant". Subsequently Segur developed the Motion Time Analysis System the first predetermined time standards.

J.H. Quick in 1934 originated the Work Factor System however following the Second World War many Variety of PTS systems were produced. Methods-Time Measurement (MTM) System came out to be the mostly used and relied upon system. The development of MTM is attributed to three Westing House Electric Corporation members, H.B. Mnyard, G.J. Stegerman and J.L. Schwab. Since they published their findings, the system was freely accessible which was the primary reason of its popularity. Several non-profit making MTM associations continue to carry out research and control the standards of training and practice into the development of MTM.

## 21.2  Definition

A predetermined time standard (PTS) is a work measurement technique whereby times established for basic human motions (classified according to the nature of the motion and the conditions under which it is made) are used to build up the time for a job at a definite level of performance.

PTS are advanced techniques aiming at defining the time needed for the performance of various operations by derivation from pre-set standards of time for various elementary motions involved in such operations and not by direct observation and measurement.

PTS systems in other words are techniques for synthesising operation times from the standard time data for basic elementary motions comprising the operation.

The nature of PTS system can be illustrated by the simple example of the work cycle involved in putting a washer on a bolt. The detailed critical analysis of this work cycle reveals the following elementary operations carried out in the sequence of listing: Operator *reaching* the washer, *grasping* the washer, *moving* the washer to the bolt, *positioning* it on the bolt and *releasing* it.

The five elementary operations reach, grasp, move, position and release in different forms of feasible combinations represent a very large number of operations. Table 18 represents the components of a basic PTS.

*Table 18. Components of a basic PTS*

| Motion | Description |
|---|---|
| REACH | Move hand to destination |
| GRASP | Secure control of object with finger |
| MOVE | Move the object |
| POSITION | Line up and engage objects |
| RELEASE | Let the object go |
| BODY MOTIONS | Leg, Trunk movement |

## 21.3   Advantages of PTS systems

In a PTS system one time is indicated for the basic elementary motions irrespective of where it is performed whereas in a stop-watch time study a sequence of motions constituting a particular operation is studied. Further the PTS systems which avoid both rating and direct observations invariably lead to more consistent standard times arrived at relative to their counterpart i.e. stop-watch time study based on direct observation. Synthesising times of the constituent basic elementary motions of an operation from standard-time tables, enables the definition of standard time for a given operation even before production begins and often while the process is still at the design stage. Such a facility facilitates modification of layout and design of workplace as well as different production aids like jigs and fixtures so that optimum productivity is ensured from the beginning without any commitment of additional expenses. An estimation of cost of production before starting of production also facilitates rational estimating, tendering and budgeting.

PTS systems are usually simple, easy to apply and are very fast particularly for operations having very short repetitive time cycles; the most common example being the assembly works in the Electronic industry.

These are the advantages of the PTS system which have resulted in a very fast rate of acceptance of the system in different industries all over the world.

## 21.4   Criticisms of PTS systems

The existence of a very large number (about 200) and of a variety of systems often results in confusion which is further increased by the fact that many of such systems are of a proprietory nature. Different systems have often a widely varying analysis and description of operations and often fail to recognise certain important aspects. These are the primary facts for criticism of the systems.

Besides, another point of criticism which is more fundamental in nature is that the assumption of Segur regarding the nature of variation associated with the times is required to perform the basic elementary motions. This criticism often arises from the rather literal interpretation which is more a misconception that these time estimates were claimed to be constant, or the contrary that the correct position is that the standards indicated in PTS tables are averages but the limits of variation of such averages though finite in nature yet are indeed small enough to be neglected in all practical situations.

Another often raised point of criticism is that PTS systems fail to recognise the effect of the influence of the immediate predecessor and follow a basic elementary motion, on the time indicated as standard time for the intervening basic elementary motion. Though the point is valid for some PTS systems more important PTS systems clearly recognise such influence and prescribe special provisions to maintain the essential correlations. As an example the MTM system can be cited where main classes of motions were further subdivided and the linking was being established through creation of special definitions and the rules of application.

Direction of motion like downwards, upwards, clockwise, anti-clockwise, away from body and near to body are also expected to influence the duration of *basic* motions; non-consideration of such influence also form a point of criticism insofar as the PTS systems are concerned. But this criticism is countered by the argument that rarely any operation comprises motions taking place in only one set of direction like downwards or anti-clockwise or away from the body. In fact more commonly the motions are seen to have complementary sets of directions of motion thereby justifying the utilisation of average values and ignoring the influence of direction.

The last point of criticism is centred around the misconception that PTS systems eliminate the need of a stop-watch since Machine Time, process time and waiting time are not amenable to treatment by PTS system. The fact remains that even with a most advanced PTS system it would not be possible to obtain 100% coverage of all the operations carried out in any organisation. For non-repetitive jobs and batch production operations application of any PTS systems entails a prohibitive level of cost.

## 21.5   Different forms of PTS systems

The successful exploitation of PTS systems is possible if the concerned work study man is well-versed, trained and familiarised with the basics of the systems. Exposure to different form of PTS systems enables a work study man to identify the differences in the context of levels and scope of application of data, motion classification and time units, which in turn, enables him to make the most judicious choice depending on the situation and nature of the operation to be studied.

The data levels pertinent to official international MTM systems: MTM-1, 2 and 3 are given in Figure 99.

*Figure 99. PTS data levels: basic motions*

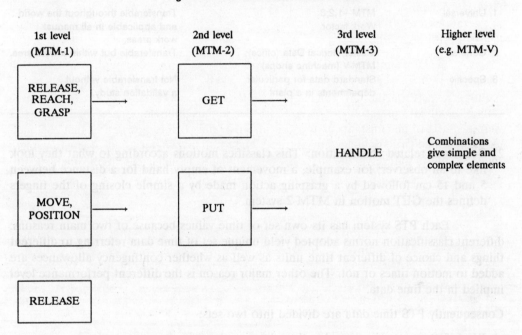

| 1st level (MTM-1) | 2nd level (MTM-2) | 3rd level (MTM-3) | Higher level (e.g. MTM-V) |

RELEASE, REACH, GRASP → GET →

MOVE, POSITION → PUT →

RELEASE

HANDLE

Combinations give simple and complex elements

Transitions from the first level to second level and to third level represent synthesis of motions while the third level represents a complete work cycle. However, beyond the third level there are as yet no clear-cut rules and hence classification vary according to the work area for which the data are intended.

The universality of the application field of PTS systems vary widely. Table 19 represents a possible set of categorisation.

As already stated PTS systems provide information about manual work cycles in terms of basic human motions. There are two main sets of criteria adopted for classification of these motions:

1) Object related classification: Employed in a majority of PTS systems (like work factor, Dimensional Motion Times MTM-1) and virtually in all the data systems relating to

main occupational groups or specifically designed for use within a plant. In such systems reference may be made to the characteristics of parts (like size e.g. grasping a 6 × 6 × 6 mm object) or the nature of the surrounding conditions (like whether the object is jumbled with other objects or relative orientation with respect to a reference). However, the classification is not entirely object related since motions such as Release Load or Disengage may have behavioural definitions.

Table 19. Scope of application of PTS system

| Degree | PTS system | Scope of application |
|---|---|---|
| 1. Universal | MTM—1,2,3<br>Work factor | Transferable throughout the world and applicable in all manual work areas. |
| 2. General | Master clerical Data (office);<br>MTM-V (machine shops) | Transferable but within a work area. |
| 3. Specific | Standard data for particular departments in a plant | Not transferable without a validation study. |

2) Behaviour related classification: This classifies motions according to what they look like to an observer; for example, a movement of empty hand for a distance between 5 and 15 cm followed by a grasping action made by a simple closing of the fingers defines the GET motion in MTM-2 system.

Each PTS system has its own set of time values because of two main reasons: different classification norms adopted yield unique set of time data referring to different things and choice of different time units as well as whether contingency allowances are added to motion times or not. The other major reason is the different performance level implied in the time data.

Consequently PTS time data are divided into two sets:

a) Work factor systems expressing time data in minutes.

b) MTM systems expressing time data in time measurement units (tmu) representing one hundred-thousandth of an hour or about one twenty-eighth of a second.

The MTM time values, derived mainly from the count of the number of frames of the film occupied by the corresponding industrial operation, were standardised by either 'Westinghouse' or 'Levelling' system and as such represent the time taken by an experienced worker of average skill, working with average effort and consistency under average conditions.

However, there are some important properties of PTS systems which are difficult to establish like the precision and accuracy of the time data, speed of application, methods description capability and learning time. The absence of sufficiently reliable, detailed information and to some extent the lack of agreed design criteria hamper comparison of these properties.

## 21.6   Use of PTS system

The mostly accepted and applicable PTS system is MTM-2. The motions and their codes as in MTM-2 system are given hereunder:

| Category | Code |
|----------|------|
| GET | GA |
|  | GB |
|  | GC |
| PUT | PA |
|  | PB |
|  | PC |
| REGRASP | R |
| APPLY PRESSURE | A |
| EYE ACTION | E |
| FOOT MOTION | F |
| STEP | S |
| BEND AND ARISE | B |
| WEIGHT FACTORS | GW |
|  | PW |
| CRANK | C |

The MTM-2 system provides time standards ranging from 3 to 61 tmu. These are shown on the data card depicted in Table 20.

*Table 20. MTM-2 data card*

| Code | Time in tmu | | | | | |
|------|-----|-----|-----|-----|-----|-----|
|  | GA | GB | GC | PA | PB | PC |
| – 5 | 3 | 7 | 14 | 3 | 10 | 21 |
| –15 | 6 | 10 | 19 | 6 | 15 | 26 |
| –30 | 9 | 14 | 23 | 11 | 19 | 30 |
| –45 | 13 | 18 | 27 | 15 | 24 | 36 |
| –80 | 17 | 23 | 32 | 20 | 30 | 41 |

GW : 1 per kg

| A | R | E | C | S | F | B |
|---|---|---|---|---|---|---|
| 14 | 6 | 7 | 15 | 18 | 9 | 61 |

Warning: Do not attempt to use these data unless you have been trained and qualified under a scheme approved by the International MTM directorate.

• GET (G) is an action with the predominant purpose of reaching with the hand or fingers to an object, grasping the object and subsequently releasing it.

The span of GET

starts : with reaching to the object,

includes: reaching to, gaining control and subsequently releasing control of the object,

ends : when the object is released.

Selection of a GET is done by considering three variables:

1) Case of GET—distinguished by the grasping action employed as for example:

  a) GA—while putting the palm of the hand on the side of a box in order to push it across a table;

  b) GB—while getting an easy-to-handle object such as a 1 inch cube which is lying by itself;

  c) GC—while getting the corner of a page of the book in order to turn it over.

The choice of a particular case can be made by the following decision modes:

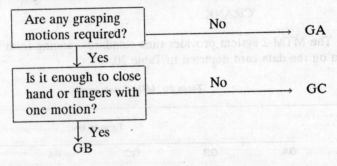

2) Distance reached estimated from the path of travel of the hand, less any body action and distinguished through codified five class intervals indicated hereunder:

| cm | | Code |
|---|---|---|
| Over | Not over | |
| 0.0 | 5.0 | − 5 |
| 5.0 | 15.0 | −15 |
| 15.0 | 30.0 | −30 |
| 30.0 | 45.0 | −45 |
| 45.0 | — | −80 |

3) Weight of the object or its resistance to motion indicated by GET WEIGHT (GW) representing the action required for the muscles of the hand and arm to take up the weight of the object or to overcome the resistance to motion.

The scope of GET WEIGHT

Starts : with the grasp on the object completed;

includes: muscular force necessary to gain full control of the weight of the object;

ends : when the object is sufficiently under control permitting movement of the object.

GET WEIGHT occurs after the fingers have closed on the object in the preceding GET. It must be accomplished before any actual movement can take place. For weight or resistance magnitude less than 2 kg per hand no time is assigned for GW. When weight or resistance exceeds 2 kg per hand, 1 tmu is assigned for every kg including the first two.

• PUT (p) is an action with the predominant purpose of moving an object to a destination with the hand or fingers.

The scope of PUT

starts : with an object grasped and kept under control at the initial place;

includes : all transporting and correcting motions necessary to place an object;

ends : with the object still under control at the initial place.

Selection of a PUT is done by considering three variables:

1) Case of PUT—distinguished by the correcting motions employed as for example:

a) PA—tossing aside an object;

b) PB—putting a 12 mm ball in a 15 mm diameter hole; and,

c) PC—inserting a Yale or Similar Key in a lock.

Cases of PUT are judged by the following decision model:

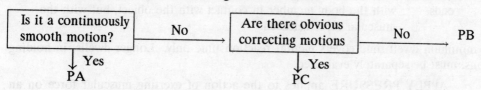

2) Distance moved— distinguished by the distance traversed and classified as well as codified in the same way as in the case of GET. An additional PUT will be allowed in cases where there is an engagement of parts following a correction and the distance exceeds 2.5 cm.

3) Weight of the object or its resistance to motion—indicated by PUT WEIGHT whose scope:

|          |   |                                                                                                                                                           |
|----------|---|-----------------------------------------------------------------------------------------------------------------------------------------------------------|
| Starts   | : | When the move begins;                                                                                                                                      |
| includes | : | the additional time, over and above the move time in PUT, to compensate for the differences in time required in moving heavy and light objects over the same distance; |
| ends     | : | when the move ends.                                                                                                                                        |

Similar to GW, PW is assigned when weight or resistance to movement exceeds magnitude of 2 kg per hand. However, between 2 kg and 5 kg, 1 tmu is allowed and coded PW 5; between 5 kg and 10 kg 2 tmu are allowed and coded PW 10 and so on.

REGRASP (R) is a hand action with the purpose of changing the grasp on the object. Its scope:

|          |   |                                                         |
|----------|---|---------------------------------------------------------|
| Starts   | : | with the object in hand;                                |
| includes | : | digital and hand muscular readjustment on an object;    |
| ends     | : | with the object in a new location in the hand.          |

A single REGRASP consists of not more than three fractional movements.

Digital and muscular readjustments while performing an APPLY PRESSURE, are included in APPLY PRESSURE. A REGRASP should not be assigned in combination with APPLY PRESSURE.

If the hand relinquishes control and then secures another grasp on the object, the action will be a GET, not a REGRASP.

An example of R is changing the grasp on a pen in order to get into the position for writing.

•   APPLY PRESSURE (A) is an action with the purpose of exerting muscular force on an object. Its scope:

|          |   |                                                                                                                                 |
|----------|---|---------------------------------------------------------------------------------------------------------------------------------|
| Starts   | : | with the body member in contact with the object;                                                                                |
| includes | : | the application of controlled increasing muscular force, a minimum reaction time to permit the reversal of force and subsequent releasing of muscular force; |
| ends     | : | with the body member in contact with the object, but with the muscular force released.                                          |

The minimum dwell time covers mental reaction time only. Longer dwells, in holding actions, must be separately evaluated.

APPLY PRESSURE applies to the action of exerting muscular force on an object to achieve control, to restrain or to overcome resistance to motion. The object is not displaced more than 6 mm during the action of APPLY PRESSURE.

APPLY PRESSURE, which can be performed by any body member, is recognised by a noticeable hesitation while force is applied.

An example of A is the final tightening action made with a screwdriver or a spanner.

- EYE ACTION (E) is an action with the purpose of either recognising a readily distinguishable characteristic of an object or shifting the aim of the axis of vision to a new viewing area. Its scope:

|  |  |
|---|---|
| Starts | : when other actions must cease because a characteristic of an object must be recognised; |
| includes: | either muscular readjustment of the lens of the eyes and the mental processes required to recognise a distinguishable characteristic of an object or the eye motion performed to shift the aim of the axis of vision to a new viewing area; |
| ends | : when other actions start again. |

A single eye focus covers an area 10 cm in diameter at 40 cm from the eyes. Recognition time included is sufficient only for a simple binary decision.

An example of E is the action of determining whether a coin is showing head or tail.

- FOOT MOTION (F) is a short foot or leg motion when the purpose is not to move the trunk. Its scope:

|  |  |
|---|---|
| Starts | : with the foot or leg at rest; |
| includes: | a motion not exceeding 30 cm that is pivoted at the hip, knee or instep; |
| ends | : with the foot in a new location. |

An example of F is depressing a foot pedal.

- STEP MOTION (S) is either a leg motion with the purpose of moving the body or a leg motion longer than 30 cm. Its scope:

|  |  |
|---|---|
| Starts | : with a leg at rest; |
| includes : | either a motion of the leg where the purpose is to achieve a displacement of the trunk or a leg motion longer than 30 cm; |
| ends | : with the leg at a new location. |

An example of S is making a single Step to the side extending the reach of the arm further.

STEP or FOOT MOTION is judged by the following decision model:

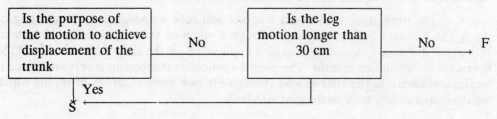

The number of times the foot hits the floor is counted to evaluate walking.

• BEND AND ARISE (B) is a lowering of the trunk followed by a rise. Its scope:

Starts : with the motion of the trunk forward from an upright posture;

includes : movement of the trunk and other body members to achieve a vertical change in body position to permit the hands to reach down or below the knees and the subsequent arise from this position;

ends : with the body in an upright posture.

The criterion for BEND AND ARISE is whether the worker is able to reach below the knees, not whether he actually does so.

Kneeling on both knees should be analysed as 2B.

• CRANK (C) is a motion with the purpose of moving an object in a circular path of more than half a revolution with the hand or finger. Its scope:

Starts : with the hand on the object;

includes: all transporting motions necessary to move an object in a circular path;

ends : with the hand on the object when one revolution is completed.

Two primary variables are considered while considering the CRANK motion:

(1) the number of revolution, and (2) weight or resistance.

The number of revolutions should be rounded to the nearest whole number. The time value of 15 tmu per revolution may be used for any crank diameter and is applicable to continuous as well as intermittent cranking. CRANK motion applies to motion in circular path irrespective of whether the cranking axis is perpendicular to the plane of rotation.

The time for moving an object depends on the weight or resistance. The rules of adding GW and PW to PUT motions also apply to CRANK. PW applies to each revolution, whether continuous or intermittent while GW is applied one only to continuous series revolutions but to each revolution where these are intermittent.

No correcting motions as applied to PUT are included in CRANK. If corrective motions occur in putting the object at the intended place an extra PUT must be allowed.

An example of C is turning a hand wheel through one revolution.

The work study man would however will have an adequate understanding of the MTM-2 system only after going through a vigorous training, practice of at least two weeks duration followed by a guided application on the shop floor with an MTM instructor for about two months. The guided application component is very vital since it creates confidence in the trainee who through his own analysis arrives at results which are compared closely with established standards.

## 21.7  Applications of PTS systems

PTS systems can be applied either through direct observation of the motions used by the worker or through mental visualisation of the motions required to accomplish a new or alternative work method.

The overall approach adopted while applying PTS system like MTM-2 in its direct observation mode is very similar to that adopted in making a time study (Chapter 16). Any person experienced in such procedural sequence like selecting the job, approaching the worker, recording the job information, breaking down the job into elements, ascertaining allowances and ultimately compiling total job times naturally becomes well equipped to become a PTS practitioner. The major difference of approach is that at the point in the total time study procedure where the observer is ready to time and rate the work cycle, will instead make an MTM-2 analysis and then enter the motion times on his analysis sheet from the MTM-2 data card. The calculation of allowances, completion of documentation and issuing job times are done in much the same way as in a time study. The study summary sheet (Figure 73) and short cycle study form (Figure 72) can be adopted to summarise the information from the MTM-2 analysis sheets.

The choice of worker continues to play the same vital role as in the case of time study. A cooperative, good-average worker is the ideal one whilst exceptionally fast or abnormally slow performances are difficult subjects for PTS practitioners also. These are difficult subjects for PTS practitioners also. The superskilled worker takes help of combined and overlapped motions while an abnormally slow worker will rely on separate one handed motions and the average worker will perform smoothly and simultaneously. The rules and motion combination tallies of the MTM system and other systems such as Work Factor, do provide information for adjusting the observed motion pattern to that applicable to the good average worker. However, an experienced PTS practitioner can effectively and reliably study fast workers and bring out clues so that training of workers can be arranged to help them to pick up the combining and overlapping skill of such workers and similarly the difficulties of slow workers can also be identified by observing the areas of their slow work and getting them equipped with techniques for overcoming such difficulties.

The workplace layout should be accurately drawn to scale so that pertinent distance parameters are observable which are significant in any PTS system.

In PTS systems the division of operations into work elements follows the same principle as in time study but the breakdown can be made to finer levels, if required, since there are no associated difficulty for timing short elements.

If desired the break points can also be changed easily without calling for retiming the cycle. Such flexibility is demonstrated in Table 21 which shows a very common work cycle—that of fitting a nut and washer on a stud. For example, if a change of method/eliminates the need for a washer, the appropriate motions (GC 30, PC 30, PA 5) and time (56 tmu) can be removed. Finger turns can also be readily separated from spanner turns and indeed, from the fitting actions and subsequent turns.

There is no problem of rating with a PTS system like MTM-2, since the times have been rated one for all. The individual motion element times are just to be added and transferred to the study summary sheet.

Figure 100. Base assembly

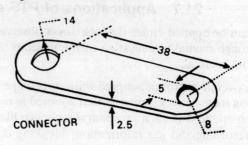

MEASUREMENTS
IN MILLIMETRES

14

38

5

2.5

8

**CONNECTOR**

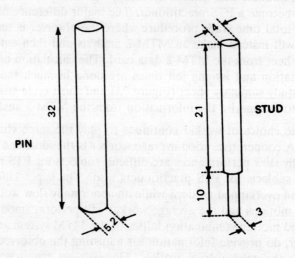

32

**PIN**

5.2

4

21

10

3

**STUD**

3.4

7

22

35

**BLOCK**

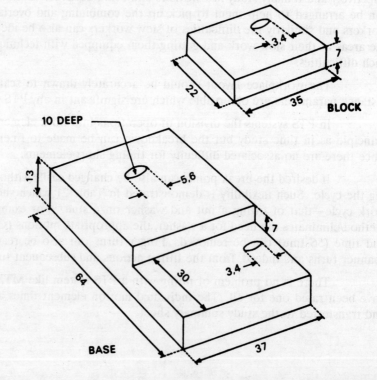

**10 DEEP**

13

5.6

7

64

30

3.4

**BASE**

37

Table 21. Fitting a nut and washer on a stud

| Element | Code | tmu | Description |
|---|---|---|---|
| Fit washer | GC 30 | 23 | Washer |
| | PC 30 | 30 | To stud |
| | PA 5 | 3 | On stud |
| Fit nut and turn | GB 15 | 10 | Nut |
| down by hand | PC 15 | 26 | To Stud |
| | 2PA 5 | 6 | Engage threads |
| | 6GB 5 | 42 | Turn down nut |
| | 6PA 5 | 18 | — |
| Tighten nut using | GB 30 | 23 | Spanner |
| spanner | PC 30 | 30 | To nut |
| | PA 15 | 6 | Turn nut |
| | A | 14 | Tighten |
| | | 231 | |

Relaxation and other allowances are added in exactly the same way as for time study, in order to give the total job time.

An example of applying PTS system in situations when the work study man does not have the opportunity of observing the work cycle is given in Figures 100 to 102. The technique of visualisation is used in these cases.

Figure 101. Base assembly workplace layout

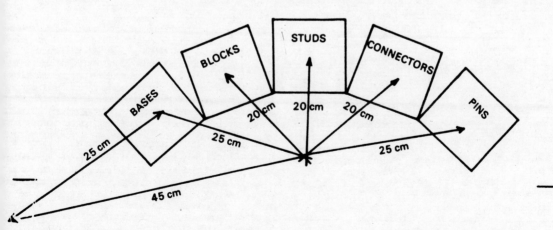

The ability to visualise motion patterns depends on the study man's intelligence and skill. In designing work methods it can be helpful to use a methods laboratory (Chapter 11). However, while performing motion analysis by a work study man, it must be borne in mind that his own performance of the motions under analysis will be falling far short of those performances which will be achieved by the regular shop floor workers. Besides when trying with the new methods the learning effect must also be given due consideration. To avoid such controllable errors motion combination possibilities expected of the average experienced worker must be used.

The nature and value of PTS systems should now be reasonably clear. If a work study man intends to become a specialist, for example in MTM, he will need full training in MTM-1 and 2 and the other techniques artlined in this book. In a more general case, where he will probably undertake work study and other jobs (line production planning and control) and MTM-2 training may be quite sufficient.

However PTS systems represent fine precision tools and should be used only after the other broader and simpler techniques have been utilised and their benefits fully absorbed.

*Figure 102. MTM-2 analysis sheet, base assembly*

| JOB DESCRIPTION: | | | REF.: | |
|---|---|---|---|---|
| *Assemble base* | | | SHEET No. *1*     of     *1* | |
| *(see sketches of parts and layout)* | | | ANALYST: | |
| | | | DATE: | |
| LEFT-HAND DESCRIPTION | LH | TMU | RH | RIGHT-HAND DESCRIPTION |
| *Get base from box* | GC30 | 23 | G— | *Get pin from box* |
| | | 14 | GC5 | |
| *Put base on bench* | PA30 | 30 | PC30 | *Locate pin to base* |
| *Get block from box* | GC30 | 23 | G— | *Get stud from box* |
| | | 14 | GC5 | |
| *Move block stud* | P— | 30 | PC30 | *Locate stud through block* |
| *Assist location* | P— | 26 | PC15 | *Fit assembly to base* |
| | | 23 | GC30 | *Get connector from box* |
| *Assist location* | GB— | 30 | PC30 | *Locate to stud* |
| *Locate to pin* | PC5 | 21 | | |
| *Pick up assembly* | GB15 | 10 | | |
| *Place on conveyor* | PA80 | 20 | | |
| | | 264 | | |

*Table 22. Methods-Time Measurement application data in tmu*
*(Based metric weights and measures)*

**OFFICIAL INTERNATIONAL MTM-1 DATA**
ⒸINTERNATIONAL MTM DIRECTORATE
AND
MTM ASSOCIATION FOR
STANDARDS AND RESEARCH
*Tables reproduced by kind permission of the International MTM Directorate.*

**TABLE I —REACH —R**

| Distance (cm) | Time (tmu) | | | | Hand in motion | | Case and description |
|---|---|---|---|---|---|---|---|
| | A | B | C or D | E | A | B | |
| 2 or less | 2.0 | 2.0 | 2.0 | 2.0 | 1.6 | 1.6 | A Reach to object in fixed |
| 4 | 3.4 | 3.4 | 5.1 | 3.2 | 3.0 | 2.4 | location, or to object in |
| 6 | 4.5 | 4.5 | 6.5 | 4.4 | 3.9 | 3.1 | other hand or on which |
| 8 | 5.5 | 5.5 | 7.5 | 5.5 | 4.6 | 3.7 | other hand rests. |
| 10 | 6.1 | 6.3 | 8.4 | 6.8 | 4.9 | 4.3 | |
| 12 | 6.4 | 7.4 | 9.1 | 7.3 | 5.2 | 4.8 | |
| 14 | 6.8 | 8.2 | 9.7 | 7.8 | 5.5 | 5.4 | B Reach to single object in |
| 16 | 7.1 | 8.8 | 10.3 | 8.2 | 5.8 | 5.9 | location which may vary |
| 18 | 7.5 | 9.4 | 10.8 | 8.7 | 6.1 | 6.5 | slightly from cycle to cycle. |
| 20 | 7.8 | 10.0 | 11.4 | 9.2 | 6.5 | 7.1 | |
| 22 | 8.1 | 10.5 | 11.9 | 9.7 | 6.8 | 7.7 | C Reach to object jumbled |
| 24 | 8.5 | 11.1 | 12.5 | 10.2 | 7.1 | 8.2 | with other objects in a |
| 26 | 8.8 | 11.7 | 13.0 | 10.7 | 7.4 | 8.8 | group so that search and |
| 28 | 9.2 | 12.2 | 13.6 | 11.2 | 7.7 | 9.4 | select occur. |
| 30 | 9.5 | 12.8 | 14.1 | 11.7 | 8.0 | 9.9 | |
| 35 | 10.4 | 14.2 | 15.5 | 12.9 | 8.8 | 11.4 | |
| 40 | 11.3 | 15.6 | 16.8 | 14.1 | 9.6 | 12.8 | D Reach to a very small object |
| 45 | 12.1 | 17.0 | 18.2 | 15.3 | 10.4 | 14.2 | or where accurate grasp is |
| 50 | 13.0 | 18.4 | 19.6 | 16.5 | 11.2 | 15.7 | required. |
| 55 | 13.9 | 19.8 | 20.9 | 17.8 | 12.0 | 17.1 | |
| 60 | 14.7 | 21.2 | 22.3 | 19.0 | 12.8 | 18.5 | E Reach to indefinite location |
| 65 | 15.6 | 22.6 | 23.6 | 20.2 | 13.5 | 19.9 | to get hand in position for |
| 70 | 16.5 | 24.1 | 25.0 | 21.4 | 14.3 | 21.4 | body balance or next |
| 75 | 17.3 | 25.5 | 26.4 | 22.6 | 15.1 | 22.8 | motion or out of way. |
| 80 | 18.2 | 26.9 | 27.7 | 23.9 | 15.9 | 24.2 | |

## TABLE II —MOVE —M

| Distance (cm) | Time (tmu) | | | | Wt allowance | | | Case and description |
|---|---|---|---|---|---|---|---|---|
| | A | B | C | Hand in motion B | Wt (kg) up to | Static con-stant (tmu) | Dyn-amic fac-tor | |
| 2 or less | 2.0 | 2.0 | 2.0 | 1.7 | | | | |
| 4 | 3.1 | 4.0 | 4.5 | 2.8 | 1 | 0 | 1.00 | |
| 6 | 4.1 | 5.0 | 5.8 | 3.1 | | | | |
| 8 | 5.1 | 5.9 | 6.9 | 3.7 | 2 | 1.6 | 1.04 | A Move object against stop or to other hand. |
| 10 | 6.0 | 6.8 | 7.9 | 4.3 | | | | |
| 12 | 6.9 | 7.7 | 8.8 | 4.9 | 4 | 2.8 | 1.07 | |
| 14 | 7.7 | 8.5 | 9.8 | 5.4 | | | | |
| 16 | 8.3 | 9.2 | 10.5 | 6.0 | 6 | 4.3 | 1.12 | |
| 18 | 9.0 | 9.8 | 11.1 | 6.5 | | | | |
| 20 | 9.6 | 10.5 | 11.7 | 7.1 | 8 | 5.8 | 1.17 | |
| 22 | 10.2 | 11.2 | 12.4 | 7.6 | | | | |
| 24 | 10.8 | 11.8 | 13.0 | 8.2 | 10 | 7.3 | 1.22 | B Move object to ap-proximate or indef-inite location. |
| 26 | 11.5 | 12.3 | 13.7 | 8.7 | | | | |
| 28 | 12.1 | 12.8 | 14.4 | 9.3 | 12 | 8.8 | 1.27 | |
| 30 | 12.7 | 13.3 | 15.1 | 9.8 | | | | |
| 35 | 14.3 | 14.5 | 16.8 | 11.2 | 14 | 10.4 | 1.32 | |
| 40 | 15.8 | 15.6 | 18.5 | 12.6 | | | | |
| 45 | 17.4 | 16.8 | 20.1 | 14.0 | 16 | 11.9 | 1.36 | |
| 50 | 19.0 | 18.0 | 21.8 | 15.4 | | | | |
| 55 | 20.5 | 19.2 | 23.5 | 16.8 | 18 | 13.4 | 1.41 | C Move object to ex-act location. |
| 60 | 22.1 | 20.4 | 25.2 | 18.2 | | | | |
| 65 | 23.6 | 21.6 | 26.9 | 19.5 | 20 | 14.9 | 1.46 | |
| 70 | 25.2 | 22.8 | 28.6 | 20.9 | | | | |
| 75 | 26.7 | 24.0 | 30.3 | 22.3 | 22 | 16.4 | 1.51 | |
| 80 | 28.3 | 25.2 | 32.0 | 23.7 | | | | |

## TABLE IIIA —TURN —T

| Weight | Time (tmu) for degrees turned | | | | | | | | | | |
|---|---|---|---|---|---|---|---|---|---|---|---|
| | 30° | 45° | 60° | 75° | 90° | 105° | 120° | 135° | 150° | 165° | 180° |
| Small 0 to 1 kg | 2.8 | 3.5 | 4.1 | 4.8 | 5.4 | 6.1 | 6.8 | 7.4 | 8.1 | 8.7 | 9.4 |
| Medium 1 to 5 kg | 4.4 | 5.5 | 6.5 | 7.5 | 8.5 | 9.6 | 10.6 | 11.6 | 12.7 | 13.7 | 14.8 |
| Large 5.1 to 16 kg | 8.4 | 10.5 | 12.3 | 14.4 | 16.2 | 18.3 | 20.4 | 22.2 | 24.3 | 26.1 | 28.2 |

## TABLE IIIB —APPLY PRESSURE —AP

| Full cylce | | | Components | | |
|---|---|---|---|---|---|
| Symbol | tmu | Description | Symbol | tmu | Description |
| APA | 10.6 | AF+DM+RLF | AF | 3.4 | Apply force |
| APB | 16.2 | APA+G2 | DM | 4.2 | Dwell minimum |
| | | | RLF | 3.0 | Release force |

## TABLE IV —GRASP —G

| Case | Time (tmu) | Description |
|---|---|---|
| 1A | 2.0 | Pick Up Grasp—Small, medium, or large object by itself, easily grasped. |
| 1B | 3.5 | Very small object or object lying close against a flat surface. |
| 1C1 | 7.3 | Interference with grasp on bottom and one side of nearly cylindrical object. Diameter larger than 12 mm. |
| 1C2 | 8.7 | Interference with grasp on bottom and one side of nearly cylindrical object. Diameter 6 to 12 mm. |
| 1C3 | 10.8 | Interference with grasp on bottom and one side of nearly cylindrical object. Diameter less than 6 mm. |
| 2 | 5.6 | Regrasp. |
| 3 | 5.6 | Transfer Grasp. |
| 4A | 7.3 | Object jumbled with other objects so search and select occur.  Larger than 25×25×25 mm. |
| 4B | 9.1 | Object jumbled with other objects so search and select occur. 6×6×3 mm. to 25×25×25 mm. |
| 4C | 12.9 | Object jumbled with other objects so search and select occur. Smaller than 6×6×3 mm. |
| 5 | 0 | Contact, sliding or hook grasp. |

## TABLE V —POSITION* —P

| Class of fit | | Symmetry | Easy to handle | Difficult to handle |
|---|---|---|---|---|
| 1 Loose | No pressure required | S | 5.6 | 11.2 |
| | | SS | 9.1 | 14.7 |
| | | NS | 10.4 | 16.0 |
| 2 Close | Light pressure required | S | 16.2 | 21.8 |
| | | SS | 19.7 | 25.3 |
| | | NS | 21.0 | 26.6 |
| 3 Exact | Heavy pressure required | S | 43.0 | 48.6 |
| | | SS | 46.5 | 52.1 |
| | | NS | 47.8 | 53.4 |

*Distance moved too engage—max. 25 mm.

## TABLE VI —RELEASE —RL

| Case | Time (tmu) | Description |
|---|---|---|
| 1 | 2.0 | Normal release performed by opening fingers as independent motion |
| 2 | 0 | Contact release |

## TABLE VII —DISENGAGE —D

| Class of fit | Easy to handle | Difficult to handle |
|---|---|---|
| 1 Loose—Very slight effort, blends with subsequent move. | 4.0 | 5.7 |
| 2 Close—Normal effort, slight recoil. | 7.5 | 11.8 |
| 3 Tight—Considerable effort, hand recoils markedly. | 22.9 | 34.7 |

## TABLE VIII —EYE TRAVEL AND EYE FOCUS —ET AND EF

Eye travel time = 15.2 x $T/D$ tmu, with a maximum value of 20 tmu.

where $T$ = the distance between points from and to which the eye travels.

$D$ = the perpendicular distance from the eye to the line of travel $T$.

Eye focus time = 7.3 tmu.

## TABLE IX —BODY, LEG AND FOOT MOTIONS

| Description | Symbol | Distance | Time (tmu) |
|---|---|---|---|
| Foot motion—Hinged at ankle. | FM | Up to 10 cm. | 8.5 |
| With heavy pressure. | FMP | | 19.1 |
| Leg or foreleg motion. | LM- | Up to 15 cm. | 7.1 |
| | | Each add'l cm. | 0.5 |
| Sidestep Case 1—Complete when leading leg contacts floor. | SS-C1 | Less than 30 cm. | Use REACH or MOVE time |
| | | 30 cm | 17.0 |
| | | Each add'l cm. | 0.2 |
| Case 2—Lagging leg must contact floor before next motion can be made. | SS-C2 | Up to 30 cm. | 34.1 |
| | | Each add'l cm. | 0.4 |
| Bend, stoop, or kneel on one knee. | B.S. KOK | | 29.0 |
| Arise. | AB.AS. AKOK | | 31.9 |
| Kneel on floor—Both knees. | KBK | | 69.4 |
| Arise. | AKBK | | 76.7 |
| Sit. | SIT | | 34.7 |
| Stand from sitting position. | STD | | 43.4 |
| Turn body 45 to 90 degrees— | | | |
| Case 1—Complete when leading leg contacts floor. | TBC1 | | 18.6 |
| Case 2—Lagging leg must contact floor before next motion can be made. | TBC2 | | 37.2 |
| Walk. | W-M | Per metre | 17.4 |
| Walk. | W-P | Per pace | 15.0 |
| Walk—Obstructed. | W-PO | "      " | 17.0 |

# Chapter 22

# Standard data

## 22.1  Introduction

Though the spectrum of operations carried out in any plant is rather wide, several common elements do exist in many operations. Walking is one such common example which is found to be involved in diverse activities such as painting and handling. While timing such activities the work study man has to time the common element walking again and again and hence his job can be made much easier if standard time for walking is easily derivable from the set of data. Besides, reducing monotony, effort and cost, such an availability would ensure a higher level of consistency in time estimations.

So the advantages of building up a standard data bank for various elements which occur repeatedly at the workplace, are rather obvious. The coverage of such data bank is made somewhat wider than time estimates for a totally new job which can be easily derived without carrying out a time study for the same.

## 22.2  Major considerations

It is however very difficult, time consuming and immensely costly even to visualise a standard data bank which would include all the possible elements making up any and every job. Therefore in practice it is always better, pragmatic and rational to restrict the number of jobs for which the standard data are desired—normally to one or a few departments of a plant, or to all the processes involved in manufacturing of a certain component or product.

The reliability of the data bank could naturally be increased if as many common elements were grouped together for analysis, and if a sufficient amount of accumulated or collected data on each element has been taken into consideration. For example, the time taken to move a sheet of given size will vary depending on whether it is a solid sheet (of metal) or an elastic one (of rubber). The weight also becomes an important factor. The times taken to move an iron sheet, a foam sheet or a cupboard will be different. Consequently, the description of the element should be precise and all the significant factors affecting the timing will also have to be indicated.

Another basic consideration concerns the way or the technique by which the time data are observed like stop-watch readings or PTS systems. The first alternative may be more acceptable to the factory personnel at least in some cases and also may be cheaper at times. However, there may be certain elements for which it is not possible to have on record enough readings to enable reliable data to be desired. Such a situation may result in a very long lead time sometimes even running into years. But on the other hand, microscopic systems like MTM should only be adapted after acquiring sufficient experience in application of such systems.

But ultimately any standard data bank must have to be built up with due cognizance of the users needs. They are indeed invaluable for a variety of purposes like production planning and control, cost estimation, incentive payments and budgetary control. However, the "level of confidence" in the developed data base which can be tolerated by those who use standard data for these purposes varies considerably: for example, the requirements for production planning allow for much greater potential deviation in the standards than the requirements for individual bonus schemes. Since it is not possible to produce a different set of data for each user, it is necessary to build a data system that produces the maximum benefit for each user at the same time.

## 22.3  Developing the standard data

Development of standard data can be made through the following steps:

1) Deciding about coverage: The coverage should normally be restricted to one department or a few departments or work areas or to a limited range of processes within a plant in which several similar elements, performed by the same method, are involved in carrying out the jobs.

2) Breaking the jobs into elements: As many common elements as possible should be attempted to be identified from the range of the jobs being considered. Let the case of a worker in a fruit packing plant who works at the end of the operation be involved in removing a carton of fruit from a conveyor, stencilling the name of the customer on the carton and carrying it to a nearby skid. Though the breakdown can be carried out in many ways, the following break up may result in identification of elements which may be found in some other works elsewhere:

   a) Lifting the carton from the conveyor and positioning it on the table.
   b) Positioning a stencil on the carton.
   c) Applying a 10 cm brush and fur to stencil the credentials of the customer.
   d) Lifting the carton.
   e) Walking with the carton.
   f) Placing on the skid.

The elements "lifting and positioning of carton" and "Walking with the carton" may occur in various other jobs in the plant but not necessarily in the same manner. The carton may also vary in size and weight depending on the size and type of fruit involved. These are factors which will have influence on the time for these elements. Furthermore, the distance covered while "walking with the carton" may also vary but it is always advisable to collect the necessary information of all such elements towards building up of the standard data bank.

3) Deciding on the type of reading: The nature of the job and the cost of applying each system are the major determining factors. In the event of accepting stop-watch time study,

sufficient time must be allowed to collect the readings necessary to produce statistically reliable data.

4) Determining the influencing factors: Identifying and classifying of the factors likely to affect the time. For example, if the time for the walking performed by a worker is being measured, the readings observed will have variations within themselves. This variation is caused by several factors which are classified either as major or minor factors. A possible list of factors is given hereunder:

*Activity*

Restricted walking starting at dead point and ending at a dead stop.

*Factors influencing the time*

| *Major* | *Minor* |
|---|---|
| Distance covered | Physical make-up of worker |
| | Temperature |
| | Humidity |
| | Lighting |
| | External alteration |
| | Variation due to time study man. |

The above classification makes it obvious that the time for walking will be affected mainly by the distance covered while other factors will exert a small influence and may cause slight variations from reading to reading.

5) Measuring the time taken to perform the activity from actual observations while using macroscopic systems: The time study man can choose arbitrary distances and time the worker for each distance. If it is found that in a majority of the cases the worker walks either 10, 20, 30 or 40 metres, readings for these distances can be timed and entered in standard tables. But rarely such ideal situations are confronted and hence for a majority of occasions a relationship exists between time and distance covered as indicated by the curve representing the discrete distances travelled and corresponding timings.

As an example let the data in Table 23 represent the readings observed for different frequented distances.

The curve showing the line of best fit is given in Figure 103. This curve also enables the derivation of standard times for values lying anywhere between 10 and 40 metres.

However, it is rather common to face situations which have more than one major factor influencing the time data. To help understanding this better, the case of cross-cutting wood by a motor-driven circular saw is reproduced here:

*Activity*

Cross-cutting wood of the same type by hand feed.

*Factors influencing the time*

| *Major* | *Minor* |
|---|---|
| Variation in the thickness of wood | Physical make-up of worker |
| | Temperature |
| Variation in the width of wood | Humidity |

*Table 23. Restricted walking*

| Distance (m) x | Actual time (minutes) a | Rating r | Base time (minutes) (axr =) E | Average (minutes) y |
|---|---|---|---|---|
| 10 | 0.13 | 85 | 0.1105 | |
| | 0.13 | 90 | 0.1170 | |
| | 0.13 | 85 | 0.1105 | |
| | 0.11 | 95 | 0.1045 | |
| | 0.12 | 90 | 0.1080 | |
| | 0.15 | 80 | 0.1200 | 0.1118 |
| 20 | 0.21 | 105 | 0.2205 | |
| | 0.21 | 105 | 0.2205 | |
| | 0.22 | 95 | 0.2090 | |
| | 0.22 | 100 | 0.2200 | |
| | 0.26 | 90 | 0.2080 | |
| | 0.22 | 80 | 0.1980 | 0.2127 |
| 30 | 0.29 | 110 | 0.3190 | |
| | 0.30 | 100 | 0.3000 | |
| | 0.32 | 90 | 0.2880 | |
| | 0.30 | 100 | 0.3000 | |
| | 0.30 | 100 | 0.3000 | |
| | 0.33 | 95 | 0.3135 | 0.3034 |
| 40 | 0.38 | 110 | 0.4180 | |
| | 0.37 | 110 | 0.4070 | |
| | 0.38 | 110 | 0.4180 | |
| | 0.43 | 90 | 0.3870 | |
| | 0.42 | 90 | 0.3780 | |
| | 0.37 | 110 | 0.4070 | 0.4025 |

Lighting
Method of holding wood
Degree of physical force applied
Machine in good working order
Experience of Worker

We are assuming here that we are dealing with skilled workers. After a period of time, it proves possible to calculate the base time for some, but not all, thicknesses and widths of wood. The results are shown in Table 24.

The time of cutting against width for each thickness (2, 4, 6 and 8 cm) is plotted in Figure 104. The curves are reliable to derive any cutting time for any width for the given thickness. But the curve cannot yield the data for certain combinations like 3 cm thickness and 8 cm width.

However such data could be derived by some simple manipulations. There are basically two possible types of solutions:

Figure 103. Restricted walking

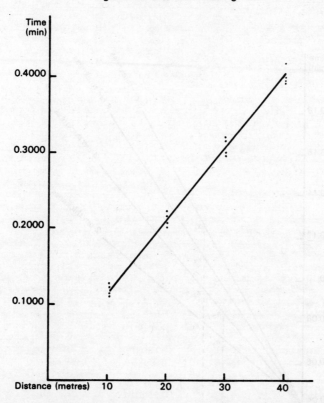

Table 24. Base times for cross-cutting wood of varying width and thickness

| Thickness (cm) | | | | | | | |
|---|---|---|---|---|---|---|---|
| 2 | | 4 | | 6 | | 8 | |
| Width (cm) | Time (minutes) | Width (cm) | Time (minutes) | Width (cm) | Time (minutes) | Width (cm) | Time (minutes) |
| 6 | 0.064 | 6 | 0.074 | 6 | 0.081 | 6 | 0.093 |
| 12 | 0.088 | 12 | 0.126 | 12 | 0.126 | 12 | 0.146 |
| 16 | 0.104 | 16 | 0.130 | 16 | | 16 | 0.181 |
| 20 | 0.120 | 20 | 0.160 | 20 | 0.180 | 20 | |

1) By calculation. A perpendicular is drawn from the point representing 8 cm width intercepting 2 cm and 4 cm thickness curves at $a_1$ and $a_2$ respectively. Then the following equation is established:

$$T = a_1(a_2 - a_1)f$$

Where $T = $ time to be calculated

$a_1 = $ time at the thickness of 2 cm (lower limit curve)

Figure 104. Base times for cross-cutting wood of varying width and thickness

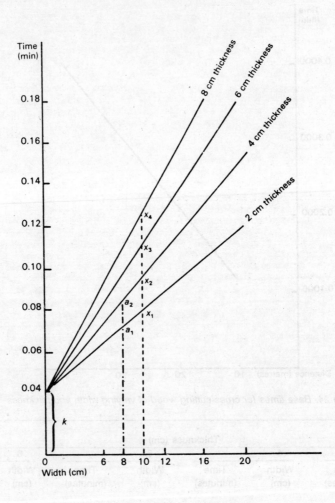

$a_2 =$ time at the thickness of 4 cm (upper limit curve)

$\quad f =$ fraction representing the required thickness in relation to $a_2$ and $a_1$

$\therefore T = 0.072 + 0.5(0.086 - 0.072) = 0.079$ minutes.

2) By graphical factor comparison. The first step in this method is to plot the curve as shown in Figure 104. Then the base cure representing one of the curves the data structure of which is complete is identified and then reproduced separately. (Figure 105).

Then any arbitrary width between 6 to 20 cm is chosen allowing for a point, say, representing 10 cm. A perpendicular is drawn from that point intercepting the four curves at $x_1$, $x_2$, $x_3$ and $x_4$ respectively.

Figure 105. Base curve for cross-cutting wood of 2 cm thickness and of varying width

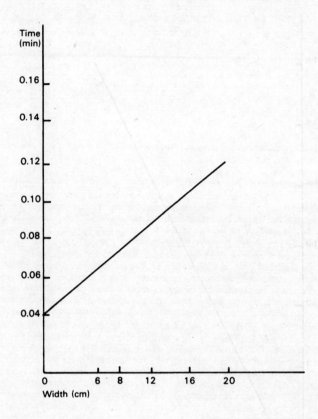

The factor curve (Figure 106) is drawn from the points calculated as follows:

| Thickness | 2 | 4 | 6 | 8 |
|---|---|---|---|---|
| Factor | $\dfrac{x_1}{x_1} = 1$ | $\dfrac{x_2}{x_1} = \dfrac{96}{80} = 1.2$ | $\dfrac{x_3}{x_1} = \dfrac{112}{80} = 1.4$ | $\dfrac{x_4}{x_1} = \dfrac{128}{80} = 1.6$ |

The required time can now be readily calculated from both the base curve and factor curve using the following equation:

$$\text{Total time} = \text{base time} \times \text{factor}$$

The time needed to cut a 3 cm thick and 8 cm wide wood is

$$T = 0.072 \times 1.1 = 0.079 \text{ minutes}$$

0.072 calculated from the base curve for 8 cm width while 1.1 is calculated from the factor curve for 3 cm thick.

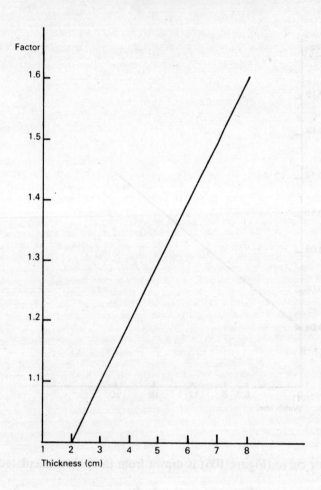

*Figure 106. Factor curve for cross-cutting wood of varying width and thickness*

It can be seen, therefore, that the data required to derive standard times can be obtained from either tables or graphs. To these data the work study man can then add any allowances in the usual way. If a firm decides that the same allowance factor is applicable to every job in a given class of work, it can then express its standard data in terms of the standard time for each element, instead of using the normal times as we did.

A word of caution is necessary here. The data collected usually cover a certain range of readings. It is not advisable to extrapolate these data for values that fall outside this range. For example, in our previous example the readings covered pieces of wood ranging from 6 to 20 cm wide and from 2 to 8 cm thick. We know what happens within this range; but there is no way of knowing whether the same type of linear relationship will continue if we go beyond this range by exceeding the widths and thicknesses actually studied and by projecting our curves beyond the points for which we have time study data.

## 22.4 Use of PTS systems to develop standard data

The method described earlier uses data derived from stop-watch time study. PTS systems can also provide such data base. In such a case the data derived for each element take into account the normal variations that are likely to arise in the execution of the job when other products, processes, equipment or materials are in use. These variations result from size, capacity, method of operation, type of tool and nature of job.

Table 25 illustrates the list of the most common elements in light engineering and assembly work, with details of their possible variations.

*Table 25. Standard data elements in light engineering and assembly work*

| General element (can be used in several departments) | Possible variations | Code |
|---|---|---|
| GET | Stillage to bench | GSB |
| | Bench to tool | GBT |
| | Stillage to tool | GST |
| | Tangled allowance | GTA |
| | Small parts to container | GSP |
| POSITION IN TOOL | Easy | PE |
| | Medium | PM |
| | Difficult | PD |
| | Complex | PC |
| CLAMP AND UNCLAMP | Fingers | CF |
| | Toggle | CT |
| | Slide | CS |
| | Air operated | CA |
| OPERATE | Close and open guard | OCG |
| | Foot pedal | OP |
| | Lever | OL |
| | Safety buttons | OSB |
| | Flypress | OFP |
| | Machine type | OMT |
| REMOVE FROM TOOL | Automatic | RA |
| | Easy | RE |
| | Medium | RM |
| | Difficult | RD |
| | Complex | RC |
| | Lever out component | RLC |
| TURN (IN) TOOL | Turn in tool | TIT |
| | Turn tool | TT |
| ASIDE | Automatic | AA |
| | Tool to bench | ATB |
| | Bench to stillage | ABS |
| | Tool to stillage | ATS |
| MISCELLANEOUS | Court parts | MCP |
| | Mark or score parts | MSP |

| | | |
|---|---|---|
| | Work area to tool | WAT |
| INSPECT OR CHECK | Component in fixture or gauge | CCF |

## Element Definitions

| | |
|---|---|
| GET | Covers picking up and moving an object or handful of objects to a destination. |
| POSITION IN TOOL | Covers positioning an objects, or handful of object, in a tool fixture, etc. or between electrodes. |
| CLAMP AND UNCLAMP | Covers all the motions necessary to close and later open a clamp of the type that operates by pressure on the object held; or to hold an object in a tool or fixture by a clamping action of the fingers. |
| OPERATE | Covers all the time and all the manual motions necessary to—<br>— close and later open a guard (OCG);<br>— grasp or contact an operating control, and later return the hand to the working area or the foot to the ground; and,<br>— operate the controls and initiate the machine cycle (OMT). |
| REMOVE FROM TOOL | Covers removing an object from a tool, fixture etc., or a part, component of fixture from under a drill or from between electrodes. |
| TURN (IN) TOOL | Occurs when two 'operate' elements follow each other, and the object must be removed from the tool, turned and repositioned in the tool; or the fixture or jig must be turned or moved in or under the tool. |
| ASIDE | Covers moving and putting down an object or handful of objects already held. |

## Word Definitions

| | |
|---|---|
| Object | Any object handled; such as parts, hand tools, sub-assemblies or complicated articles. Also any jig, fixture or other holding device. |
| Handful | The optimum number of objects which can be conveniently picked up, moved and placed as required. |
| Bench | The term 'bench' includes any table, tool pan or other storage area, convenient to the tool or workplace. |
| Stillage | A storage box or container on legs, for moving by a hand-lifting or fork-lift truck. The term 'stillage' includes a pallet, the floor or any other storage device at floor level. |
| Tool | A general term to cover any fixture, jig, electrode, press or other tool used to hold or operate on an object or objects.<br>One tool can be positioned in another —for example, a parts holding fixture under a drill or a welding electrode. |

Figure 107 illustrates a typical operation in a light engineering plant. This particular example contains one or other of the following sequences of elements:

a) get material; position in tool; operate machine; remove part; aside; or

b) get material; position in tool; position fixture in machine; operate machine; remove fixture; remove part; aside.

**Figure 107. Sequence of elements**

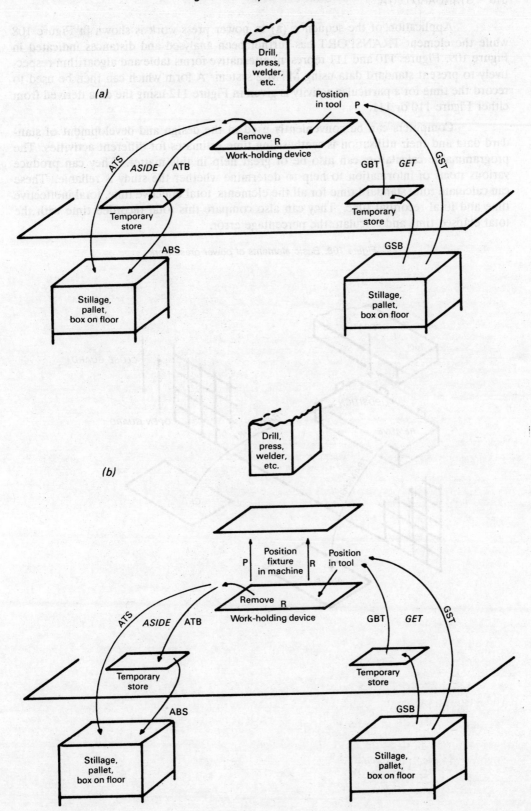

Application of the sequence (a) in power press work is shown in Figure 108 while the element TRANSPORT has further been analysed and distances indicated in Figure 109. Figures 110 and 111 represent alternative forms table and algoarithm respectively to present standard data using MTM-2 system. A form which can then be used to record the time for a particular activity is given in Figure 112 using the data derived from either Figure 110 or 111.

Computers can be conveniently used in the design and development of standard data and their utilisation is synthesising time estimates for different activities. The programmable calculators can also be of great help in this matter. They can produce various items of information to help to determine whether the study is reliable. These can calculate total standard time for all the elements, total effective time, total ineffective time and total recordial time. They can also compare this total recorded time with the total elapsed time and calculate the percentage error.

Figure 108. Basic elements of power press work

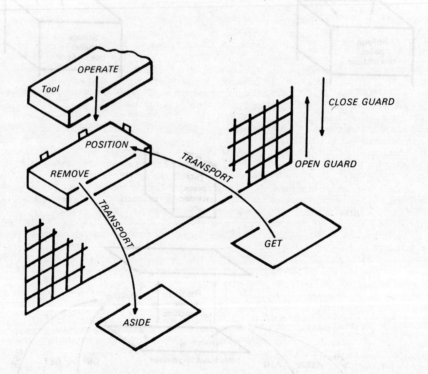

*Figure 109. Power press work: example of TRANSPORT elements and distances*

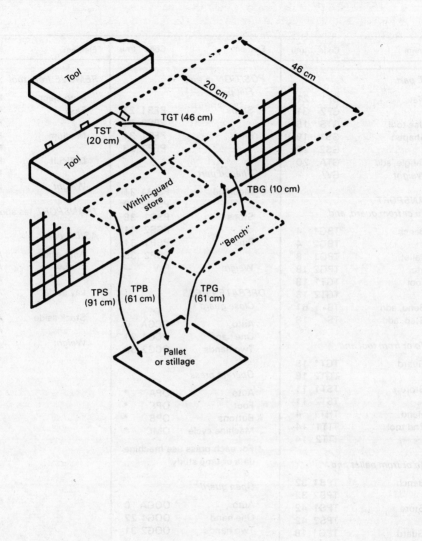

*Figure 110: Power press work: example of standard data determined by MTM-2 (tabular presentation)*

| Element | Code | tmu | Element | Code | tmu | Element | Code | tmu |
|---|---|---|---|---|---|---|---|---|
| **GET part** | | | **POSITION in tool** | | | **REMOVE from tool** | | |
| | | | *Flat part* | | | | | |
| Flat | GF1 | 21 | | | | Auto eject | RA- | 0 |
| | GF2 | 31 | Stops | PFS1 | 27 | Easy | RE1 | 17 |
| Use tool | GTS | 15 | | PFS2 | 30 | | RE2 | 17 |
| Shaped | GS1 | 19 | Pins | PFP1 | 31 | Medium | RM1 | 36 |
| | GS2 | 28 | | PFP2 | 33 | | RM2 | 52 |
| Tangle, add | GTA | 20 | | | | Difficult | RD1 | 50 |
| *Weight* | GW | — | *Shaped part* | | | | RD2 | 50 |
| | | | Moulded | PSM1 | 31 | *Weight* | GW | — |
| **TRANSPORT** | | | | PSM2 | 39 | | | |
| *To or from guard, and* | | | Stops | PSS1 | 38 | **TRANSPORT** (as above) | | |
| Bench | TBG1 | 4 | | PSS2 | 41 | **ASIDE part** | | |
| | TBG2 | 4 | Pins | PSP1 | 31 | | | |
| Pallet, | TPG1 | 18 | | PSP2 | 35 | Auto aside | AA- | 0 |
| etc. | TPG2 | 18 | *Weight* | PW | — | Throw | AT1 | 7 |
| Tool | TGT1 | 18 | | | | | AT2 | 7 |
| | TGT2 | 18 | **OPERATE PRESS** | | | Lay aside | AL1 | 10 |
| Bend, add | TB- | 61 | *Close guard* | | | | AL2 | 10 |
| Step, add | TS- | 18 | Auto | OCGA | 0 | Stack aside | AS1 | 11 |
| | | | One hand | OCG1 | 21 | | AS2 | 19 |
| *To or from tool and* | | | Two hands | OCG2 | 30 | *Weight* | — | — |
| Guard | TGT1 | 18 | | | | | | |
| | TGT2 | 18 | *Operate press* | | | | | |
| Store | TST1 | 11 | Auto | OPA | * | | | |
| | TST2 | 11 | Foot | OPF | * | | | |
| Hand | THT1 | 4 | Buttons | OPB | * | | | |
| 2nd tool | TTT1 | 14 | Machine cycle | OMC | * | | | |
| | TTT2 | 14 | * For each press use machine data or time study | | | | | |
| *To or from pallet and* | | | *Open guard* | | | | | |
| Bench | TPB1 | 32 | Auto | OOGA | 0 | | | |
| | TPB2 | 32 | One hand | OOG1 | 22 | | | |
| Store | TPS1 | 42 | Two hands | OOG2 | 31 | | | |
| | TPS2 | 42 | | | | | | |
| Guard | TPG1 | 18 | | | | | | |
| | TPG2 | 18 | | | | | | |
| *Weight* | PW | — | | | | | | |

*Note:* Last character in code indicates:   1 — One-handed
                                            2 — Two-handed

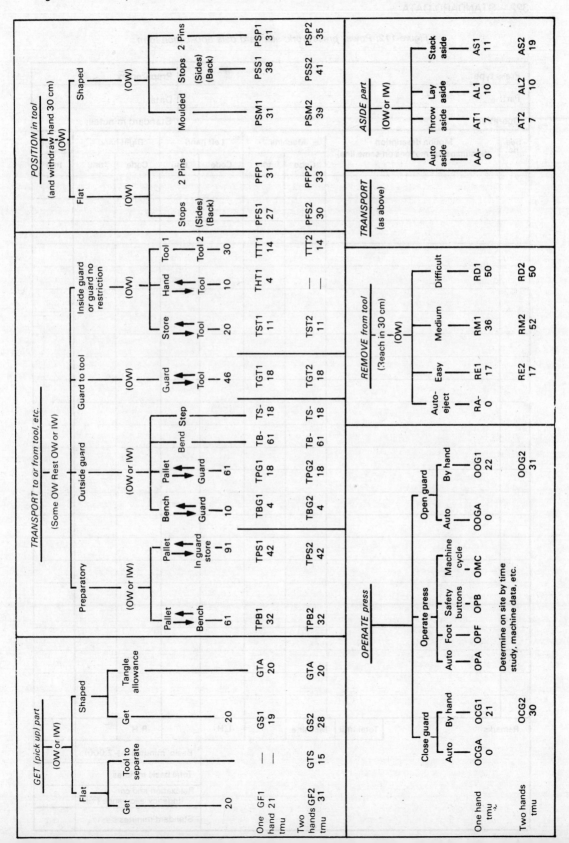

Figure 111: Power press work: example of standard data determined by MTM-2 (algorithmic presentation)

Figure 112: Power press work: standard data application form

| Press type: | | | | | | | Prepared by: | |
|---|---|---|---|---|---|---|---|---|
| Part: | | | | | | | Date: | |
| Operation: | | | | | | | Standard minutes: | |
| Seq. no. | Motion description (Simultaneous motions on same line) | Machine | | Left hand | | Right hand | | Charge |
| | | Code | tmu | Code | tmu | Code | tmu | tmu |
| | | | | | | | | |
| | | | | | | | | |
| | | | | | | | | |
| | | | | | | | | |
| | | | | | | | | |
| | | | | | | | | |
| | | | | | | | | |
| | | | | | | | | |
| | | | | | | | | |
| | | | | | | | | |
| | | | | | | | | |
| | | | | | | | | |
| | | | | | | | | |
| | | | | | | | | |
| | | | | | | | | |
| | | | | | | | | |
| | | | | | | | | |
| | | | | | | | | |
| | | | | | | | | |
| | | | | | | | | |
| | | | | | | | | |
| | | | | | | | | |
| | | | | | | | | |
| | | | | | | | | |
| | | | | | | | | |

| Remarks | | Total tmu | Machine | | L.H. | | R.H. | |
|---|---|---|---|---|---|---|---|---|
| | | | Basic minutes (÷ 2,000) | | | | | |
| | | | Total basic minutes | | | | | |
| | | | Relaxation and contingency allowance (%) | | | | | |
| | | | Standard minutes | | | | | |

# Chapter 23

# The uses of time standard

## 23.1  Introduction

In the preceding chapters of this part we have examined the techniques of Work Measurement at various levels of detail. These techniques provide us with a scientific approach for arriving at standard time for a job or an operation. The time standards, in turn, help us to evaluate the performance of the system and identify the presence of ineffective time, if any. Proper documentation of the methods for compiling the standard for a given system together with the systems characteristics under which the study is done is essential for future references. Technical set-up and Work specification are two such documents. Apart from serving as a comparison base, the time standard is useful for production planning and control, and for designing incentive schemes. These documents provide valuable aid in such cases as well. Finally, wherever Work Measurement techniques are applied, it should be backed by a proper information system. It is important to have records of operatives' time, output of work, waiting time etc., from which certain control statements are to be generated at regular intervals, to enable the management to take proper and timely actions.

The objective of this concluding chapter on Work Measurement is to examine the issues as raised in the above paragraph. Accordingly, Technical Set-up and Work Specification are discussed in the next section, followed by a presentation of the different uses of time standards. In the concluding section, we briefly discuss the organisation of the recording system associated with Work Measurement.

## 23.2  Technical set-up and work specification

Once the time standards are fixed, it is usual to prepare two documents to describe and define completely the way in which these standards have been compiled, and to maintain a record of the working conditions to which these standards refer. These two documents are known respectively as the technical set-up and the work specification. We elaborate on these in the following paragraphs.

> **Technical set-up of a section of an enterprise is a document that shows in summary form, in suitably presented tables and graphs, the main results of the study work undertaken in the section and how all the time standards which have been set have been derived.**

Thus, the technical set-up is essentially a work study document, having no reference to rates of pay, control of workers or other matters of contract between employers and employees. It contains all the information necessary to calculate fresh time standards, should the jobs or the working conditions change, insofar as these fresh standards can be compiled from the studywork already undertaken. It is thus in effect a manual from which time standards can be built up.

It will be necessary to compile a separate set-up for each technically different section of an enterprise, since the methods by which time standards are compiled will differ from section to section. Thus in a vitreous enamelling shop there would probably be one set-up for the sprayers, another for the operators of the shot-blast machines, a third for the furnace men, and so on.

Summaries of all the data on which the technical set-up is based should be attached to it, including—

    a) flow process charts showing the improved methods developed;

    b) analysis of studies sheets;

    c) relaxation allowance calculation sheets;

    d) data from predetermined time standards (PTS); and,

    e) curves and graphs relating to standard data.

The greatest care should be taken of the technical set-up and of all the original documents attached to it, since they are essential evidence in any disputes which may arise. They are also of great value in compiling time standards for similar work in the future. Technical set-up are normally filed in the work study department, where they are available to the management or to the workers' representatives whenever they may be needed.

> **A work specification is a document setting out the details of an operation or job, how it is to be performed, the layout of the work-place, particulars of machines, tools and appliances to be used, and the duties and responsibilities of the worker. The standard time or allowed time assigned to the job is normally included.**

The work specification thus represents the basic data on which the contract between an employer and an employee for the operation of an incentive scheme rests. The amount of detail necessary in a work specification varies greatly according to the nature of the operation concerned. In machine shop work in the engineering industry, where a large number of different jobs are done on machines the methods of which operation are broadly similar, general conditions governing all jobs can be established for the whole shop and only variations in detail need be specifically recorded. On the other hand, where an operation involves a whole shop or department and will run for an indefinite period substantially unchanged, as is the case in parts of the textile industry, the work specification may be lengthy and detailed.

Generally speaking, the following points should be covered by a work specification, which should, of course, embrace the standard method laid down as a result of the method study:

a) Details of the workpieces or product, including—
drawing, specification or product number and title;
material specification;
sketch, where necessary, of parts or surfaces to be treated.

b) Details of the machine or plant on which the operation is performed, including—
make, size or type, plant register number;
speeds and feeds, pulley sizes or other equivalent data;
jigs, tools and fixtures;
other equipment;
sketch of workplace layout (where not available on the method study).

c) Operation number and general description of the work covered.

d) Quality standards, including—
quality grade;
finish and/or tolerances, where applicable;
checking and gauging requirements, gauges and other inspection apparatus;
frequency of inspection

e) Grade and sex of labour, including—
direct and indirect labour;
part-time assistance by inspectors or supervisors.

f) Detailed description of all work involved, including—
repetitive elements, constant and variable;
occasional elements;
indirect work: setting up and breaking down;
cleaning, greasing, etc., and frequency with which such operations are carried out.

g) Details of time standards, including—
standard time for each element, job or operation, as appropriate; allowed time for all indirect work, with a note on how it has been assessed;
percentage relaxation allowance included in each element time;
other allowances.

h) Clerical procedure to be carried out by operatives in recording output and booking waiting time.

i) Conditions under which the time standard is issued, and any special provisos.

It may be necessary to supply copies of the work specification to the management and to the departmental and shop supervisors and, in the case of specifications affecting a large number of workers, to the workers' representatives. The manner in which the time standards are made known to the operatives depends largely on the nature of the work. If the job is one that is done only by a single worker (the one who was timed), it is usually enough for the work study man to tell him personally, in the first instance. When Work Study has been accepted, workers do not usually want lengthy explanations. What they are interested in are the targets at which they must aim in order to earn a reasonable bonus. Time standards are likely to be better understood if they are expressed in terms of a number of units to be produced in an hour to earn a certain bonus, than in the form of a number of standard units per piece. In batch production, the standard time is generally written or printed on the job card.

## 23.3 The uses of time standard

A standard, is general, gives us a performance base for a system. Time standard is no exception and as such, the output expressed in standard minutes can be used for computing the productivity of a system. The time standard serves several other purposes:

a) It is a necessary input for production planning.

b) It provides us with a base for designing incentive schemes.

c) It is useful for estimation of production cost and for standard costing and budgetary control.

We discuss the different uses in brief in the following paragraphs.

### Time standard as a performance base

Standard times are generally set down in the form of $x$ minutes per piece or $z$ minutes per ton, etc., depending on the type of system for which the standard is determined. These time values represent the output at standard performance. The time allowed for any job is not in minutes or hours of continuous work, as each unit of time contains within it an element of relaxation.

Thus, a standard minute being a measure of output, it can be used for computing the system performance as follows:

$$\text{Performance} = \frac{\text{output of work in standard minutes}}{\text{input of labour time or machine time in clock minutes}} \times 100$$

In the above form, we essentially get an expression for the productivities of the system which can be used for comparing different systems.

### Time standard as an input for production planning

One of the causes of ineffective time due to management shortcomings mentioned in Chapter 3 is failing to plan the flow of work and of orders, with the result that one order does not immediately follow another and plant and labour are not continuously employed. In order to plan a programme of work effectively, it is necessary to know precisely—

a) What is to be made or done.

b) The quantity involved.

c) What operations are necessary to carry out the work.

d) What plant, equipment and tools are needed.

e) What types of labour are needed.

f) How long each operation may be expected to take.

g) How much plant and equipment of the types necessary are available.

h) How much labour of the types necessary is available.

The information on items (a) and (b) is generally supplied by the sales office or commercial department. The information for determining items (c), (d) and (e) is supplied by process planning and method study. The information on item (f) is supplied by work measurement. The information on item (g) is supplied from plant department records or those of the department concerned. The information on item (h) is supplied from personnel office records or those of the department concerned.

Once this information is available, it is a matter of simple arithmetic to match the requirements with the available capacity. Both the requirements and the capacity available to fulfil them must be stated in terms of time. Requirements will be stated as equal to the number of operations of each type to be performed multiplied by the expected time for each operation. This must be matched against the total time available on each type of plant and with each type of labour necessary to perform the operations.

When a programme is being planned, only the actual times which the operations may be expected to take are of interest. These will depend, among other things, on whether the general conditions in the plant—including the state of labour-management relations and the system of remuneration in use are such that the workers are working at their best rate. Where this is the case and the work study application has had time to settle down, these times should be those of the average performance of the shop or department as given by the production records over a period of time. This may even apply to an individual machine or process. It is the only realistic basis for such calculations. The times are arrived at by multiplying the standard times by

$$\frac{100}{\text{Average performance}}$$

The plant and labour capacity available is expressed in man-minutes or machine minutes, due regard being paid to any time that is necessary to allot for cleaning, setting up, dismantling, change-overs, repairs, etc.

The matching of production or operational requirements to capacity in this way makes it possible to—

a) show whether there is an insufficiency of any type of plant or labour likely to hold up the programme or cause bottlenecks in the course of production and, if so, its extent;

b) show whether there is an excess of capacity in any type of plant or labour and its extent; and,

c) give estimates of delivery dates.

If the management can have such information, compiled from realistic standards of performance, available well before production is due to start, it can take steps to prevent hold-ups from occurring. Alternatively, it can start looking for work to fill up spare capacity. Without such standards it has no sure basis for doing either of those things.

## Time standard as an input to Costing and Budgetary Control

The success or failure of a firm in a competitive market may depend on the accuracy with which it is able to price its products. Unless the manufacturing time of the product is accurately known, the labour cost cannot be estimated, and many indirect costs dependent on time—such as, plant depreciation, fuel and power consumption, rent, and the salaries of staff and supervision—cannot be accurately determined. If the management can rely on the accuracy of the costing, economic prices can be fixed. If these are below those of the firm's competitors, the management can be happy in the knowledge that it is underselling them in safety; if they are above, the cutting of costs can be undertaken with more assurance than would otherwise be the case and with a knowledge of the margins available to be cut.

Standard and actual labour costs per 100 or per 1,000 standard minutes of production are frequently calculated each week from the weekly control statements. Since the actual labour cost per 100 standard minutes takes into account both direct and indirect labour costs, it is the more useful figure to use for estimating production costs. The standards can also be used as the basis of the labour budgets for budgetary control; they provide certain elements of the information necessary for the production and indirect expense budgets and, related to the sales budget, indicate the plant and labour capacity likely to be available over the period of the budget.

## Time standard as the basis for Incentive Schemes

The merits of time standard as a basis for incentive schemes lie in several features inherent in the techniques, namely—

a) The times are based on direct observation and on recording by the most accurate practicable means.

b) Enough observations are taken of all elements of work, both repetitive and occasional, to ensure that the times finally selected to make up the standard time are truly representative and that random occurrences are taken into account.

c) Full records are made and retained so as to be available for examination by either management or workers, should the occasion arise.

d) The recorded times and associated data give a factual basis to any management-labour negotiations on performance standards, as opposed to the bargaining based on opinion which must take place when times are estimated.

e) Properly applied method study followed by work measurement enables the management to guarantee the time standards with reasonable assurance that it is not exposing itself to risks of perpetuating uneconomic rates.

The basic objective of any incentive scheme is to motivate the worker for producing more. A performance base is identified for the purpose and the worker is rewarded based on his performance relative to the identified base. In the initial days, most of the schemes used output as the base, where the reward is decided based on the actual output of the worker and the standard output. The schemes that we find today are mostly multifactor schemes where over and above the output several other factors like quality, material consumption etc., are taken into account to work out a composite performance base. However, in all such composite measures, output forms the major constituent. Thus standard time serves as an useful input for designing incentive schemes.

## 23.4  Conclusion

Any work measurement application needs to be backed up by a proper information system to enable the management to take effective and timely actions. This forms a part of the job of a work study man.

The sheets on which output and performance information is summarised and reported to the management are known as control statements. In a fully developed labour control system there will probably be three different labour control statements, prepared at different intervals and for different purposes. A daily statement may be prepared each morning, separately for each section of the organisation, to indicate to the foreman or supervisor in charge of the section the results of the previous day's working. Once a week the weekly control statement will be compiled, usually on a departmental rather than a section-by-section basis. The weekly statement will go to both foreman and departmental heads. A single sheet frequently has space for the record of 13 weeks of work, a fresh line being used each week, so that the current week's results can be compared with those of earlier weeks during the same quarter. The control statement which goes to the top management is usually made up monthly, on either a departmental or a whole-works basis.

The design of labour control statements varies according to the needs of the organisation, but the usual form is divided into two parts. In the first part, the labour utilisation and effectiveness are expressed in terms of time; in the second part, the figures are translated into costs. In addition to the output (in standard minutes) and the clock minutes worked, from which the productivity of the department may be calculated, waiting time and additional allowances are analysed by causes, so that the manager can at once question and take action on any cause of excessively high waiting time, and can see the cost of it.

In any system of recording associated with work measurement and an incentive system, the minimum data given in Table 26 must be recorded and eventually transmitted to the wages and cost office. It should be noted that the application of work measurement will almost certainly entail and increase in clerical staff. The idea of this frightens many managers, who fear increases in their overhead expenses, forgetting that the increased cost is likely to be very small compared with the savings which the techniques of Work Study can make in their total costs of production or operation.

Finally, any such information system should have the following characteristics:

a) provide accurate and full information;

b) ensure that all the necessary information is recorded as a matter of routine and transmitted with the minimum delay to the central office;

*Table 26. Minimum data required for work measurement and labour control records*

| Information | Source |
|---|---|
| (1) Hours of attendance of each operative | Clock card or time sheet |
| (2) Standard time for each operation | Job card or work study office |
| (3) Times of starting and finishing each operation | Job card or work sheet (via shop clerk) |
| (4) Quantities produced | Job card or work sheet (via work checker) |
| (5) Scrap or rectification: quantities and times | Scrap note or rectification slip (via inspector and shop clerk) |
| (6) Waiting time and non-productive time | Waiting time slips or daily work sheet (via shop clerk) |

c) be simple to understand and to operate and as nearly as possible foolproof, so that all the routine work can be carried on by comparatively unskilled clerical staff;

d) be economical of staff; and,

e) be economical of paper.

Devising a system to fulfil all these requirements is a difficult task. The discussion in the foregoing paragraph provides only a general guideline and has to be followed up by further study.

# Part D

## Two integrated exercises

# Chapter 24
# Two integrated exercises

## 24.1 Introduction

The task that we set for ourselves in this book has been attained. The concepts of productivity and Work Study were introduced and their relationship established in Part A of this book. In parts B and C, we have discussed the techniques of Method Study and Work Measurement, and shown how in different situations these work study techniques can be useful in improving resource utilisation and increasing productivity. In these discussions, any improvement in productivity or resource utilisation has been specified in physical terms. While this may be sufficient in many cases, it may not be enough for some other cases. In a man-machine system, for example, a choice often exists in terms of either reducing the idle time of the worker or that of machines to zero. Behind any decision in such cases, there are always implicit assumptions on costs of idle time of machine and worker. These are not needed to be spelt out if the trade-off is obvious, and hence one can be satisfied in specifying the improvement in physical units. If the trade-off is not obvious, one needs to impute the relevant cost for each choice and finally choose the one with minimum cost as the improved method. Given the direct and often known or logically inferable relationship that exists between physical and monetary units (time and money, in this case), the extension should be straightforward. We present this as an exercise to the reader in the form of two problem situations given in the next two sections.

The first exercise given in section 1 pertains to a manpower planning problem in a department in a process industry. The concepts of labour hour and work sampling will be useful for solving the problem.

The second exercise is provided in the context of production planning in a record manufacturing set-up. The concept of cycle time in a man-machine system will be particularly useful. The reader can also test out his understanding on the use of multiple activity chart for the purpose.

## 24.2 Exercise 1: Manpower planning

D.N. Limited has been thinking of diversifying into the production of a chemical product DNPT. The technical manager of the company, Mr. M, has already visited a DNPT manufacturer abroad, and studied the manufacturing process. The process consists in first preparing Hexamin from Formalin and Ammonia Vapour. Sodium Nitrite is then charged into the Hexamin and chilled. Acid is then added, after which Ammonia is injected. A second round acid addition follows and the resulting mass is transferred to a settler where it is water washed to produce wet DNPT. The product is then dried and pulverised.

Nine types of equipment (vessels or machines) are used for the different operations. The processing times for various operations together with the equipment type used, are provided in Exhibit-1. The data pertains to the manufacturing of a 250 kg batch size of wet DNPT. The operations are semi-automatic with operations 1 to 8 requiring operator's intervention and operations 9 and 10 requiring helper's intervention.

From his experience with the company abroad, and looking at the capital sanction, Mr. M has decided to go for a 250 kg batch size with one equipment for each operation. Thus, the acid vessels (V004) which is used twice during the process, will be two in number, while the rest of the equipments will be one in number. Exhibit-2 provides the number and cost of equipments of each type, as estimated by Mr. M.

Mr. M knows that there is no harm in leaving equipment empty at certain point(s) in times in the production process. However, the compounds formed at the end of each operation, except for the material produced at the end of operation 6 (2nd phase acid addition), are unstable. Thus, no in-process is allowed in these cases. Transferring of materials from one equipment to another, at the end of each operation, is immediate, except for the transferring time to settler. The material at the end of operation 6 is found to be stable for the time required to transfer it.

The company abroad uses one operator and two helpers per shift for operating 250 kg batch size with one equipment for each operation. Mr. M. noted that there was no such thing as peak load for them and the work is uniformly distributed over the shift. The production process being continuous, the operators and helpers in a shift take their turns for different breaks, which can be taken at any point in time in the shift. Thus, there lies a possibility of only two helpers manning the department in any point in time in a shift.

Mr. M is not convinced about departments left to helpers, even for a short duration; so he has decided to provide for 2 operators and 2 helpers per shift.

Please answer the following:

1) a) Once the production process stabilises, maximum how much pulverised DNPT per week (in kg) can Mr. M expect? (The production process is continuous with 3 shifts and 7 days/week operating, and two helpers and two operators in a shift are more than sufficient.) Justify your answer.

   b) How many operators and helpers should Mr. M plan for, if each of the two operators and two helpers are to have a weekly off and the absenteeism (including leave, etc.) is 20%? (Operators and helpers are salaried staff and are not available on hourly basis.)

2) Suppose that two years hence, DNPT production at DNL has been established, and Mr. M is thinking of doubling production. He knows that there will be additional cost on equipment(s), and also the manpower requirement might increase. To assess the current idle time for the operators and helpers, he has undertaken a work sampling

study for a shift. The work sampling result is shown in Exhibit 3. Can you help Mr. M in finding out—

a) What will be the minimum additional expenditure on equipments?

b) What is the minimum cost required to carry out a work sampling study, if Mr. M desires a 5% relative accuracy for the working time of an operator and a helper? (Assume 95% confidence interval and cost of hundred observations on 1 operator and 1 helper = Rs 50.)

c) What will be an estimate of the new manpower requirement by using Exhibit-3 data on work sampling as representative?

3) Supposing, as a measure for doubling production, the Industrial Department of DN Limited have suggested that Mr. M should add a 'drier' (M003) to the current system. Can you illustrate with the help of a bar chart, a feasible schedule for the above alternative? (No constraint on operators and helpers.)

*NOTE*:
The production process is continuous and sufficient number of operators and helpers are there so that there are no process delays.

## EXHIBIT-1: OPERATION DETAILS FOR 250 KG BATCH SIZE

| | Operation | Time hours | Vessel/Machine Type |
|---|---|---|---|
| 1. | Hexamin preparation | 1 | V001 |
| 2. | Charging of Sodium Nitrite | $\frac{1}{2}$ | V002 |
| 3. | Chilling | $1\frac{1}{2}$ | V003 |
| 4. | Acid addition 1st phase | 1 | V004 |
| 5. | Ammonia injection | 1 | V005 |
| 6. | 2nd phase acid addition | 1 | V004 |
| 7. | Transferring to Settler | $\frac{1}{2}$ | M001 |
| 8. | Waterwashing | 1 | M002 |
| 9. | Drying | 8 | M003 |
| 10. | Pulverising | 1 | M004 |

## EXHIBIT-2

| | VESSEL/Machine No. to be currently used | | Cost/Vessel or machine (Rs) |
|---|---|---|---|
| 1. | V001 | 1 | 20000 |
| 2. | V002 | 1 | 10000 |
| 3. | V003 | 1 | 15000 |
| 4. | V004 | 2 | 40000 |
| 5. | V005 | 1 | 10000 |
| 6. | M001 | 1 | 5000 |
| 7. | M002 | 1 | 40000 |
| 8. | M003 | 1 | 50000 |
| 9. | M004 | 1 | 5000 |

## EXHIBIT-3: WORKSAMPLING FOR A SHIFT

| Operator/Helper Observation No. | Operator | Helper |
|---|---|---|
| 1 | 1 | 9 |
| 2 | 1 | 9 |
| 3 | (1) | 9 |
| 4 | 1 | 9 |
| 5 | 2 | (1) |
| 6 | (1) | 9 |
| 7 | 3 | 9 |
| 8 | (1) | 9 |
| 9 | 4 | (1) |
| 10 | (1) | 9 |
| 11 | 5 | 9 |
| 12 | 5 | (1) |
| 13 | (1) | 9 |
| 14 | 6 | 9 |
| 15 | (1) | (1) |
| 16 | 6 | (B) |
| 17 | (1) | 9 |
| 18 | (B) | 9 |
| 19 | 1 | (1) |
| 20 | (1) | (1) |
| 21 | (1) | 9 |
| 22 | 7 | 9 |
| 23 | 8 | (1) |
| 24 | (1) | 10 |
| 25 | (1) | 10 |

*NOTE*: i) (1) Implies Idle state.  ii) (B) Implies break.  iii) Numbers inside the matrix correspond to the operation number being performed.

## 24.3 Exercise 2: Production planning

P & Co. is a small-time manufacturer of L.P. records. Orders amounting to 4000 records/week are produced from two big companies which supply P & Co., with the necessary labels, stampers (negative of records), and the covers (for packaging) for the records to be manufactured. P & Co., buys the raw material (PVC) which is preheated to a certain temperature by an automatic Extruder, converting the granular PVC into cake form. The cakes are then used for producing records from the stampers by compression moulding. Finally, the records are put in the cover and dispatched to the respective companies.

The steps involved in the record manufacturing process is given in Exhibit-1, and the relevant time elements for each step is shown in Exhibit-2.

Normally 50 records can be produced from a single stamper, and P & Co., is accordingly supplied with stampers. The attaching of stampers is done by the press operator and the time required for the activity is negligible. The Extruder does not require any manning and cakes are made available to a press operator as and when required. Contribution/record is Rs 8 and P & Co., is allowed to operate only one shift and six days a week. Considering lunch break etc., the available time per shift is 7 hours. Presently, the company has one Press, one Extruder and one Trimming Machine. One operator has been managing the whole production. However, capacity can be augmented by hiring presses at the rate of Rs 1000 per hour and operators at the rate of Rs 10 per hour. Trimming machine is not available on such basis, and if required has to be bought at a cost of Rs 1 lakh/machine. The Extruder is currently utilised to the extent of 10 to 15% and is not a bottleneck.

## ISSUES FOR DISCUSSION

a) Given the current demand, is it worthwhile adding any type of machine and/or operator? (You would perhaps be interested in finding out the percentage utilisation of each.)

b) If the demand doubles, what alternatives are available to P & Co., for meeting the extra demand? (Please feel free to generate even the most trivial one also, i.e., avoid coupling generation of alternatives with evaluation.)

c) Which alternative among the above would you suggest, given that on any order not met, the company not only loses the contribution but also has to pay a penalty as follows:

| Shortage(% of records) | Penalty (Rs) |
|---|---|
| 0 to 50 | 0 |
| 51 to 100 | 500 |
| 101 to 500 | 1000 |
| 501 and above | 1500 |

d) If the material cost goes up by 200%, what will be its implication on the above decision? (Cost of one bag (50 kg) of PVC = Rs 100.)

### EXHIBIT-1: Steps for record manufacturing process

a) Granular PVC is preheated to a certain temperature to produce a PVC cake of 200 gm (1 cake/record) from the automatic extruder.

b) Stampers corresponding to side A and side B of a record are attached to the top and bottom half of the compression moulding press.

c) Corresponding labels are then placed at the centre of the stampers.

d) The PVC cake is then placed on the centre of the bottom half of the Press.

e) Two buttons on the body of the Press are then pressed by the operator which makes the top half of the Press slide over the bottom half.

f) The sliding activates a limit switch and sets the following automatic set of operations going:

   i) Steam passes.

   ii) Bottom half rises till the gap between the top and bottom half is very small.

   iii) Steam passes.

   iv) Water passes.

   v) Bottom half goes down to the original position.

   vi) Top half slides back to the original position marking the end of the automatic operations.

g) The earlier step results in pressing of the PVC cake into a record with some excess material (sprue) with it. To remove the sprue, the record is placed on a Trimming machine attached to the body of the Press.

h) The trimmer is switched on which results in removing the sprue.

i) The record (finished product) is then taken out from trimmer and packed.

### EXHIBIT-2: Time data

| OPERATION | TIME (Seconds) |
|---|---|
| 1. Place labels | 2 |
| 2. Take cake from Extruder and place on the press | 2 |
| 3. Press buttons | 1 |
| 4. Automatic set of operations | 30 |
| 5. Take out record with sprue and put on Trimmer | 2 |
| 6. Switch on Trimmer | 1 |
| 7. Trimming operation | 3 |
| 8. Switch off Trimmer | 1 |
| 9. Take out record from Trimmer, Take cover and pack | 9.5 |

# Part E

---

# From
# analysis to synthesis:
# new forms of
# work organisation

# Chapter 25

# Combined methods and tasks: New forms of work organisation

## 25.1 Method study and work measurement: basic tools for job design

In the preceding chapters we have thoroughly discussed modern work study techniques. Since the introduction of these techniques at the beginning of this century, work study has become an effective tool in improving the performance of enterprises. Few developments have contributed so much towards attaining that goal. Moreover, the underlying principles of these methods will, for the foreseeable future, continue to be of immense importance in the great majority of enterprises, regardless of their size or area of economic activity.

Let us briefly summarise the basic significance of systematic work study for the development of better methods of work.

### METHODS: SYSTEMATIC v. HAPHAZARD

The first rule of work study is that each task must be systematically analysed in advance and the ways of carrying it out must be thought through. If the task in question is to be carried out only once, perhaps this preliminary analysis is of no great importance—indeed, there might be no point in paying too much attention to it. But if the task is to be carried out repeatedly, we can easily see that much is to be gained by carefully scrutinising the manner in which the task is executed. Every movement that can be eliminated or improved, every time span that can be shortened will produce economies—and if each task is repeated many times, as happens with mass production or long runs, the saving of even tiny movements or of a few seconds here and there can be of crucial economic importance.

It can thus readily be seen that if systematic analyses of this kind are not carried out, preferably before production is begun, inefficiency will in effect be built into the job.

## WORK ANALYSIS: STEP-BY-STEP EXAMINATION

An important feature of work study is therefore the systematic analysis of the job, that is the division of a task into its various component parts followed by a careful examination and discussion of each part. By thus breaking down a complex problem into its underlying elements, a clearer and more readily understandable picture of the task can be obtained, and a good method of carrying it out can be deduced. In Chapter 8 we examined various methods of breaking down work processes into small parts. In the same chapter we went over the questioning technique—a method of questioning everything that is done and taking nothing for granted, with the aim of finding new alternatives, new combinations and new ideas.

## PRE-SET TIMES FOR VARIOUS MOVEMENTS

One of the most important features of modern work study is that it is possible to fix in advance, with moderate margins of error, the times necessary to carry out different movements. There are many different methods of doing this, ranging from summary estimates to highly refined PTS systems. One point that these methods have in common, however, is that they all contain a more or less established method of determining, on the basis of the characteristics of the work in question, the "normal" time that a task should require.

This process of pre-setting times for various tasks is of overwhelming importance in production management. Most important, it makes it possible to test alternative methods and combinations of methods of performing a certain job and to determine which alternative is the most time-saving. Furthermore, with the help of these systematic time guidelines, it becomes feasible to distribute work assignments among different individuals and groups in order to plan production more efficiently and to construct a foundation for discussing production-linked wages and similar incentives.

Again, this is an element of modern work study that is virtually indispensable in normal industrial activities. Without the help of work study methods and systematic time formulae, the determination of guidelines would be pure guesswork.

## THE LATEST ROLE OF WORK STUDY: FROM ANALYSIS TO SYNTHESIS

So far we have discussed the basic role of work study in the design of individual jobs and of work organisation. Before we go into more detail, it should be emphasised that the development of method study and work measurement has been continuous, so that it is now possible to apply work study to any kind of activity. Furthermore, the workers' understanding of and active involvement in work study has increased rapidly.

With this point clear in our minds, let us now turn to the question of how the basic "building blocks" of method study and work measurement can be put together in designing jobs, and how work organisation can best be shaped in other respects. We shall divide this discussion into three parts, corresponding to three organisational levels—

1. Design of individual work roles.

2. Design of group work in production.

3. Design of product-oriented organisations.

A detailed examination of these topics falls outside the scope of this introductory book, and we shall limit ourselves here to a discussion of some of their basic features.

## 25.2  Design of individual work roles

### GUIDELINES IN THE DESIGN OF JOBS: SOME EXAMPLES

In putting together an individual work role with the help of the fundamental building blocks we have been discussing (that is, the component parts of each task and the description of methods), we may adopt a number of criteria as guidelines for satisfactory job design.

Most important are the economic aspects. With the help of systematic work study the component parts of a task are put together in such a way that as little time as possible is required to carry it out. In this book we have so far confined our discussion to this point.

However, the design of individual work roles is too complex to be effected with the aid of a single criterion—that is, what appears on paper to be the shortest time needed to carry out a task. In practice, numerous different factors must be considered.

Some of these are purely practical considerations, such as the need for different types of machinery, the nature of the different components of each job, and so on. For example, if it takes ten minutes to carry out a particular component part of the task and if this component part is repeated 1,000 times within a 50-man work group, it is easy to see that the results of this study must be combined with other information about the work situation in order to arrive at a reasonable division of the task among the various members of the group. This example is given merely to indicate the problem, which we shall not examine here. There is, however, one special group of factors that we must look at more closely: namely, the worker's needs and preferences, his experience of the work and his reaction to different kinds of work organisation. This is a new and important dimension, since it implies the need to adapt work design to the individual's wishes and capacities, to create jobs in industry that offer a reasonable challenge, and to provide the worker with the chance of a working climate that offers some degree of satisfaction. The reader will no doubt recall that this point was made earlier, in Chapter 5. Here we can identify three important factors that can lead to increased job satisfaction—

(1) A moderate amount of variety in the work done.

(2) Decoupling of man/machine processes, that is, freedom from being tied to a machine during the entire working day.

(3) The opportunity to integrate various service and auxiliary tasks into a production job.

These three topics will be treated separately below.

### Variety at work

If work is to be done well, there must be a reasonable correlation between the job and the person doing that job. A job that consists of only a few simple movements and takes only a few seconds to do can certainly be easy to learn. At first sight, it may seem that this is an efficient way of organising the work. But this type of job is hardly efficient from a more practical viewpoint. It will rapidly become monotonous and tiring, and such extreme specialisation requires long runs, plus a degree of structural stability and production volume that is not often found in reality. It is much better to create work roles that display a reasonable amount of variety, that require something from the worker in terms of learning and that are adapted to reality in terms of the true length of runs, a stable product mix and infrequent production disturbances.

There is no complete, clear answer to the question of how a task cycle that gives just the right amount of variety should be designed. However, a study of the following factors offers some guidance in bringing about improvements:

the basic structure of the technical system;

the pattern of the physical load;

the information content of the task;

the balance between physical and intellectual task components;

the demand for learning and the need for individual development opportunities.

In many production technologies the basic structure of the technical system is the determining factor. For example, on a motor car assembly line the length and content of the job cycle are wholly dictated by the technical system. If 500 cars are to be produced in 500 minutes, each operative has one minute in which to do his job. There is nothing that can be done about it. In other words the job cycle can be changed only if the concept of the technical system itself (i.e. the assembly line as a working arrangement) is changed. We shall come back to this question of assembly system design later.

But the fixed-speed assembly line is not the only technical system that prevents the introduction of a time cycle of reasonable length. Short-cycle man/machine operations, such as those carried out with eccentric shaft presses, offer another example of the need to reshape the entire technical system in order to apply time cycles that are of a comfortable length for the operative. This also will be discussed later.

It should be emphasised that variety in the time cycle is primarily a subjective concept and therefore cannot be precisely defined, either technically or mathematically. However, it is more or less closely related to other factors such as—

length of the time cycle;

size of the run;

frequency of recurrence of a product (that is, the time that passes before the same product is worked on again);

amount and distribution, in repetitive jobs, of non-repetitive tasks;

differences in work structure and job content between different series.

*Example.* In an enterprise manufacturing electrical circuit breakers, two alternatives for the organisation of the work were identified. The first would require that assembly be done in four separate jobs, each carried out at a specially built and specially equipped work station. At the last of these work stations the assembly work is completed and a control check is made. In this type of arrangement the cycles are about ten seconds in length. Variations within cycles are virtually non-existent.

The second alternative would require that the entire circuit breaker assembly be done at each of the work stations (i.e. one job at each work station). In order to arrive at this solution, the materials supply system would have to be completely reorganised. By planning the work in this way the cycle is lengthened to 40 seconds. In addition, opportunities for varying the cycles increase markedly.

After an analysis of the practical consequences of the two choices at the workplace, the second alternative was chosen. The decision is significant, since it exemplifies the efforts that have been made in recent years to limit monotony in jobs and to achieve a practical balance of working conditions.

One important point in an analysis of this kind is the fact that people are different. At any one time the people at the same workplace will present quite different characteristics. And if we study the same person at different times during his working life, we shall find significant differences in his performance. This is an important, indeed fundamental element in the design of individual work roles. Jobs should be different, and should present different degrees of difficulty to those who execute them. Thus different people can find a work role and a level of difficulty that match their own aptitudes and preferences. In addition, an individual can begin working in a particular job that has a particular level of difficulty, and can then move steadily to more challenging jobs as he develops further.

## Decoupling man/machine systems

The rigidity of the links on a worker in a man/machine system may be due to several factors. The person can be tied to the workplace in a geographical sense—it may be impossible for him to be absent from his station for even a short time. He can also be tied by the method—it may be impossible to vary the order in which operations are carried out. And he can be tied in terms of time—he may be required to carry out certain operations at fixed times.

The degree of rigidity with which he is tied can be "planned"—that is, the man and the machine are consciously and deliberately tied together in a man/machine system—but in many cases the rigidity is quite "unplanned". In some cases this unplanned rigidity arises from a fault in the technical system; the operational stability of the machines may be so poor that the machines must be continuously tended, usually with only simple movements. Unplanned rigidity can, however, be reduced through the use of more operationally reliable technology.

Three different types of solution may be offered for this problem of rigid man/machine links—

(1) Complete decoupling through increased mechanisation.

(2) Use of technical auxiliary equipment to free the operative from the machine.

(3) Decoupling through contact and co-operation among operatives.

Let us examine more closely each of these three choices.

### Complete decoupling through mechanisation

Decoupling of this kind requires heavy capital investment. Therefore, production processes that are to be handled in this way must be characterised by mass production, extremely short cycles and severe rigidity and monotony. In such cases mechanisation means the complete elimination of all human intervention.

### Technical auxiliary equipment for the operator

This principle can be put into effect by establishing buffers and magazines in an integrated man/machine system in order to reduce dependence relationships between men and machines. (A buffer is a waiting point located between two consecutive operations in the production flow; a magazine is a point of accumulation located within an operation and providing automatic feeding of material to the machine.) The key is to create processes that can accept variations in the speed at which different sections of the line move.

Both buffers and magazines are characterised by an "accumulation of products for continued processing" which can be completely identical in their technical design.

Since buffers and magazines are placed at different points in the man/machine system, their characteristics as accumulators of time are influenced by different types of time gaps in the process.

A buffer makes it possible to accumulate:

*(a)* the waiting times created when two operatives on opposite sides of the buffer work at different speeds; and

*(b)* the waiting times created because the quantities of work done at two stations are not absolutely identical.

A magazine makes it possible to accumulate:

*(a)* waiting times created because an operative works at a different speed from the over-all speed of the technical process; and

*(b)* waiting times created because an operative is forced to wait while a machine does its part of the work.

### Decoupling through contact and co-operation

Finally, decoupling can be achieved if, through job rotation and mutual co-operation and in agreement with the management, workers are able to interchange tasks and assignments.

## Integration of production and auxiliary tasks

In the design of individual work roles it can often be advantageous to include various service and auxiliary tasks in production jobs. This leads to greater variety for the individual in his job.

Auxiliary tasks that are most often combined in this way are:

maintenance of machines and tools;

setting-up of machines;

handling of materials near the work station;

inventory work;

quality control.

Let us discuss some of these auxiliary tasks further.

When we speak of maintenance in production positions, we are referring to measures that can be taken to reduce the number and extent of production errors. Maintenance can include a regular inspection of the system in order to find errors and take remedial measures. Maintenance can also include repairs of parts so as to make it possible to achieve the established precision norms required in production. In addition, it can include a statistical follow-up in order to improve the capacity utilisation of equipment.

The possibility of including machine setting-up and similar preparatory functions in the ordinary operative's role depends on a number of factors, among which are the following:

degree of difficulty and time available for the setting-up operation;

frequency of setting-up operations;

degree of rigidity in other production tasks;

need for special auxiliary equipment to undertake this work.

*Example.* A metalworking enterprise conducts its operations with the help of advanced computer-controlled equipment. In one department the operative was trained to programme the computer equipment himself. He was thus able to handle the traditional job as well as the programming of the machine tool's computer equipment. He therefore works both as a programmer and as a machine operator. This example shows that even moderately difficult and specialised tasks can sometimes be integrated into a normal production job.

Regarding the possible integration of material-handling work near the work station, the following factors are some of the more decisive:

character of the product;

volume of materials to be handled;

design of the transport system;

degree of rigidity in the production operation.

These are some examples showing how direct production jobs can be supplemented with various auxiliary and service tasks. There are no simple, standard solutions in this area; each case must be examined in the light of its special characteristics. However, the guiding principle in making these decisions is that a practical and smoothly functioning arrangement must be feasible, that jobs can be broadened sufficiently to include everyday variations and that they must not be excessively monotonous.

## 25.3   Design of group work in production

### ADVANTAGES OF GROUP WORK

Once individual jobs have been designed, the next logical step is to co-ordinate these roles. One method of co-ordination that has attracted increasing interest in recent years is the tying together of individual jobs into work groups. Organisational descriptions of a complete work group specify which roles are included in the group and the principles according to which these roles should be co-ordinated. Group work in production can have many advantages. Here we shall touch only on some of the more important of them.

The most important advantage is the way in which objectives are established and the results measured. In this connection it must be borne in mind that it is much easier to formulate appropriate objectives for a group than for an individual job, and this is an important advantage.

Another advantage is that the leeway for variations in the individual's activities increases and that a stronger sense of participation in the larger process can be experienced than when each person is tied to a limited individual task. People working in a group have a better chance to co-operate continuously in improving methods and eliminating unnecessary work. Attitudes can change as team spirit develops.

A further merit of group organisation is that the organisation's capacity to adapt itself to change increases. An enterprise is in a state of continuous change. The management alone cannot completely control, manage and follow up this process of adaptation to change; the organisation itself must possess a strong built-in capacity for self-adaptation.

These are some of the most important reasons why ideas of group work in production have been gaining ground in the design of work organisation. But group work is not suitable everywhere. In certain types of production systems it is an excellent concept, while in others it is completely unworkable. Let us look at some models of production systems and see how group work might fit with specific working conditions.[1]

## SEVEN PRODUCTION SYSTEM MODELS: WHERE DOES GROUP WORK FIT?

We shall divide these production systems schematically into seven main types, and then use this classification to discuss where group production is most suitable as an organisational concept. We may refer to these seven models as follows:

(1) The machine-paced line.

(2) The man-paced line.

(3) The automated process.

(4) The concentrated operation.

(5) The diversified line group.

(6) The service group.

(7) The construction group.

Let us study briefly the requisite characteristics for group work in each of these categories.

### The machine-paced line

This type of arrangement is most often found in situations where material

---

[1] These models are taken from Hans Lindestad and Jan-Peder Norstedt: *Autonomous groups and payment by result* (Stockholm, Swedish Employers' Confederation, 1973). For further details see also George Kanawaty (ed.): *Managing and developing new forms of work organisation* (Geneva, ILO, 2nd ed., 1981).

handling is an important factor and where the material-handling function occupies a dominant role. The classical example of this type is the motor car factory's final assembly on a fixed-speed assembly line.

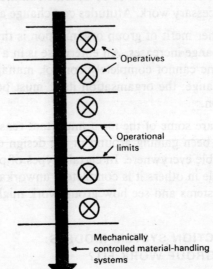

Figure 113.   Machine-paced line

Operatives

Operational limits

Mechanically controlled material-handling systems

In this type of production system a high degree of mechanised handling is chosen. The flow of materials and the organisation of work are therefore completely under the control of the technical system. Until only a few years ago this was the only assembly arrangement used in situations where a high volume of materials was the rule. The disadvantage of this system is that the individual's work role is strictly limited and that the work pace is completely controlled by the technical system. In systems of this type, where operatives are tightly tied to a short task cycle, no genuine group work is possible. Consequently, the most important disadvantage of such production systems is the way in which operatives experience their work. Other disadvantages include the extreme sensitivity of such lines to disturbances. These production chains are only as strong as their weakest link, and it requires only a small influenza epidemic in the region where the factory is located to upset the whole system. Moreover, it is difficult to make changes in such production lines.

The advantages are short through-put times, the efficient utilisation of space, machines and auxiliary equipment and, consequently, the efficient operation that is achieved through the extreme division of work and specialisation. However, these advantages apply only when the production system is in operation.

During recent years a considerable number of attempts to "loosen up" the assembly line have been made with the help of different innovative arrangements—a point to which we shall return later.

### The man-paced line

If we imagine an assembly line from which we have removed the mechanised control and flow speed and introduced some inventories between work stations, we have a type of functional arrangement that is common in many companies (in the clothing and metalworking industries, for example).

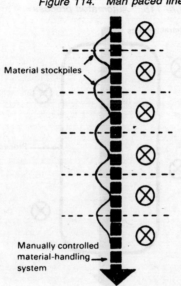

Figure 114.   Man paced line

In this sort of production system the control is less rigorous and the existence of buffers makes it possible to adapt the individual work pace in a completely different way from work on an assembly line. In such a system work organisation based on production groups is an excellent arrangement. Within a group made up of individual work roles, operatives can help each other, take care of work disturbances, even out peaks and valleys of work flows and strive for a good common work result.

### The automated process

If it were possible to mechanise all the manually executed tasks on a conventional assembly line, the result would be a kind of process line where the individual's work would be concerned primarily with supervision and control. Process lines of this type are extensively used, particularly in the steel, chemical and paper and pulp industries.

On a process line the possibilities of creating meaningful group work are often excellent. Operatives rely greatly on one another and possess a common goal. Working together to attain this goal is a clear-cut necessity. One factor that may sometimes make group co-operation difficult is an excessive distance between group members. A key question in this type of production system is the relationship of direct production tasks and maintenance tasks executed in the work organisation. The higher the degree of mechanisation, the fewer production workers there are; but the number of maintenance workers normally increases at almost the same rate as the number of production workers decreases.

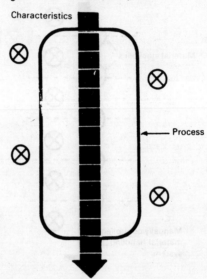

*Figure 115.    Automated process*

### The concentrated operation (functional layout)

A constant element in the three types of system that we have discussed up to now is the grouping together of production equipment along the production flow so that different types of machines are placed in the correct order along the direction of flow. However, if we group the machines in such a way that all machines of a certain type are concentrated in one department, all machines of another type in another department, and so on, we obtain a concentration of each type of operation in one place (this is the "functional layout" referred to earlier in the book). In this layout the product to be worked is sent through the various departments in turn—the drilling department, the turning department, the milling department, and so forth.

This type of concentrated operation often occurs in batch production, where series are short and the products varied.

In this type of production system it is extremely difficult to organise meaningful group work. In everyday reality each individual is bound to his own individual job and work station. Genuine group work with spontaneous interaction between different roles and role occupants is virtually impossible to bring about.

Figure 116.   *Concentrated operation*

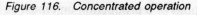

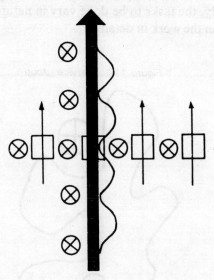

## The diversified line group

In many cases the conditions affecting production are such that neither highly developed line grouping nor an advanced degree of operation grouping is suitable. Instead, an intermediate type is chosen—what we may call the "diversified line group". Production is concentrated in an arrangement that is primarily flow-oriented, but in order that it may carry out many combinations of tasks, some critical operational stages are repeated two or more times. In this way a system is obtained that can, with a high degree of efficiency, combine the capacity of the flow group to accept and channel a large volume of materials with the capacity of the functional layout to execute all conceivable production assignments.

In this type of production system, group work is often an excellent organisational concept. With this arrangement the division of work between various individuals must be adapted continuously to varying conditions. But this cannot be done entirely by the management, and a substantial proportion must occur spontaneously at the initiative of the members of the group. In a group organisation the capacity for such spontaneous self-adaptation can gradually be generated.

## The service group

Conditions within service-producing organisations differ in several respects from the types of activity we have discussed earlier. Various forms of services are produced in large sectors, such as commerce, transportation, hotels and restaurants and motor vehicle repair shops. But service functions also occur in manufacturing industry, a good example being repair and maintenance activities.

The service functions of a production unit must be highly adaptable to varying demands. Generally, the tasks to be done vary in nature. The work load is uneven and it is difficult to plan the work in detail.

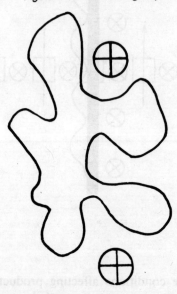

Figure 117.   Service group

Group organisation is a good concept in this type of situation also. The work group can itself handle much of the variation that shows up in the inflow of tasks, in routine work planning and in other circumstances that tend to vary.

### The construction group

For the final type in our classification, let us see how construction operations are carried out. In this case the product itself is the centre for the whole organisation, which is built up around the construction object itself. Work organisations of this type are also found in industry, for example in manufacturing very large products (e.g. turbines, ships, process machinery).

In production work of this type, group work is not only a good idea: it is the only conceivable type of work organisation. Moreover, the work is varied, and the spontaneous adaptation of the division of work and planning is such an essential feature that flexible group organisation is the only possible solution.

We have now briefly examined the possibilities of group work in different types of production system. We have seen that group work is more suitable in some cases than in others.

*Figure 118.   Construction group*

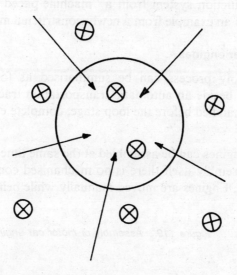

One of the lines of development that has been particularly advocated in discussions about group work in production is the degree to which groups can be organised along the direction of production flow. Grouping of this type makes it possible to direct the group's interests and strivings toward a good common production result. We might look rather more closely at the possibilities of organising such groups, either in assembly work or in machine shops. Our purpose in taking up these examples for special discussion is not to provide ready-made solutions but to point out a line of development that nowadays is assuming particular importance.

## FLOW GROUPS IN ASSEMBLY WORK:
## SOME TRENDS AND EXAMPLES

In assembly work, flow groups have always been the most natural arrangement. Let us take final assembly of a motor car, for example. When this arrangement was first conceived it was quite natural to introduce an assembly system that moved beside a materials inventory, with the different components being assembled on the car as it moved past. This is an extreme example of flow orientation in assembly work. The flow of materials was completely decisive in arranging the work.

But an arrangement of this type can also have its disadvantages. The work is strictly controlled and the cycle time is normally very short.

At subsequent stages of development, efforts were made to introduce buffers in the production line in order to create greater freedom in different parts of the production system. This placed new demands on the system, and various technical solutions were advocated to separate the different links in the chain from each other.

With reference to our previous discussion of different production system models, we may say that the introduction of buffer arrangements in motor car assembly changes the production system from a "machine-paced line" to a "man-paced line". The following is an example from a newly constructed motor car engine factory.

### Assembly of motor car engines

The assembly process can be summarised as follows. Seven assembly groups are organised beside an automatic transportation track loop. Except for certain steps which are handled before the loop stage, complete engines are assembled in each group.

Up to six engines can be assembled at the same time within each production group. During the assembly itself there is no mechanised control of the flow as in a moving assembly line. Engines are moved manually while being assembled. When an

*Figure 119. Assembly of motor car engines*

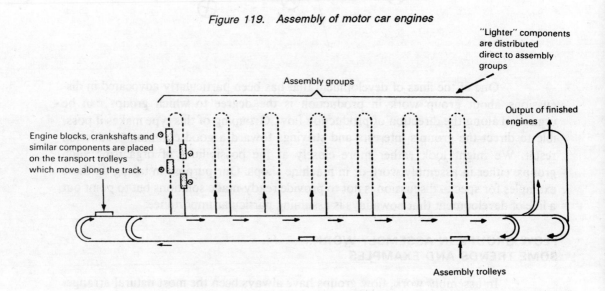

"Lighter" components are distributed direct to assembly groups

Assembly groups

Output of finished engines

Engine blocks, crankshafts and similar components are placed on the transport trolleys which move along the track

Assembly trolleys

engine has been completely assembled in a group, it is transported automatically to a testing station which is common to all groups. At the same time it is automatically registered that an engine has left the group and a new assembly trolley is moved forward to that group on the transport track.

The advantages and disadvantages of this type of assembly process, as compared with the traditional assembly line, are as follows:

(1) This arrangement is more flexible and less susceptible to interruptions and fluctuations in the production flow.

(2) It offers good possibilities for job expansion and a more stimulating kind of group

work. Each of the small loops contains a production group, a "gang" whose members co-operate closely in everyday tasks and themselves take care of such chores as the adaptation of work to changing conditions. One of the seven groups is a training group. In this group there is a fairly strict and extensive division of tasks based on detailed instructions. In the other groups the division of work is made on the basis of the abilities of individual members. There is therefore an opportunity to adapt the design of jobs within the group to the workers' knowledge and experience.

(3) It is not necessary to carry out an extensive and costly reconstruction of the line every time the production volume has to be increased or decreased. Capacity can be expanded to a certain extent by varying the numbers of members in the groups, up to six. Further increases in capacity can be achieved by increasing the number of groups.

(4) Job design is better adapted to the individual and should therefore lead to better recruiting possibilities, reduced personnel turnover and less absenteeism.

(5) The new arrangement requires greater floor space and higher goods-in-process inventories than a moving assembly line.

(6) Capital investment is somewhat higher for the new arrangement.

(7) Work efficiency (primarily as regards speed of movement) is lower than on a moving assembly line, because of the lower degree of specialisation and the fragmentation of work assignments.

This example illustrates not only how buffer arrangements can be introduced between different jobs and different capacities for work of different individuals but also how different parts of an assembly line—or an entire line—can be rearranged in a parallel pattern. The assembly of the engines is carried out at a number of stations, with an entire engine being assembled at each station.

The nature of parallel production operations is made clear in figure 136.

The most important advantages offered by the parallel arrangement of an assembly operation (or parts of an assembly operation) are as follows:

(1) Production reliability—it is naturally less likely that several subsystems will all be simultaneously affected by disturbances than that one large system will be so affected.

(2) Flexibility—it is easier to handle different product models, as well as changes in production volume, in a parallel system.

(3) Work content and work organisation—the possibility of creating tasks with a richer content, and of finding natural dividing lines between groups, is considerably greater. Opportunities for production groups to accept responsibility for quality and the division of work, for example, are also greater.

*Figure 120.   Line grouping and parallel grouping*

Line grouping

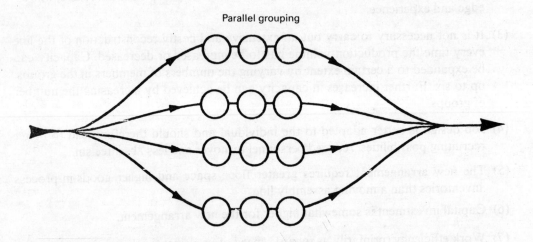

Parallel grouping

## Flow-oriented machine groups in batch production

In a traditional layout in batch production, machines and personnel are grouped in departments, with each department carrying out its own separate function. For example, one department may handle turning, another drilling, a third milling, and so on. The advantage of this arrangement is that it results in great flexibility and a high degree of utilisation of machine capacity. A major disadvantage is that the volume of goods-in-process, and therefore the amount of working capital tied up in these goods, is always substantial. Moreover, the work in a plant of this kind is highly fragmented. It is difficult for an individual or a group of individuals to see the connection between their own work roles and the over-all activity of the company. It is therefore difficult for individuals and groups to participate actively in work planning and in attaining the established goals of the company.

During recent years, interest has grown in finding ways of grouping machinery and equipment around flow-oriented groups in batch production, that is, groups formed around the manufacture of entire products or complex product components. We shall discuss these trends briefly here.

What is a **flow-oriented group?** Figure 137 illustrates the basic principle.

With the help of a standard classification method, we have selected an assortment of different components, such as axles and flanges. In each of these groups there are subgroups that resemble each other as regards the types of work required.

Machines, personnel and other resources needed for the components—from metal supplies to finished parts—are collected in one unit. Through the choice of suitable components, methods and equipment we can create a simple flow pattern.

With this manufacturing arrangement through-put times, and therefore also the working capital tied up in the system, can both be reduced. Production can be carried out with a minimum of supplies of materials on hand—this applies particularly to the work stations themselves. The lower the supply of materials on hand, the shorter and surer the through-put times become.

In a functional organisation, each operative's task at "his" machine and the job planned for the machine are fixed in advance. A flow-oriented group is a machine group for the finished manufacture of a mix of components. It contains more machines or work stations than there are operatives, and each operative should preferably master several types of job. This means that all the members of the group must be able to work relatively independently. The group members themselves have the responsibility for dividing the work between them and seeing that material flows through the group as it should. Thus the work of a flow group relies heavily on teamwork and co-operation.

Unlike a functional grouping of machines, a flow group makes heavy demands on individuals. But a flow group also makes possible the creation of more attractive work roles for group members, because—

(1) They have a better over-all view of their contribution to the larger production process.

(2) They have more variety in their work because they can move between various tasks.

(3) They have the chance of being trained for new jobs.

(4) They have increased contact with their colleagues at work as well as with the management.

*Example.* In figure 138, a flow group has been created for the manufacture of pump axles in a metalworking company. In this group approximately 150 types of axle are produced; however, these are based on about ten general methods, of which the most widely used account for about 65 articles.

The simplest components are manufactured from pre-cut metal pieces during a single trip through the group. The most complicated components must go through the group three times. Operatives can easily return parts to the incoming station with the help of roller conveyor tracks. Two men work in this group; their work is delineated by the shape of the conveyor.

However, flow-oriented manufacturing in short series requires certain definite conditions and cannot be used in all situations. For example, a systematic structuring of the product mix must be made, to make it possible to channel certain main types of product in a homogeneous flow. Moreover, production must be of such

*Figure 121. Schematic diagram of a flow-oriented group*

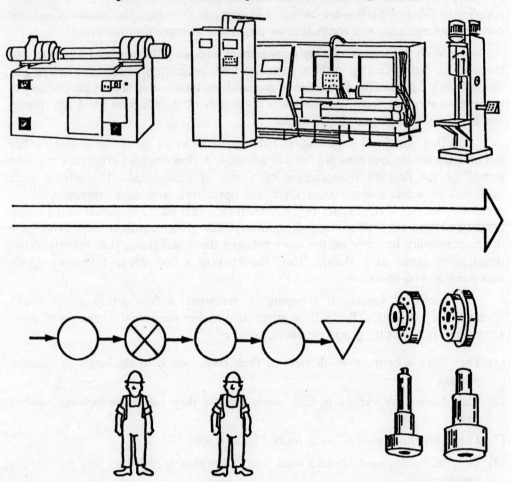

a nature that an "unbroken flow principle" can be applied. If it is necessary to break the material flow within the flow group at a certain operational step and to send components outside the group for working, the planning will naturally become substantially more complicated.

A key issue in the formation of flow-oriented groups is the degree of utilisation of equipment that can be attained, especially in the case of more expensive production machinery. Here it is necessary to weigh machine costs against the costs of tying up capital in everyday work. Recently, the clear trend is towards a recognition of the fact that tying up capital in goods-in-process inventories has reached such proportions that the order of priorities has had to be modified in favour of the use of flow groups.

A further factor of decisive importance is of course the stability of the product mix. Flow grouping of machinery has to be based on the assumption that it is

*Figure 122.  Flow group for the manufacture of pump axles*

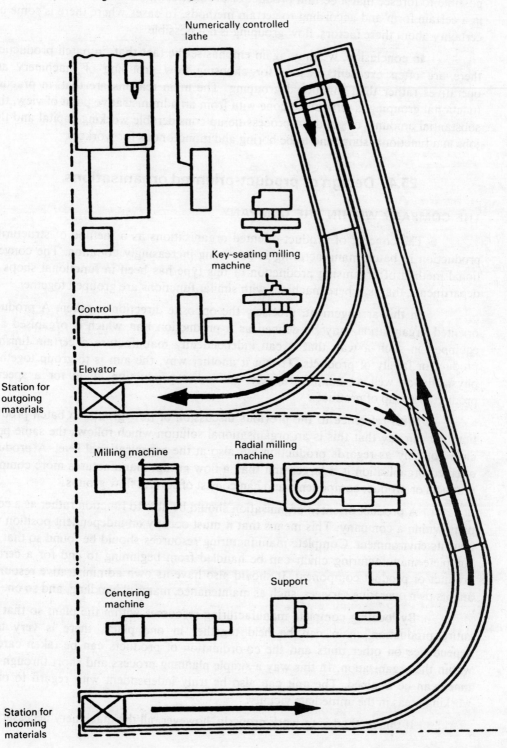

Numerically controlled
lathe

Key-seating milling
machine

Control

Elevator

Station for
outgoing
materials

Milling machine

Radial milling
machine

Support

Centering
machine

Station for
incoming
materials

possible to foresee that a certain product or product component will be manufactured in a certain form and according to certain methods. In cases where there is some uncertainty about these factors, flow grouping is not possible.

In conclusion, we may again emphasise the fact that, in batch production, there are often excellent reasons for choosing flow grouping of machinery and operatives rather than functional grouping. The main reasons are that, in practice, functional grouping is difficult to cope with from an administrative point of view, that substantial amounts of goods-in-process tie up considerable working capital and that jobs in a functional shop tend to be boring and monotonous for workers.

## 25.4   Design of product-oriented organisations

### THE COMPANY WITHIN THE COMPANY

The concept of product-oriented organisations as a method of structuring production in batch manufacturing is becoming increasingly common. The conventional method of organising production of this type has been in functional shops or departments, that is, where machines with similar functions are grouped together.

In this arrangement, precisely the opposite direction is taken. A product-oriented organisation may be defined as a production unit which is organised and equipped in such a way that it can independently manufacture a certain finished product or family of products. To put it another way, the aim is to group together, physically as well as administratively, the entire production chain for a specific product or group of products.

With reference to the previous discussion of flow groups in batch production, we can say that this is an organisational solution which follows the same principle not only as regards production but also at the organisational level. A product-oriented organisation is a larger unit than a flow group, manufactures more complex products or product components and can consist of several flow groups.

A product-oriented organisation should be able to function rather as a company within a company. This means that it must occupy an independent position vis-a-vis its environment. Complete manufacturing resources should be found so that the complete manufacturing chain can be handled from beginning to end for a certain product or product component. It should also have its own administrative resources and its own operating services, such as maintenance, material handling, and so on.

By locating complete manufacturing resources within the plant so that the entire production chain can be held together in one place, there is very little dependence on other units and the co-ordination of products can be taken care of within the organisation. In this way a simple planning process and short through-put times can be attained. The unit can also be truly independent with regard to other working areas in the immediate vicinity.

If this method is to work properly, however, all the machinery necessary carry out the complete production operation must be available. In general, the capacity of utilisation of most machines will be lower than in a functional shop. The possible machine utilisation will thus be a key factor in examining the feasibility of this organisational concept, and should be weighed against its other advantages, especially as regards lower working capital tied up in inventories and simpler administration.

## FLOW PATTERNS IN A PRODUCT-ORIENTED ORGANISATION: AN EXAMPLE

By definition, the product-oriented organisation refers to a certain flow of production. Within the unit itself, however, this flow can be more or less divided, and machine grouping can vary from very pronounced line grouping to a more operationally grouped functional arrangement. Let us look at two examples of the organisation of a product shop.

In the first example, a heat exchanger unit, a systematic attempt has been made to build the production structure on the basis of flow groups. It proved possible to do so for the main part of the manufacturing process despite the fact that it is heavily influenced by customer orders and that batches are small. Figure 139 shows how an attempt was made to come as near as possible to a "straight-line" arrangement. This simplifies material handling and gives all operatives a good over-all view of the manufacturing process.

However, it will be seen that the flow is divided between two areas in the manufacturing chain. There is a materials buffer for assembly, and there is also a buffer between the pressing of plates and finishing of products (see figure 139). The reason for this is to achieve reasonable batch sizes and to reduce change-over times in production.

In our second example, relating to the manufacture of electric motors, figure 141 shows a product-oriented organisation consisting of a number of flow groups in which different components are manufactured. Among the principles on which the arrangement was based are the following:

(1) Manufacture of components in units from raw materials, each in its own component flow or flow group.
(2) Co-ordination of component flow directly with the main flow without material buffers or interim inventories.
(3) Completion of main flow with delivery of finished motors.

This arrangement of the flow means that the quantity of goods-in-process is very small, and the through-put times from the first operation to the finished motor is only two or three days. Furthermore, no interim inventories are needed for assembly.

Figure 123. Layout for a heat exchanger unit

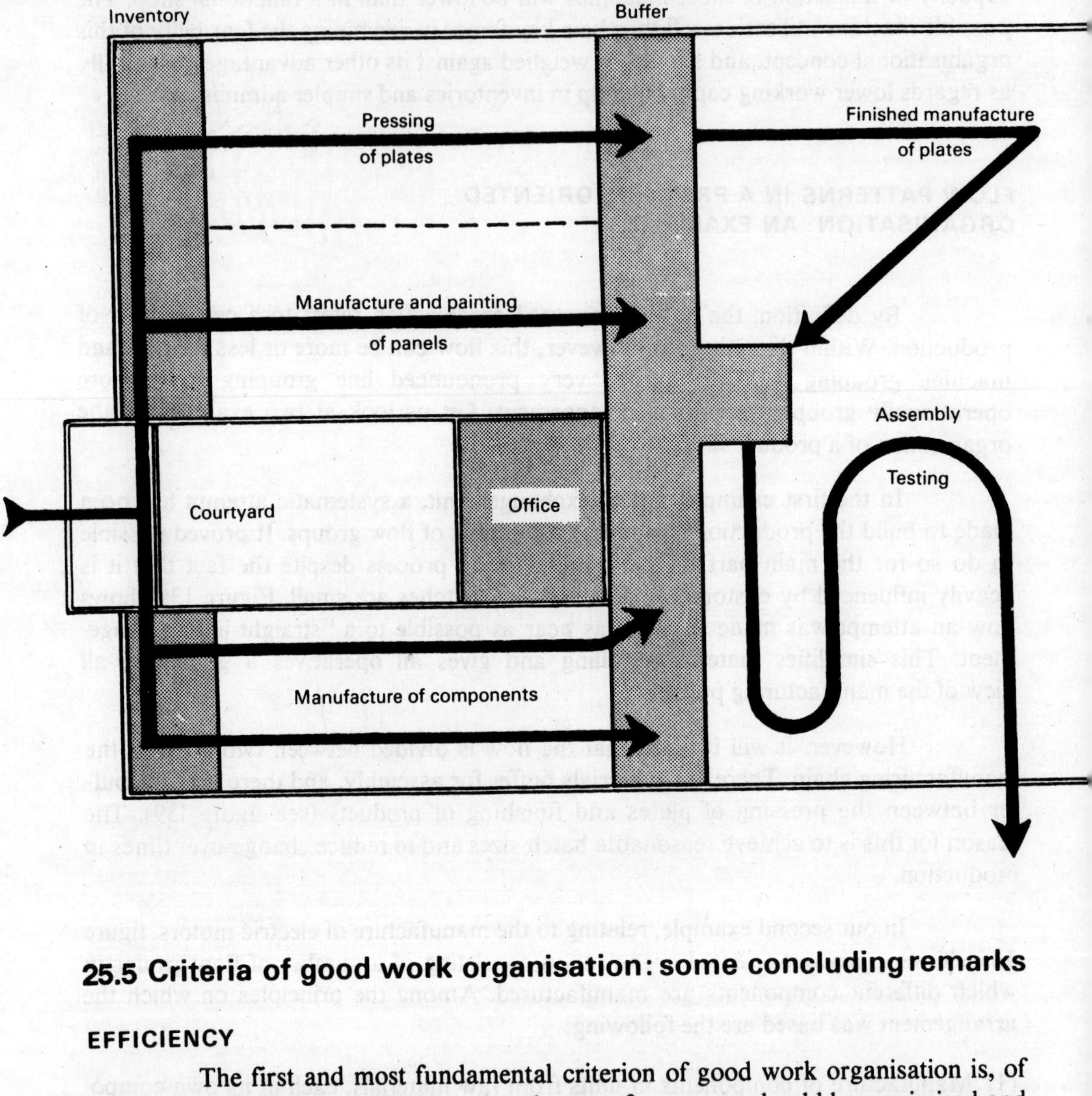

## 25.5 Criteria of good work organisation: some concluding remarks

### EFFICIENCY

The first and most fundamental criterion of good work organisation is, of course, that it should be effective—that the use of resources should be maximised and that the largest possible output should be obtained from the smallest possible input. The various chapters of this book have dealt extensively with this criterion, because this factor will always be of fundamental significance—in all types of technology, in all stages of development and at every workplace.

Naturally, there are situations in which considerations other than those of a purely economic nature are of paramount importance. If, for example, there are evident safety or health risks at a workplace, and if additional investment is required to

Figure 124.   Some examples of the building of buffer stock in manufacturing operations

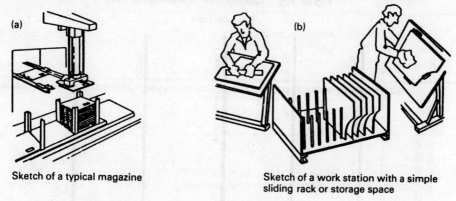

(a)

**Sketch of a typical magazine**

(b)

**Sketch of a work station with a simple
sliding rack or storage space**

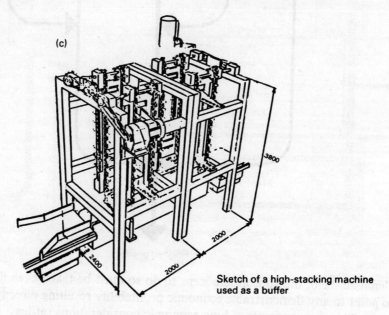

(c)

**Sketch of a high-stacking machine
used as a buffer**

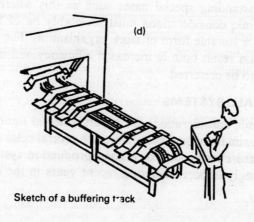

(d)

**Sketch of a buffering track**

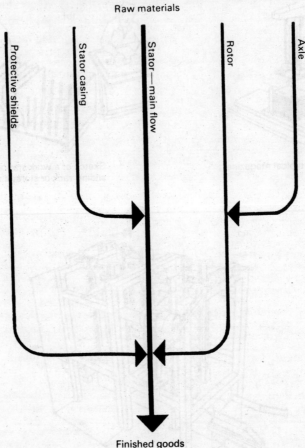

Figure 125.   Manufacture of electric motors

eliminate them, the appropriate steps to do so must be taken even if it is not possible to point to any demonstrable economic profitability resulting directly from such measures. This is an example of how economic considerations (at least in the short term) have to give way to other factors.

But, notwithstanding special cases such as this where particular circumstances obtain, economic considerations must inevitably be of fundamental importance in the choice of a suitable form of work organisation. The organisational principles and solutions that result both in increased efficiency and in better jobs for the workers are naturally to be preferred.

## AUTONOMY OF SMALL SYSTEMS

Even if economic considerations are of fundamental significance and must be carefully analysed in each individual case, there are several rules of thumb, or general lines of thinking, for the construction of a good production system—guidelines that have become increasingly important during recent years in the development of new forms of work organisation but in which precise calculations of short-term profitability are difficult, if not impossible. Nevertheless, there has been so much emphasis on these guidelines that we take special note of them here; but we must also stress that they stand somewhat apart from the basic economic factors.

The first of these criteria for constructing good production systems is the search for greater independence for small systems in company organisation. By this we mean production systems that consist of moderately large production units and can function with a relatively high degree of independence within the larger company. The underlying intention is to create a production arrangement that emphasises local independence within smaller units. Breaking down the company into these smaller units reduces the need for co-ordination, and therefore management problems too become simpler to deal with.

The decentralisation that results from this type of production arrangement is also of great value in stimulating local initiative and in increasing the ability to adapt to the changing conditions and needs that arise in different parts of the company. It has also been shown that workers are often more satisfied and more involved in their work if they are members of smaller and more independent production units.

If we wish to create production systems based on this principle, four points are particularly significant—

(1) The possibility of dividing up larger systems into smaller systems.

(2) The possibility of arranging finished manufacturing units into smaller units so that the need for contacts with adjacent units is reduced.

(3) The possibility of arranging for self-sufficiency as regards production resources, operational service, and so on.

(4) The possibility of arranging for less direct management control from high levels, so that the independence of the smaller units is not eroded too much by control from the upper levels of the hierarchy.

## STABILITY OF THE PRODUCTION SYSTEM

One further rule of thumb or criterion of a good production system which has received increasing interest in recent years is the desire to arrange for stable production activity with a minimum of disturbance. The following requirements in particular arise in this connection:

(1) A simple flow pattern, so that as far as possible the workers have an over-all view and that it becomes easier to plan the work.

(2) An operationally reliable technology with an optimum level of mechanisation, so that technical disturbances are held within reasonable limits.

(3) A disturbance-resistant work arrangement, so that all production stages that are critical for production are organised in parallel and that those that are particularly sensitive to disturbance are surrounded with buffers of different kinds.

## ATTRACTIVE JOBS

It is important to be able to offer people jobs that they find attractive and in which they can feel personally involved. Personal aspirations vary from individual to individual and from situation to situation, and depend not only on a person's ambitions and desires but also on his or her abilities, knowledge and capacity to develop. A production organisation must therefore offer a variety of jobs, so that the desires of as many people as possible can be satisfied and so that a particular individual can progress from simple jobs to more complex work roles.

Among the factors that should be considered in any endeavour to create sufficiently attractive jobs are the following:

(1) The creation of jobs with different degrees of difficulty through flow orientation, different degrees of subdivision of work and different degrees of integration of auxiliary tasks. Variations of this kind make it possible to offer to different individuals at different times jobs that correspond to their abilities and wishes.

(2) The creation of individual jobs and group arrangements that bring about a degree of independence in work, through finished manufacturing of entire products, self-sufficiency of production service functions and buffering vis-à-vis adjacent systems. This independence is of value both in terms of the production results obtained and for the way the work is experienced by individuals in the group.

(3) The design of a work organisation that is suitable for teamwork, as a result of flow grouping and similar arrangements that are compatible not only with more attractive jobs and work situations but also with greater efficiency.

(4) Provision of over-all views from inside the organisation. In order for a person to find his work attractive, he must also be able to view the larger context of which his work is a part. It is also important that he should be involved, if possible, in the design of his work and be able to feel some sense of "belonging" with his group of fellow workers and with the over-all production process in which he performs his function.

## GOOD WORKING ENVIRONMENTS

An important criterion of a good job is the quality of the working environment. In Chapter 6 we indicated the basic factors that have to be considered with respect to safety at the workplace.

In addition, however, a working environment should also be pleasant to work in—in other words, it should be so designed that it becomes easier to adopt ergonomically correct working positions.

## CONCLUSION

We have briefly touched on some of the trends leading towards new forms of work organisation. We have given some principles and general guidelines. We have provided some examples and emphasised certain current lines of development. Finally, we have given some criteria to be borne in mind when designing good working environments.

It is important, however, to stress the fact that there are no standard solutions to these problems. Our aim has been merely to put forward a few ideas, tendencies and general indications of solutions to problems. It must be remembered that the best solution to a problem can be found only in the specific circumstances of the particular case—when the actual conditions are known, when local values are considered and when the persons involved are able to find their own solutions.

# Part F
# Appendices

# 1. Glossary of terms used

## A. Work study

**Activity Sampling**

See *Work Sampling*.

**Basic Time**

The time for carrying out an element of work at standard rating, i.e.—

$$\frac{\text{Observed Time} \times \text{Observed Rating}}{\text{Standard Rating}}$$

**Break Point**

The instant at which one element in a work cycle ends and another begins.

**Check Time**

The time intervals between the start of a time study and the start of the first element observed, and between the finish of the last element observed and the finish of the study.

**Chronocyclegraph**

A cyclegraph in which the light source is suitably interrupted so that the path appears as a series of pear-shaped dots, the pointed end indicating the direction of movement and the spacing indicating the speed of movement.

**Contingency Allowance**

A small allowance of time which may be included in a standard time to meet legitimate and expected items of work or delays, the precise measurement of which is uneconomical because of their infrequent or irregular occurrence.

**Cumulative Timing**

See *Timing*.

**Cyclegraph**

A record of a path of movement, usually traced by a continuous source of light on a photograph, preferably stereoscopic.

**Cycle Time**

The total time taken to complete the elements constituting the work cycle.

**Elapsed Time**

The total time from the start to the finish of a time study.

Element

The portion of an operation selected for convenience of observation, measurement and analysis.

Operation Element

The portion in which the basic time-consuming operation can be recognised.

Finger Element

A movement carried out during a cycle which, after analysis, is not found to be a necessary part of the job.

Operating Element

A manual act having a higher time value than that of the other elements which, is not a part of the operation.

Genuine Period (also Basic Time)

A value per automatic action which the worker driven machine performance.

Manual Element

An element in which the worker's action governs the...

Machine Element

An element in which the worker's action governs the part of the machine which may be performed whilst the worker is otherwise engaged.

Repetitive Element

An element which occurs in every work cycle of an operation.

Therblig

The amount of time which the basic time-values to attend characteristics of the product, equipment, materials, temperature, weight, quality, etc.

Tension

The difference of base-time from observed times.

Fatigue Allowance

An addition to the relaxation allowance intended to allow for the physiological and psychological effects of carrying out specified work under specified conditions.

Flow Analysis

The reinforcement examination of a cine film of an operation to determine the area of action of the surface during each exposure.

Flow Diagram

A diagram or model, substantially to scale, which shows the location of specific activities carried out and the routes followed by workers, materials or equipment in their execution.

Flow Process Chart

A process chart setting out the sequence of the flow of a product or a procedure by recording...

**Element**

 A distinct part of a specified job selected for convenience of observation, measurement and analysis.

### Constant Element

An element for which the basic time remains constant whenever it is performed.

### Foreign Element

An element observed during a study which, after analysis, is not found to be a necessary part of the job.

### Governing Element

An element occupying a longer time than that of any other element which is being performed concurrently.

### Machine Element

An element automatically performed by a power-driven machine (or process).

### Manual Element

An element performed by a worker.

### Occasional Element

An element which does not occur in every work cycle of the job, but which may occur at regular or irregular intervals.

### Repetitive Element

An element which occurs in every work cycle of the job.

### Variable Element

An element for which the basic time varies in relation to some characteristics of the product, equipment or process, e.g. dimensions, weight, quality, etc.

**Extension**

 The calculation of basic time from observed time.

**Fatigue Allowance**

 A subdivision of the relaxation allowance intended to cater for the physiological and psychological effects of carrying out specified work under specified conditions.

**Film Analysis**

 The frame-by-frame examination of a ciné film of an operation to determine the state of activity of the subject during each exposure.

**Flow Diagram**

 A diagram or model, substantially to scale, which shows the location of specific activities carried out and the routes followed by workers, materials or equipment in their execution.

**Flow Process Chart**

 A process chart setting out the sequence of the flow of a product or a procedure by recording all events under review using the appropriate process chart symbols.

### Equipment Type Flow Process Chart

A flow process chart which records how the equipment is used.

### Man Type Flow Process Chart

A flow process chart which records what the worker does.

### Material Type Flow Process Chart

A flow process chart which records how material is handled or treated.

## Flyback Timing

See *Timing*.

## Idle Time

That part of attendance time when the worker has work available but for various reasons does not do it.

## Ineffective Time

That portion of the elapsed time, excluding the check time, spent on any activity which is not a specified part of a job.

## Inside Work

Elements which can be performed by a worker within the machine- (or process-)controlled time.

## Interference Allowance

An allowance of time for production unavoidably lost through synchronisation of stoppages on two or more machines (or processes) attended by one worker. Similar circumstances arise in team work.

## Interference Time

The time when the machine (or process) is idle awaiting attention, while the worker attends to another machine (or process). Similar circumstances arise in team work.

## Job Breakdown

A listing of the content of a job by elements.

## Load Factor

The proportion of the over-all cycle time required by the worker to carry out the necessary work at standard performance, during a machine- (or process-)controlled cycle.

## Machine Ancillary Time

The time when a machine is temporarily out of productive use owing to change-overs, setting, cleaning, etc.

## Machine Available Time

The time during which a machine could work based on attendance time—i.e. working day or week plus overtime.

**Machine Capacity**

The potential volume of a machine, usually expressed in physical units capable of being produced in any convenient unit of time, e.g. tons per week, pieces per hour, etc.

**Machine-Controlled Time**

The time taken to complete that part of the work cycle which is determined only by technical factors peculiar to the machine.

**Machine Down Time**

The time during which a machine cannot be operated on production or ancillary work owing to breakdown, maintenance requirements, or for other similar reasons.

**Machine Effective Utilisation Index**

The ratio of:  Machine Running Time at Standard
   to:  Machine Available Time.

**Machine Efficiency Index**

The ratio of:  Machine Running Time at Standard
   to:  Machine Running Time.

**Machine Hour**

The running of a machine or piece of plant for one hour.

**Machine Idle Time**

The time during which a machine is available for production or ancillary work but is not used owing to shortage of work, materials or workers, including the time that the plant is out of balance.

**Machine Interference**

The queuing of machines (or processes) for attention—e.g. when one worker is responsible for attending to more than one machine. Similar circumstances arise in team work where random delays at any point may affect the output of the team.

**Machine Maximum Time**

The maximum possible time during which a machine or group of machines could work within a given period, e.g. 168 hours in one week or 24 hours in one day.

**Machine Running Time**

The time during which a machine is actually operating, i.e. the machine available time *less* any machine down time, machine idle time, or machine ancillary time.

**Machine Running Time at Standard**

The running time that should be incurred in producing the output if the machine is working under optimum conditions.

**Machine Utilisation Index**

The ratio of:  Machine Running Time.
   to:  Machine Available Time.

**Man-Hour**

The labour of one man for one hour.

**Memomotion Photography**

A form of time-lapse photography which records activity by a ciné camera adapted to take pictures at longer intervals than normal. The time intervals usually lie between ½ and 4 seconds.

**Method Study**

The systematic recording and critical examination of existing and proposed ways of doing work, as a means of developing and applying easier and more effective methods and reducing costs.

**Methods-Time Measurement (MTM)**

A system of Predetermined Time Standards (q.v.).

**Micromotion Study**

The critical examination of a simo chart prepared by a frame-by-frame examination of a cine film of an operation.

**Multiple Activity Chart**

A chart on which the activities of more than one subject (worker, machine or item of equipment) are each recorded on a common time scale to show their inter-relationship.

**Multiple Machine Work**

Work which requires the worker to attend two or more machines (of similar or different kinds) running simultaneously.

**Observed Time**

The time taken to perform an element or combination of elements obtained by means of direct measurement.

**Outline Process Chart**

A process chart giving an over-all picture by recording in sequence only the main operations and inspections.

**Outside Work**

Elements which must necessarily be performed by a worker outside the machine- (or process-)controlled time.

**Personal Needs Allowance**

A subdivision of the relaxation allowance intended to cater for attention to personal needs.

**Plant and Machine Control**

The procedures and means by which efficiency and utilisation of units of plant and machinery are planned and checked.

**Policy Allowance**

An increment, other than bonus increment, applied to standard time (or to some constituent part of it, e.g. work content) to provide a satisfactory level of earnings for a specified level of performance under exceptional circumstances.

**Predetermined Time Standards (PTS)**

A work measurement technique whereby times established for basic human motions (classified according to the nature of the motion and the conditions under which it is made) are used to build up the time for a job at a defined level of performance.

### Primary Questions

The first stage of the questioning technique, which queries the fundamental purpose, place, sequence, person and means of every activity recorded, and seeks a reason for each reply.

### Principles of Motion Economy

Characteristics which, when incorporated in the methods adopted, make for easier working.

### Process Charts

Charts in which a sequence of events is portrayed diagrammatically by means of a set of process chart symbols to help a person to visualise a process as a means of examining and improving it.

### Process-Controlled Time

The time taken to complete that part of the work cycle which is determined only by technical factors peculiar to the process.

### Qualified Worker

One who is accepted as having the necessary physical attributes, who possesses the required intelligence and education, and who has acquired the necessary skill and knowledge to carry out the work in hand to satisfactory standards of safety, quantity and quality.

### Questioning Technique

The means by which the critical examination is conducted, each activity being subjected in turn to a systematic and progressive series of questions.

### Random Observation Method

See *Work Sampling*.

### Rating

(1)     The assessment of the worker's rate of working relative to the observer's concept of the rate corresponding to standard pace.

(2)     The numerical value or symbol used to denote the rate of working.

  *(a)* Loose rating: an inaccurate rating which is too high.
  *(b)* Tight rating: an inaccurate rating which is too low.
  *(c)* Inconsistent ratings: a mixture of loose, tight and accurate ratings.
  *(d)* Flat ratings: a set of ratings in which the observer has underestimated the variations in the worker's rate of working.
  *(e)* Steep ratings: a set of ratings in which the observer has overestimated the variations in the worker's rate of working.

### Rating Scale

The series of numerical indices given to various rates of working. The scale is linear.

### Ratio-Delay Study

See *Work Sampling*.

### Relaxation Allowance

An addition to the basic time intended to provide the worker with the opportunity to recover from the physiological and psychological effects of carrying out specified work under specified conditions and to allow attention to personal needs. The amount of the allowance will depend on the nature of the job.

### Representative Worker

A worker whose skill and performance is the average of a group under consideration. He is not necessarily a qualified worker.

### Restricted Work

Work in which the output of the worker is limited by factors outside the control of the worker.

### Secondary Questions

The second stage of the questioning technique, during which the answers to the primary questions are subjected to further query to determine whether possible alternatives of place, sequence, persons and/or means are practicable and preferable as a means of improvement upon the existing method.

### Selected Time

The time chosen as being representative of a group of times for an element or group of elements. These times may be either observed or basic and should be denoted as selected observed or selected basic times.

### Flyback Timing

A method in which the hands of the stop-watch are returned to zero at the end of each element and are allowed to restart immediately, the time for the element being obtained directly.

### Tool Allowance

An allowance of time, which may be included in a standard time, to cover adjustment and maintenance of tools.

### Travel Chart

A tabular record for presenting quantitative data about the movements of workers, materials or equipment between any number of places over any given period of time.

### Two-Handed Process Chart

A process chart in which the activities of a worker's hands (or limbs) are recorded in their relationship to one another.

### Unoccupied Time

The periods during machine- (or process-)controlled time when a worker is engaged neither on inside work nor in taking authorised rest, the time for carrying out the work being calculated at a defined performance.

### Unoccupied Time Allowance

An allowance made to a worker when there is unoccupied time during machine- (or process-) controlled time.

### Unrestricted Work

Work in which the output of the worker is limited only by factors within the control of the worker.

**Work Content**

Basic time + relaxation allowance + any allowance for additional work—e.g. that part of contingency allowance which represents work.

**Work Cycle**

The sequence of elements which are required to perform a job or yield a unit of production. The sequence may sometimes include occasional elements.

**Work Factor**

A system of Predetermined Time Standards (q.v.).

**Work Measurement**

The application of techniques designed to establish the time for a qualified worker to carry out a specified job at a defined level of performance.

**Work Sampling**

A method of finding the percentage occurrence of a certain activity by statistical sampling and random observations. (Work sampling is also known as ratio-delay study; observation ratio study; snap-reading method; random observation method; and activity sampling.)

**Work Specification**

A document setting out the details of an operation or job, how it is to be performed, the layout of the workplace, particulars of machines, tools and appliances to be used, and the duties and responsibilities of the worker. The standard time or allowed time assigned to the job is normally included.

**Flyback Timing**

A method in which the hands of the stop-watch are returned to zero at the end of each element and are allowed to restart immediately, the time for the element being obtained directly.

**Tool Allowance**

An allowance of time, which may be included in a standard time, to cover adjustment and maintenance of tools.

**Travel Chart**

A tabular record for presenting quantitative data about the movements of workers, materials or equipment between any number of places over any given period of time.

**Two-Handed Process Chart**

A process chart in which the activities of a worker's hands (or limbs) are recorded in their relationship to one another.

**Unoccupied Time**

The periods during machine- (or process-)controlled time when a worker is engaged neither on inside work nor in taking authorised rest, the time for carrying out the work being calculated at a defined performance.

**Unoccupied Time Allowance**

An allowance made to a worker when there is unoccupied time during machine- (or process-)controlled time.

**Unrestricted Work**

Work in which the output of the worker is limited only by factors within the control of the worker.

**Work Content**

Basic time + relaxation allowance + any allowance for additional work—e.g. that part of contingency allowance which represents work.

**Work Cycle**

The sequence of elements which are required to perform a job or yield a unit of production. The sequence may sometimes include occasional elements.

**Work Factor**

A system of Predetermined Time Standards (q.v.).

**Work Measurement**

The application of techniques designed to establish the time for a qualified worker to carry out a specified job at a defined level of performance.

**Work Sampling**

A method of finding the percentage occurrence of a certain activity by statistical sampling and random observations. (Work sampling is also known as ratio-delay study; observation ratio study; snap-reading method; random observation method; and activity sampling.)

**Work Specification**

A document setting out the details of an operation or job, how it is to be performed, the layout of the workplace, particulars of machines, tools and appliances to be used, and the duties and responsibilities of the worker. The standard time or allowed time assigned to the job is normally included.

**Work Study**

A generic term for those techniques, particularly method study and work measurement, which are used in the examination of human work in all its contexts, and which lead systematically to the investigation of all the factors which affect the efficiency and economy of the situation being reviewed, in order to effect improvement.

## B. Plant layout

**Factory Flow Analysis**

Part of production flow analysis (q.v.). A technique which uses networks to study and simplify the flow of materials between departments.

**Fixture**

A device for holding parts which would otherwise have to be held in one hand while the other worked on them.

**Group Analysis**

Part of production flow analysis (q.v.). A technique used to determine the best division of the machines in a machining department into groups and the best division of the parts made into families.

### Group Layout

A layout in which a set of machines, chosen so that it can carry out the complete processing of a given family of products, is laid out together in one area.

### Jig

A device which holds parts in an exact position and guides the tool that works on them.

### Line Analysis

Part of production flow analysis (q.v.). A technique used to study the flow of materials between the machines in a group, in order to find the best arrangement for their layout.

### Line Layout

A layout in which the machines are set out in a line in their sequence of use, with materials flowing along the line.

### Plant Layout

The arrangement of the desired machinery and equipment of a plant, established or contemplated, in the way which will permit the easiest flow of materials, at the lowest cost and with the minimum of handling, in processing the product from the receipt of raw materials to the dispatch of the finished product.

### Process Layout

A layout in which all machines or processes of the same type are grouped together.

### Product Layout

A layout in which all machines or processes concerned in the manufacture of the same product or range of products are grouped together.

### Production Flow Analysis

A technique used to study the flow of materials in a factory and to find the best division into groups and families for group layout (q.v.).

### Tooling Analysis

Part of production flow analysis (q.v.). A technique used to find the sequence for loading parts on a machine which will give minimum setting time.

### Workplace Layout

A convenient term used to describe the space and the arrangement of facilities and conditions provided for a worker in the performance of a specified job.

## C.  Management

### Budgetary Control

A means of controlling the activities of an enterprise by carefully forecasting the level of each activity and converting the estimate into monetary terms. The actual cost of or revenue from each activity is checked against the estimates.

**Incentive Scheme**

Any system of remuneration in which the amount earned is dependent on the results obtained, thereby offering the employee an incentive to achieve better results.

**Inspection**

The application of tests with the aid of measuring appliances to discover whether a given item or product is within specified limits of variability.

**Maintenance** *(in the management sense)*

The systematic inspection, servicing and repair of plant, equipment and buildings with a view to preventing breakdowns while in use.

**Market Research**

The gathering, recording and analysing of all facts about problems relating to the transfer and marketing of specified goods and services from producer to consumer.

**Marketing Policy**

The policy of an enterprise regarding the marketing of its products or services. It includes questions relating to the range of goods or services to be offered, markets to be entered, price ranges, selling methods, distribution and sales promotion, and the appropriate policy mix to be followed by management with regard to the marketing of its products.

**Material Control**

Procedures and means by which the correct quantity and quality of materials and components are made available to meet production plans.

**Operator Training**

The systematic training or retraining of workers in manual skills with a view to ensuring sound and uniform working methods.

**Personnel Policy**

The policy of an enterprise towards its employees. It embraces methods of selection, recruitment, training, remuneration, welfare services, consultation, relations with unions, social security and all other matters in which the attitude of the employer can affect the quality of working life and well-being of those employed.

**Process Planning**

The detailed planning of the processes of manufacture necessary to convert raw material into finished products before commencing operation. The term originated in the engineering industry.

**Process Research**

Research into the nature and characteristics of a given production process.

**Product Development**

The stage, usually between design and large-scale production, during which units of the product are tested and studied with a view to improving performance, ease of manufacture and market appeal.

**Product Research**

Research into the nature and characteristics of a product or potential product in relation to the functions it has to or may have to perform.

### Production Control

The planning, direction and control of the supply of materials and processing activities of an enterprise.

### Production Planning

The planning of the physical means of production. It is concerned with process planning, with the design of tooling, with the layout of plant and equipment and with the handling of materials and tools in the workshop. Work study is a major technique in production planning.

### Productivity

The ratio of output to input.

### Progressing

Systematic control procedures designed to ensure that the programmes and orders issued by production control are carried out.

### Quality Control

The function of management which controls the quality of products. It includes inspection and other procedures and means (including sampling methods based on statistical principles) of maintaining the quality of products.

### Standard Costing

A system of costing in which standard costs are estimated in advance; the actual costs incurred are compared with the standards and any variance is analysed for causes.

### Standardisation

The development and application of a standard for a particular product or type of component or range of products or components or a given procedure.

### Value Analysis

The systematised investigation of the product and its manufacture to reduce cost and improve value.

### Variety Reduction

The systematic reduction of the number of varieties of products made and materials, parts and tools used in a factory.

# 2. Check-list of questions which may be of use in applying the questioning sequence in method study

Most of the questions listed below apply generally to method study investigations. They amplify the questioning procedure described in Chapter 8, and may be of service in suggesting to studymen aspects of the method which might otherwise be overlooked. The questions are listed under the following headings:

A. Operations          G. Work Organisation
B. Design              H. Workplace Layout
C. Inspection Requirements   I. Tools and Equipment
D. Material Handling    J. Working Conditions
E. Process Analysis     K. Job Enrichment
F. Material

## A. Operations

1. What is the purpose of the operation?

2. Is the result obtained by the operation necessary?
   If so, what makes it necessary?

3. Is the operation necessary because the previous operation was not performed correctly?

4. Is the operation instituted to correct a condition that has now been corrected otherwise?

5. If the operation is being carried out to improve appearance, does the additional cost give extra saleability?

6. Can the purpose of the operation be obtained in another way?

7. Can the material supplier perform the operation more economically?

8. Is the operation being performed to satisfy the requirements of all users of the product, or is it made necessary by the requirements of one or two customers only?

9. Does a subsequent operation eliminate the necessity for this operation?

10. Is the operation being performed as a result of habit?

11. Was the operation established to reduce the cost of a previous operation, or a subsequent operation?

12. Was the operation added by the sales department as a special feature?

13. Can the part be purchased at a lower cost?

14. Would adding a further operation make other operations easier to perform?

15. Is there another way to perform the operation and still maintain the same results?

16. If the operation has been established to correct a subsequent difficulty, is it possible that the corrective operation is more costly than the difficulty itself?

17.    Have conditions changed since the operation was added to the process?

18.    Could the operation be combined with a previous or a subsequent operation?

## B. Design

1.    Can the design be changed to simplify or eliminate the operation?

2.    Is the design of the part suitable for good manufacturing practice?

3.    Can equivalent results be obtained by changing the design and thus reducing cost?

4.    Can a standard part be substituted?

5.    Would a change in design mean increased saleability, an increased market?

6.    Can a standard part be converted to do the job?

7.    Is it possible to improve the appearance of the article without interfering with its utility?

8.    Would an additional cost caused by improved appearance and greater utility be offset by increased business?

9.    Has the article the best possible appearance and utility on the market at the price?

10.    Has value analysis been used?

## C. Inspection Requirements

1.    What are the inspection requirements for this operation?

2.    Does everybody involved know exactly what the requirements are?

3.    What are the inspection details of the previous and following operations?

4.    Will changing the requirements of this operation make it easier to perform?

5.    Will changing the requirements of the previous operation make this operation easier?

6.    Are tolerance, allowance, finish and other standards really necessary?

7.    Can standards be raised to improve quality without unnecessary cost?

8.    Will lowering standards reduce cost considerably?

9.    Can the finished quality of the product be improved in any way above the present standard?

10.    How do standards for this operation/product compare with standards for similar items?

11.    Can the quality be improved by using new processes?

12.    Are the same standards necessary for all customers?

13.    Will a change in standards and inspection requirements increase or decrease the defective work and expense in the operation, shop or field?

14.    Are the tolerances used in actual practice the same as those shown on the drawing?

15.    Has an agreement been reached by all concerned as to what constitutes acceptable quality?

16.    What are the main causes of rejections for this part?

17.    Is the quality standard definitely fixed, or is it a matter of individual judgement?

## D. Material Handling

1.    Is the time spent in bringing material to the work station and in removing it large in proportion to the time used to handle it at the work station?

2.    If not, could material handling be done by the operatives to provide a rest through change of occupation?

3.   Should hand, electric or fork-lift trucks be used?

4.   Should special racks, containers or pallets be designed to permit the handling of material with ease and without damage?

5.   Where should incoming and outgoing materials be located in the work area?

6.   Is a conveyor justified, and if so, what type would best be suited for the job?

7.   Can the work stations for progressive steps of the operation be moved closer together and the material-handling problem overcome by gravity chutes?

8.   Can material be pushed from operative to operative along the bench?

9.   Can material be dispatched from a central point by means of a conveyor?

10.   Is the size of the container suitable for the amount of material transported?

11.   Can material be brought to a central inspection point by means of a conveyor?

12.   Could the operative inspect his own work?

13.   Can a container be designed to make material more accessible?

14.   Could a container be placed at the work station without removing the material?

15.   Can an electric or air hoist or any other lifting device be used with advantage?

16.   If an overhead travelling crane is used, is the service prompt and accurate?

17.   Can a tractor-trailer train be used? Could this or an individual railway replace a conveyor?

18.   Can gravity be utilised by starting the first operation at a higher level?

19.   Can chutes be used to catch material and convey it to containers?

20.   Would flow process charts assist in solving the flow and handling problem?

21.   Is the store efficiently located?

22.   Are truck loading and unloading stations located centrally?

23.   Can conveyors be used for floor-to-floor transportation?

24.   Can waist-high portable material containers be used at the work stations?

25.   Can a finished part be easily disposed of?

26.   Would a turntable eliminate walking?

27.   Can incoming raw material be delivered at the first work station to save double handling?

28.   Could operations be combined at one work station to save double handling?

29.   Would a container of standard size eliminate weighing?

30.   Would a hydraulic lift eliminate a crane service?

31.   Could the operative deliver parts to the next work station when he disposes of them?

32.   Are containers uniform to permit stacking and eliminate excessive use of floor space?

33.   Could material be bought in a more convenient size for handling?

34.   Would signals, i.e. lights, bells, etc., notifying men that more material is required, save delay?

35.   Would better scheduling eliminate bottlenecks?

36.   Would better planning eliminate crane bottlenecks?

37.   Can the location of stores and stockpiles be altered to reduce handling and transportation?

### E. Process Analysis

1. Can the operation being analysed be combined with another operation? Can it be eliminated?

2. Can it be broken up and the various parts of the operation added to other operations?

3. Can a part of the operation being performed be completed more effectively as a separate operation?

4. Is the sequence of operations the best possible, or would changing the sequence improve the operation?

5. Could the operation be done in another department to save the cost of handling?

6. Should a concise study of the operation be made by means of a flow process chart?

7. If the operation is changed, what effect will it have on the other operations? On the finished product?

8. If a different method of producing the part can be used, will it justify all the work and activity involved?

9. Can the operation and inspection be combined?

10. Is the job inspected at its most critical point, or when it is completed?

11. Will a patrol form of inspection eliminate waste, scrap and expense?

12. Are there other similar parts which could be made using the same method, tooling and set-up?

### F. Material

1. Is the material being used really suitable for the job?

2. Could a less expensive material be substituted and still do the job?

3. Could a lighter-gauge material be used?

4. Is the material purchased in a condition suitable for use?

5. Could the supplier perform additional work on the material that would improve usage and decrease waste?

6. Is the material sufficiently clean?

7. Is the material bought in amounts and sizes that give the greatest utilisation and limit scrap, offcuts and short ends?

8. Is the material used to the best possible advantage during cutting, processing?

9. Are materials used in connection with the process—oils, water, acids, paint, gas, compressed air, electricity—suitable, and is their use controlled and economised?

10. How does the cost of material compare with the cost of labour?

11. Can the design be changed to eliminate excessive loss and scrap material?

12. Can the number of materials used be reduced by standardisation?

13. Could the part be made from scrap material or offcuts?

14. Can newly developed materials—plastics, hardboard, etc.—be used?

15. Is the supplier of the material performing operations on it which are not necessary for the process?

16. Can extruded materials be used?

17. If the material was of a more consistent grade, could better control of the process be established?

18. Can a fabricated part be substituted instead of a casting to save pattern costs?

19. Is the activity low enough to warrant this?

20. Is the material free from sharp edges and burrs?

21. What effect does storage have on material?

22. Could a more careful inspection of incoming materials decrease difficulties now being encountered in the shop?

23. Could sampling inspection combined with supplier rating reduce inspection costs and delays?

24. Could the part be made more economically from offcuts in some other gauge of material?

## G. Work Organisation

1. How is the job assigned to the operative?

2. Are things so well controlled that the operative is never without a job to do?

3. How is the operative given instructions?

4. How is material obtained?

5. How are drawings and tools issued?

6. Is there a control on time? If so, how are the starting and finishing times of the job checked?

7. Are there many possibilities for delays at the drawing-room, tool-room and store-room and at the clerk's office?

8. Does the layout of the work area prove to be effective, and can it be improved?

9. Is the material properly positioned?

10. If the operation is being performed continually, how much time is wasted at the start and end of the shift by preliminary operations and cleaning up?

11. How is the amount of finished material counted?

12. Is there a definite check between pieces recorded and pieces paid for?

13. Can automatic counters be used?

14. What clerical work is required from operatives for filling in time cards, material requisitions and the like?

15. How is defective work handled?

16. How is the issue and servicing of tools organised?

17. Are adequate records kept on the performance of operatives?

18. Are new employees properly introduced to their surroundings and do they receive sufficient instruction?

19. When workers do not reach a standard of performance, are the details investigated?

20. Are suggestions from workers encouraged?

21. Do the workers really understand the incentive plan under which they work?

## H. Workplace Layout

1. Does the plant layout aid efficient material handling?

2. Does the plant layout allow efficient maintenance?

3. Does the plant layout provide adequate safety?

4. Is the plant layout convenient for setting-up?

5.    Does the plant layout help social interaction between the operatives?

6.    Are materials conveniently placed at the workplace?

7.    Are tools pre-positioned to save mental delay?

8.    Are adequate working surfaces provided for subsidiary operations, e.g. inspection and deburring?

9.    Are facilities provided for the removal and storage of swarf and scrap?

10.   Is adequate provision made for the comfort of the operative, e.g. fan, duckboard or chairs?

11.   Is the lighting adequate for the job?

12.   Has provision been made for the storage of tools and gauges?

13.   Has provision been made for the storage of the operatives' personal belongings?

## I. Tools and Equipment

1.    Can a jig be designed that can be used for more than one job?

2.    Is the volume sufficient to justify highly developed specialised tools and fixtures?

3.    Can a magazine feed be used?

4.    Could the jig be made of lighter material, or so designed with economy of material to allow easier handling?

5.    Are there other fixtures available that can be adapted to this job?

6.    Is the design of the jig correct?

7.    Would lower-cost tooling decrease quality?

8.    Is the jig designed to allow maximum motion economy?

9.    Can the part be quickly inserted and removed from the jig?

10.   Would a quick-acting, cam-actuated mechanism be desirable for tightening the jig, clamp or vice?

11.   Can ejectors be installed on the fixture for automatically removing the part when the fixture is opened?

12.   Are all operatives provided with the same tools?

13.   If accurate work is necessary, are proper gauges and other measuring instruments provided?

14.   Is the wooden equipment in use in good condition and are work benches free from splinters?

15.   Would a special bench or desk designed to eliminate stooping, bending and reaching reduce fatigue?

16.   Is pre-setting possible?

17.   Can universal tooling be used?

18.   Can setting time be reduced?

19.   How is material supply replenished?

20.   Can a hand or foot air-jet be supplied to the operative and applied with advantage?

21.   Could jigs be used?

22.   Could guides or bullet-nosed pins be used to position the part?

23.   What must be done to complete the operation and put away all the equipment?

**J. Working Conditions**

1.  Is the light even and sufficient at all times?

2.  Has glare been eliminated from the workplace?

3.  Is the proper temperature for comfort provided at all times; if not, can fans or heaters be used?

4.  Would installation of air-conditioning equipment be justified?

5.  Can noise levels be reduced?

6.  Can fumes, smoke and dirt be removed by exhaust systems?

7.  If concrete floors are used, are duckboards or matting provided to make standing more comfortable?

8.  Can a chair be provided?

9.  Are drinking fountains with cool water provided and are they located nearby?

10. Has due consideration been given to safety factors?

11. Is the floor safe, smooth but not slippery?

12. Has the operative been taught to work safely?

13. Is the clothing suitable from a safety standpoint?

14. Does the plant present a neat and orderly appearance at all times?

15. How thoroughly is the workplace cleaned?

16. Is the plant unduly cold in winter, or stuffy in summer, especially on the first morning of the week?

17. Are dangerous processes adequately guarded?

**K. Job Enrichment**

1.  Is the job boring or monotonous?

2.  Can the operation be made more interesting?

3.  Can the operation be combined with previous or subsequent operations to enlarge it?

4.  What is the cycle time?

5.  Can the operative do his own setting?

6.  Can he do his own inspection?

7.  Can he deburr his own work?

8.  Can he service his own tools?

9.  Can he be given a batch of tasks and do his own scheduling?

10. Can he make the complete part?

11. Is job rotation possible and desirable?

12. Can group layout be used?

13. Are flexible working hours possible and desirable?

14. Is the operation machine paced?

15. Can buffer stock be provided to allow variations in work pace?

16. Does the operative receive regular information about his performance?

# 3. Example of tables used to calculate relaxation allowances

This appendix is based on information supplied by Peter Steele and Partners (United Kingdom). Similar tables have been developed by various institutions, such as REFA (Federal Republic of Germany), and by other consulting firms.

Relaxation allowances may be determined by means of the tables of comparative strains and the points conversion table reproduced in this appendix. The analysis should proceed as follows:

(1) For the element of work under consideration, determine the severity of the strain imposed under each sub-heading of the table of strains below, by reference to the tables of comparative strains.

(2) Allocate points as indicated and determine the total points for the strains imposed by the performance of the element of work.

(3) Read off from the points conversion table the appropriate relaxation allowance.

*Table 1.   Points allocated for various strains: summary*

| Type of strain | Severity | | |
| --- | --- | --- | --- |
| | Low | Medium | High |
| **A.** *Physical strains resulting from nature of work* | | | |
| 1. Average force exerted | 0-85 | 0-113 | 0-149 |
| 2. Posture | 0-5 | 6-11 | 12-16 |
| 3. Vibration | 0-4 | 5-10 | 11-15 |
| 4. Short cycle | 0-3 | 4-6 | 7-10 |
| 5. Restrictive clothing | 0-4 | 5-12 | 13-20 |
| **B.** *Mental strains* | | | |
| 1. Concentration/anxiety | 0-4 | 5-10 | 11-16 |
| 2. Monotony | 0-2 | 3-7 | 8-10 |
| 3. Eye strain | 0-5 | 6-11 | 12-20 |
| 4. Noise | 0-2 | 3-7 | 8-10 |
| **C.** *Physical or mental strains resulting from nature of working conditions* | | | |
| 1. Temperature | | | |
| Low humidity | 0-5 | 6-11 | 12-16 |
| Medium humidity | 0-5 | 6-14 | 15-26 |
| High humidity | 0-6 | 7-17 | 18-36 |

| Type of strain | Severity | | |
|---|---|---|---|
| | Low | Medium | High |
| 2. Ventilation | 0-3 | 4-9 | 10-15 |
| 3. Fumes | 0-3 | 4-8 | 9-12 |
| 4. Dust | 0-3 | 4-8 | 9-12 |
| 5. Dirt | 0-2 | 3-6 | 7-10 |
| 6. Wet | 0-2 | 3-6 | 7-10 |

*Note*: Allocate points for each strain independently, irrespective of what has been allowed for other strains. If any strain occurs for only a proportion of the time, allocate a similar proportion of the points:

e.g. High concentration: 16 points, 25 per cent of the time;
  Low concentration: 4 points, 75 per cent of the time.
  Allocate 16 × 0.25 = 4 points plus 4 × 0.75 = 3 points, which gives a total of 4 + 3 = 7 points.

# TABLES OF COMPARATIVE STRAINS

## A. Physical strains resulting from the nature of the work

### i. AVERAGE FORCE EXERTED (FACTOR A.1)

Consider the whole of the element or period for which the relaxation allowance is required and determine the **average** force exerted.

*Example:*

Lift and carry a weight of 40 lb. (time 12 seconds) and return empty-handed (time 8 seconds). In this example, if the relaxation allowance is to apply to the full 20 seconds, the "average force exerted" should be calculated as follows:

$$\left(40 \times \frac{12}{20}\right) + \left(0 \times \frac{8}{20}\right) = 24 \text{ lb.}$$

The number of points allocated for the average force exerted will depend upon the type of stress involved. Stresses are classified as follows:

*(a)* Medium stress

   (i) Where the work is primarily concerned with carrying or supporting loads;

   (ii) shovelling, swinging hammers and other rhythmical movements.
This category covers most operations.

*(b)* Low stress

   (i) Where the weight of the body is transferred in order to exert force, e.g. foot-pedal operation; pressing an article, with the body, against a buff;

   (ii) supporting or carrying well balanced loads strapped to the body or hung from the shoulders; arms and hands free.

*(c)* High stress

   (i) Where the work is primarily concerned with lifting;

   (ii) exerting the force by continued use of certain muscles of fingers and arms;

   (iii) lifting or supporting loads in awkward attitudes, manipulation of heavy weights into awkward positions;

(iv) operations in hot conditions, hot metalworking, etc.

Relaxation allowances should be awarded in this category only after every endeavour has been made to improve facilities which will make the physical task lighter.

*Table 2. Medium stress: points for average force exerted*

| lb. | 0 | 1 | 2 | 3 | 4 | 5 | 6 | 7 | 8 | 9 |
|---|---|---|---|---|---|---|---|---|---|---|
| 0 | 0 | 0 | 0 | 0 | 3 | 6 | 8 | 10 | 12 | 14 |
| 10 | 15 | 16 | 17 | 18 | 19 | 20 | 21 | 22 | 23 | 24 |
| 20 | 25 | 26 | 27 | 28 | 29 | 30 | 31 | 32 | 32 | 33 |
| 30 | 34 | 35 | 36 | 37 | 38 | 39 | 39 | 40 | 41 | 41 |
| 40 | 42 | 43 | 44 | 45 | 46 | 46 | 47 | 48 | 49 | 50 |
| 50 | 50 | 51 | 51 | 52 | 53 | 54 | 54 | 55 | 56 | 56 |
| 60 | 57 | 58 | 59 | 59 | 60 | 61 | 61 | 62 | 63 | 64 |
| 70 | 64 | 65 | 65 | 66 | 67 | 68 | 69 | 70 | 70 | 71 |
| 80 | 72 | 72 | 72 | 73 | 73 | 74 | 74 | 75 | 76 | 76 |
| 90 | 77 | 78 | 79 | 79 | 80 | 80 | 81 | 82 | 82 | 83 |
| 100 | 84 | 85 | 86 | 86 | 87 | 88 | 88 | 88 | 89 | 90 |
| 110 | 91 | 92 | 93 | 94 | 95 | 95 | 96 | 96 | 97 | 97 |
| 120 | 97 | 98 | 98 | 98 | 99 | 99 | 99 | 100 | 100 | 100 |
| 130 | 101 | 101 | 102 | 102 | 103 | 104 | 105 | 106 | 107 | 108 |
| 140 | 109 | 109 | 109 | 110 | 110 | 111 | 112 | 112 | 112 | 113 |

*Table 3. Low stress: points for average force exerted*

| lb. | 0 | 1 | 2 | 3 | 4 | 5 | 6 | 7 | 8 | 9 |
|---|---|---|---|---|---|---|---|---|---|---|
| 0 | 0 | 0 | 0 | 0 | 3 | 6 | 7 | 8 | 9 | 10 |
| 10 | 11 | 12 | 13 | 14 | 14 | 15 | 16 | 16 | 17 | 18 |
| 20 | 19 | 19 | 20 | 21 | 22 | 22 | 23 | 23 | 24 | 25 |
| 30 | 26 | 26 | 27 | 27 | 28 | 28 | 29 | 30 | 31 | 31 |
| 40 | 32 | 32 | 33 | 34 | 34 | 35 | 35 | 36 | 36 | 37 |
| 50 | 38 | 38 | 39 | 39 | 40 | 41 | 41 | 42 | 42 | 43 |
| 60 | 43 | 43 | 44 | 44 | 45 | 46 | 46 | 47 | 47 | 48 |
| 70 | 48 | 49 | 50 | 50 | 50 | 51 | 51 | 52 | 52 | 53 |
| 80 | 54 | 54 | 54 | 55 | 55 | 56 | 56 | 57 | 58 | 58 |
| 90 | 58 | 59 | 59 | 60 | 60 | 60 | 61 | 62 | 62 | 63 |
| 100 | 63 | 63 | 64 | 65 | 65 | 66 | 66 | 66 | 67 | 67 |
| 110 | 68 | 68 | 68 | 69 | 69 | 70 | 71 | 71 | 71 | 72 |
| 120 | 72 | 73 | 73 | 73 | 74 | 74 | 75 | 75 | 76 | 76 |
| 130 | 77 | 77 | 77 | 78 | 78 | 78 | 79 | 80 | 80 | 81 |
| 140 | 81 | 82 | 82 | 82 | 83 | 83 | 84 | 84 | 84 | 85 |

### Table 4.   High stress: points for average force exerted

| lb. | 0 | 1 | 2 | 3 | 3-4 | 4 | 5 | 6 | 7 | 8 | 9 |
|-----|---|---|---|---|-----|---|---|---|---|---|---|
| 0 | 0 | 0 | 0 | 3 | 6 | 8 | 11 | 13 | 15 | 17 | 18 |
| 10 | 20 | 21 | 22 | 24 | | 25 | 27 | 28 | 29 | 30 | 32 |
| 20 | 33 | 34 | 35 | 37 | | 38 | 39 | 40 | 41 | 43 | 44 |
| 30 | 45 | 46 | 47 | 48 | | 49 | 50 | 51 | 52 | 54 | 55 |
| 40 | 56 | 57 | 58 | 59 | | 60 | 61 | 62 | 63 | 64 | 65 |
| 50 | 66 | 67 | 68 | 69 | | 70 | 71 | 72 | 73 | 74 | 75 |
| 60 | 76 | 76 | 77 | 78 | | 79 | 80 | 81 | 82 | 83 | 84 |
| 70 | 85 | 86 | 87 | 88 | | 88 | 89 | 90 | 91 | 92 | 93 |
| 80 | 94 | 94 | 95 | 96 | | 97 | 98 | 99 | 100 | 101 | 101 |
| 90 | 102 | 103 | 104 | 105 | | 105 | 106 | 107 | 108 | 109 | 110 |
| 100 | 110 | 111 | 112 | 113 | | 114 | 115 | 115 | 116 | 117 | 118 |
| 110 | 119 | 119 | 120 | 121 | | 122 | 123 | 124 | 124 | 125 | 126 |
| 120 | 127 | 128 | 128 | 129 | | 130 | 130 | 131 | 132 | 133 | 134 |
| 130 | 135 | 136 | 136 | 137 | | 137 | 138 | 139 | 140 | 141 | 142 |
| 140 | 142 | 143 | 143 | 144 | | 145 | 146 | 147 | 148 | 148 | 149 |

A study should be made of the elements in relation to low, medium and high stress conditions. The points to be allocated, according to the type of stress and the average force applied, are set out in tables II to IV.

*Example:* If the weight carried is 25 lb.—

(i) determine the type of stress involved (medium, low or high);

(ii) in the left-hand column of the table for the type of stress (table II, II or IV), find the line for 20 lb.;

(iii) on this line, move across the table to the right, to column 5;

(iv) read off the points allocation for 25 lb. carried, which is—

table II, medium stress: 30 points;

table III, low stress: 22 points;

table IV, high stress: 39 points.

## 2. POSTURE (FACTOR A.2)

Consider whether the worker is sitting, standing, stooping or in a cramped position and whether a load is handled easily or awkwardly.

| | Points |
|---|---|
| Sitting easily | 0 |
| Sitting awkwardly, or mixture of sitting and standing | 2 |
| Standing or walking freely | 4 |
| Ascending or descending stairs unladen | 5 |
| Standing or walking with a load | 6 |

| | |
|---|---|
| Climbing up or down ladders, or some bending, lifting, stretching or throwing | 8 |
| Awkward lifting, shovelling ballast to container | 10 |
| Constant bending, lifting, stretching or throwing | 12 |
| Coalmining with pickaxes, lying in a low seam | 16 |

## 3. VIBRATION (FACTOR A.3)

Consider the impact of the vibration on the body, limbs or hands and the addition to mental effort due to it, or to a series of jars or shocks.

|  | Points |
|---|---|
| Shovelling light materials | 1 |
| Power sewing-machine | |
| Power press or guillotine if operative is holding the material | 2 |
| Cross-cut sawing | |
| Shovelling ballast | |
| Portable power drill operated by one hand | 4 |
| Pickaxing | 6 |
| Power drill (two hands) | 8 |
| Road drill on concrete | 15 |

## 4. SHORT CYCLE (HIGHLY REPETITIVE) (FACTOR A.4)

In highly repetitive work, if a series of very short elements form a cycle which is continuously repeated for a long period, award points as indicated below, to compensate for the lack of opportunity to vary the muscles used during the work.

| Average cycle time (centiminutes) | Points |
|---|---|
| 16-17 | 1 |
| 15 | 2 |
| 13-14 | 3 |
| 12 | 4 |
| 10-11 | 5 |
| 8-9 | 6 |
| 7 | 7 |
| 6 | 8 |
| 5 | 9 |
| Less than 5 | 10 |

## 5. RESTRICTIVE CLOTHING (FACTOR A.5)

Consider the weight of the protective clothing in relation to effort and movement. Consider also whether ventilation and breathing are affected.

|  | Points |
|---|---|
| Thin rubber (surgeon's) gloves | 1 |
| Household rubber gloves | |
| Rubber boots | 2 |

| | |
|---|---:|
| Grinder's goggles | 3 |
| Industrial rubber or leather gloves | 5 |
| Face mask (e.g. for paint-spraying) | 8 |
| Asbestos suit or tarpaulin coat | 15 |
| Restrictive protective clothing and respirator | 20 |

# B. Mental strains

## 1. CONCENTRATION/ANXIETY (FACTOR B.1)

Consider what would happen if the operative relaxed his attention, the responsibility carried, the need for exact timing of movements, and the accuracy or precision required.

*Points*

| | |
|---|---:|
| Routine simple assembly } <br> Shovelling ballast } | 0 |
| Routine packing, labourer washing vehicles } <br> Wheeling trolley down clear gangway } | 1 |
| Feed press tool; hand clear of press } <br> Topping up battery } | 2 |
| Painting walls | 3 |
| Assembling small and simple batches, performed without much thinking } <br> Sewing-machine work, automatically guided } | 4 |
| Assembling warehouse orders by trolley } <br> Simple inspection } | 5 |
| Load/unload press tool, hand feed into machine } <br> Spray-painting metalwork } | 6 |
| Adding up figures } <br> Inspecting detailed components } | 7 |
| Buffing and polishing | 8 |
| Guiding work by hand on sewing-machine <br> Packing assorted chocolates, memorising pattern and selecting accordingly } <br> Assembly work too complex to become automatic <br> Welding parts held in jig | 10 |
| Driving a motor bus in heavy traffic or fog } <br> Marking out in detail with high accuracy } | 15 |

## 2. MONOTONY (FACTOR B.2)

Consider the degree of mental stimulation and if there is companionship, competitive spirit, music, etc.

*Points*

| | |
|---|---:|
| Two men on jobbing work | 0 |

| | | |
|---|---|---|
| Cleaning own shoes for half an hour on one's own | | 3 |
| Operative on repetitive work | } | 5 |
| Operative working alone on non-repetitive work | | |
| Routine inspection | | 6 |
| Adding similar columns of figures | | 8 |
| One operative working alone on highly repetitive work | | 11 |

## 3. EYE STRAIN (FACTOR B.3)

Consider the lighting conditions, glare, flicker, illumination, colour and closeness of work and for how long the strain is endured.

| | | Points |
|---|---|---|
| Normal factory work | | 0 |
| Inspection of easily visible faults | } | 2 |
| Sorting distinctively coloured articles by colour | | |
| Factory work in poor lighting | | |
| Intermittent inspection for detailed faults | } | 4 |
| Grading apples | | |
| Reading a newspaper in a motor bus | | 8 |
| Arc-welding using mask | } | 10 |
| Continuous visual inspection, e.g. cloth from a loom | | |
| Engraving using an eyeglass | | 14 |

## 4. NOISE (FACTOR B.4)

Consider whether the noise affects concentration, is a steady hum or a background noise, is regular or occurs unexpectedly, is irritating or soothing. (Noise has been described as "a loud sound made by somebody else".)

| | | Points |
|---|---|---|
| Work in a quiet office, no distracting noise | } | 0 |
| Light assembly factory | | |
| Work in a city office with continual traffic noise outside | | 1 |
| Light machine shop | } | 2 |
| Office or assembly shop where noise is a distraction | | |
| Woodworking machine shop | | 4 |
| Operating steam hammer in forge | | 5 |
| Rivetting in a shipyard | | 9 |
| Road drilling | | 10 |

# C. Physical or mental strains resulting from the nature of the working conditions

## 1. TEMPERATURE AND HUMIDITY (FACTOR C.1)

Consider the general conditions of atmospheric temperature and humidity and classify as indicated below. Select points according to average temperature within the ranges shown.

| Humidity (per cent) | Temperature | | |
|---|---|---|---|
| | Up to 75°F | 76° to 90°F | Over 90°F |
| Up to 75 | 0 | 6- 9 | 12-16 |
| 76-85 | 1-3 | 8-12 | 15-26 |
| Over 85 | 4-6 | 12-17 | 20-36 |

## 2. VENTILATION (FACTOR C.2)

Consider the quality and freshness of the air and its circulation by air-conditioning or natural draught.

| | Points |
|---|---|
| Offices<br>Factories with "office-type" conditions | 0 |
| Workshop with reasonable ventilation but some draught | 1 |
| Draughty workshops | 3 |
| Working in sewer | 14 |

## 3. FUMES (FACTOR C.3)

Consider the nature and concentration of the fumes: whether toxic or injurious to health; irritating to eyes, nose, throat or skin; disagreeable odour.

| | Points |
|---|---|
| Lathe turning with coolants | 0 |
| Emulsion paint<br>Gas cutting<br>Soldering with resin | 1 |
| Motor vehicle exhaust in small commercial garage | 5 |
| Cellulose painting | 6 |
| Moulder procuring metal and filling mould | 10 |

## 4. DUST (FACTOR C.4)

Consider the volume and nature of the dust.

|  | Points |
|---|---|
| Office | |
| Normal light assembly operations } | 0 |
| Press shop | |
| Grinding or buffing operations with good extraction | 1 |
| Sawing wood | 2 |
| Emptying ashes | 4 |
| Linishing weld | 6 |
| Running coke from hoppers into skips or trucks | 10 |
| Unloading cement | 11 |
| Demolishing building | 12 |

## 5. DIRT (FACTOR C.5)

Consider the nature of the work and the general discomfort caused by its dirty nature. This allowance covers "washing time" where this is paid for (i.e. where operatives are allowed three minutes or five minutes for washing, etc.). Do **not** allow both points and time.

|  | Points |
|---|---|
| Office work | |
| Normal assembly operations } | 0 |
| Office duplicators | 1 |
| Dustman | 2 |
| Stripping internal combustion engine | 4 |
| Work under old motor vehicle | 5 |
| Unloading bags of cement | 7 |
| Coalminer | |
| Chimney-sweep with brushes } | 10 |

## 6. WET (FACTOR C.6)

Consider the cumulative effect of exposure to this condition over a long period.

|  | Points |
|---|---|
| Normal factory operations | 0 |
| Outdoor workers, e.g. postman | 1 |
| Working continuously in the damp | 2 |
| Rubbing down walls with wet pumice block | 4 |
| Continuous handling of wet articles | 5 |
| Laundry wash-house, wet work, steamy, floor running with water, hands wet | 10 |

# POINTS CONVERSION TABLE

Table 5.   *Percentage relaxation allowance for total points allocated*

| Points | 0 | 1 | 2 | 3 | 4 | 5 | 6 | 7 | 8 | 9 |
|--------|----|----|----|----|----|----|----|----|----|----|
| 0 | 10 | 10 | 10 | 10 | 10 | 10 | 10 | 11 | 11 | 11 |
| 10 | 11 | 11 | 11 | 11 | 11 | 12 | 12 | 12 | 12 | 12 |
| 20 | 13 | 13 | 13 | 13 | 14 | 14 | 14 | 14 | 15 | 15 |
| 30 | 15 | 16 | 16 | 16 | 17 | 17 | 17 | 18 | 18 | 18 |
| 40 | 19 | 19 | 20 | 20 | 21 | 21 | 22 | 22 | 23 | 23 |
| 50 | 24 | 24 | 25 | 26 | 26 | 27 | 27 | 28 | 28 | 29 |
| 60 | 30 | 30 | 31 | 32 | 32 | 33 | 34 | 34 | 35 | 36 |
| 70 | 37 | 37 | 38 | 39 | 40 | 40 | 41 | 42 | 43 | 44 |
| 80 | 45 | 46 | 47 | 48 | 48 | 49 | 50 | 51 | 52 | 53 |
| 90 | 54 | 55 | 56 | 57 | 58 | 59 | 60 | 61 | 62 | 63 |
| 100 | 64 | 65 | 66 | 68 | 69 | 70 | 71 | 72 | 73 | 74 |
| 110 | 75 | 77 | 78 | 79 | 80 | 82 | 83 | 84 | 85 | 87 |
| 120 | 88 | 89 | 91 | 92 | 93 | 95 | 96 | 97 | 99 | 100 |
| 130 | 101 | 103 | 105 | 106 | 107 | 109 | 110 | 112 | 113 | 115 |
| 140 | 116 | 118 | 119 | 121 | 122 | 123 | 125 | 126 | 128 | 130 |

*Example:* If the total number of points allocated for the various strains is 37:

   (i) in the left-hand column of table V, find the line for 30;

   (ii) on this line, move across the table to the right, to column 7;

   (iii) read off the relaxation allowance for 37 points, which is 18 per cent.

# EXAMPLES OF CALCULATION OF RELAXATION ALLOWANCES

1. *Power press operation.* As press guard opens automatically, reach in with left hand, grasp piece-part, and disengage it. With left hand move piece-part to tote bin, while right hand places new blank in press tool. Withdraw right hand, while left hand closes guard. Operate press with foot. Simultaneously, with right hand reach to tote bin, grasp blank and orient it in hand, move blank near guard and wait for guard to open.

   On 20-ton press. Maximum reach 50 cm (20 in.). Posture somewhat unnatural; seated at machine. Noisy department, adequate lighting.

2. *Carry 50 lb. sack up stairs.* Lift sack on to bench 90 cm (3 ft.) high; transfer to shoulder, carry up stairs, drop sack on floor. Dusty conditions.

3. *Pack chocolates* in three layers of 4 lb. box, according to pattern for each layer, average 160 chocolates. Operative sits in front of straight shelves bearing 11 kinds of chocolates in trays or tins; he must pack the chocolates according to a memorised pattern for each layer. Air-conditioned, good light.

Table 6.    Calculation of relaxation allowances: examples

| Type of strain | Job | | | | | |
|---|---|---|---|---|---|---|
| | Power press operation | | Carrying 50 lb. sack | | Packing chocolates | |
| | Stress | Points | Stress | Points | Stress | Points |
| **A.  Physical strains** | | | | | | |
| 1.  Average force (lb.) | — | — | M | 50 | — | — |
| 2.  Posture | L | 4 | M | 6 | L | 2 |
| 3.  Vibration | L | 2 | L | — | — | — |
| 4.  Short cycle | H | 10 | L | — | — | — |
| 5.  Restrictive clothing | — | — | — | — | — | — |
| **B.  Mental strains** | | | | | | |
| 1.  Concentration/anxiety | M | 6 | L | 1 | H | 10 |
| 2.  Monotony | M | 6 | L | 1 | L | 2 |
| 3.  Eye strain | L | 3 | — | — | L | 2 |
| 4.  Noise | M | 4 | L | — | L | 1 |
| **C.  Working conditions** | | | | | | |
| 1.  Temperature/humidity | — | — | L/L | 1 | L/L | 3 |
| 2.  Ventilation | — | — | — | — | — | — |
| 3.  Fumes | — | — | — | — | — | — |
| 4.  Dust | — | — | H | 9 | — | — |
| 5.  Dirt | M | 3 | L | — | — | — |
| 6.  Wet | — | — | L | — | — | — |
| Total points | | 38 | | 68 | | 20 |
| Relaxation allowance, including tea breaks (per cent) | | 18 | | 35 | | 13 |

# 4. Conversion factors

| (1) | (2) | To convert column (1) into column (2), multiply by |
|---|---|---|
| **Length** | | |
| Inches | Feet | 0.083 |
| Inches | Centimetres | 2.540 |
| Feet | Yards | 0.333 |
| Feet | Metres | 0.305 |
| Yards | Feet | 3 |
| Yards | Metres | 0.914 |
| Poles | Yards | 5.502 |
| Poles | Metres | 5.029 |
| Furlongs | Miles | 0.125 |
| Furlongs | Kilometres | 0.201 |
| Miles | Yards | 1,760 |
| Miles | Kilometres | 1.609 |
| Fathoms | Feet | 6 |
| Fathoms | Metres | 1.829 |
| Centimetres | Inches | 0.394 |
| Metres | Feet | 3.281 |
| Metres | Yards | 1.094 |
| Metres | Poles | 0.199 |
| Metres | Fathoms | 0.547 |
| Kilometres | Furlongs | 4.975 |
| Kilometres | Miles | 0.62[1] |
| **Area** | | |
| Square inches | Square feet | 0.0069 |
| Square inches | Square centimetres | 6.452 |
| Square feet | Square yards | 0.111 |
| Square feet | Square metres | 0.093 |
| Square yards | Square feet | 9 |
| Square yards | Square metres | 0.836 |
| Acres | Square feet | 43,560 |
| Acres | Square miles | 0.0016 |
| Acres | Square metres | 4047 |
| Acres | Hectares | 0.405 |
| Square miles | Square feet | 27,878,400 |
| Square miles | Square kilometres | 2.590 |

| (1) | (2) | To convert column (1) into column (2), multiply by |
|---|---|---|
| Square miles | Hectares | 259.2 |
| Square miles | Acres | 640 |
| Square centimetres | Square inches | 0.155 |
| Square metres | Square feet | 10.754 |
| Square metres | Square yards | 1.196 |
| Square metres | Acres | 0.0025 |
| Square kilometres | Hectares | 100 |
| Square kilometres | Acres | 247.105 |
| Square kilometres | Square miles | 0.386 |
| Hectares | Acres | 2.471 |
| Hectares | Square kilometres | $10^{-2}$ |
| Hectares | Square miles | 0.0039 |

*Volume*

| | | |
|---|---|---|
| Cubic inches | Cubic feet | $5.787 \times 10^{-4}$ |
| Cubic inches | Cubic centimetres | 16.387 |
| Cubic feet | Cubic yards | 0.037 |
| Cubic feet | Cubic metres | 0.028 |
| Cubic yards | Cubic feet | 27 |
| Cubic yards | Cubic metres | 0.765 |
| Cubic centimetres | Cubic feet | $3.53 \times 10^{-5}$ |
| Cubic centimetres | Cubic inches | 0.061 |
| Cubic metres | Cubic yards | 1.308 |
| Cubic metres | Cubic feet | 35.315 |

*Liquid measure*

| | | |
|---|---|---|
| Fluid ounces (Imperial) | Fluid ounces (US) | 0.961 |
| Fluid ounces (Imperial) | Millilitres | 28.413 |
| Fluid ounces (US) | Fluid ounces (Imperial) | 1.041 |
| Fluid ounces (US) | Millilitres | 29.574 |
| Pints (Imperial) | Pints (US) | 1.201 |
| Pints (Imperial) | Quarts | 0.5 |
| Pints (Imperial) | Gallons (Imperial) | 0.125 |
| Pints (Imperial) | Litres | 0.568 |
| Pints (US) | Pints (Imperial) | 0.833 |
| Pints (US) | Litres | 0.473 |
| Gills | Pints | 0.25 |
| Gills | Litres | 0.142 |
| Gallons (Imperial) | Gallons (US) | 1.201 |
| Gallons (Imperial) | Litres | 4.546 |
| Gallons (US) | Gallons (Imperial) | 0.833 |
| Gallons (US) | Litres | 3.785 |
| Cubic centimetres | Litres | $10^{-3}$ |
| Litres | Pints (Imperial) | 1.760 |
| Litres | Pints (US) | 2.113 |

| (1) | (2) | To convert column (1) into column (2), multiply by |
| --- | --- | --- |
| *Weight* | | |
| Grains (avdp.) | Grains (troy) | 1.003 |
| | Grams | 0.0648 |
| Grains (troy) | Grains (avdp.) | 0.996 |
| | Grams | 0.0648 |
| Pennyweight (troy) | Grains (troy) | 24 |
| | Grams | 1.555 |
| Ounces (avdp.) | Ounces (troy) | 0.9115 |
| | Pounds | 0.0625 |
| | Grams | 28.350 |
| Ounces (troy) | Ounces (avdp.) | 1.097 |
| | Grams | 31.104 |
| Pounds (avdp.) | Pounds (troy) | 1.215 |
| | Ounces (avdp.) | 16 |
| | Kilograms | 0.454 |
| Pounds (troy) | Pounds (avdp.) | 0.823 |
| | Ounces (troy) | 12 |
| | Kilograms | 0.373 |
| Stones | Pounds (avdp.) | 14 |
| | Grams | 6350.297 |
| Tons (short) | Pounds (avdp.) | 2000. |
| | Kilograms | 907.185 |
| Tons (long) | Pounds (avdp.) | 2240. |
| | Kilograms | 1016.047 |
| Grams | Ounces (avdp.) | 0.035 |
| | Ounces (troy) | 0.032 |
| Kilograms | Pounds (avdp.) | 2.205 |
| | Tons (short) | 0.0011 |
| | Tons (long) | 0.00098 |

# 5. Selected bibliography

Aguren, S.; Hansson, R.; Karlsson, K. G.: *The Volvo Kalmar plant: The impact of new design on work organisation* (Stockholm, Rationalisation Council–Swedish Employers' Confederation–Swedish Trade Union Confederation, 1976).

Alford, L. P.; Bangs, J. R.: *Production handbook* (New York, Ronald Press, 2nd ed., 1964).

Allenspach, Heinz: *Flexible working hours* (Geneva, ILO, 1975).

Arscott, P. E.; Armstrong, M.: *An employer's guide to health and safety management: A handbook for industry* (London, Engineering Employers' Federation, 1976).

Ashcroft, H.: "The productivity of several machines under the care of one operator", in *Journal of the Royal Statistical Society* (London), Series B, Vol. XII, 1950, pp. 145-151.

Barnes, Ralph M.: *Work sampling* (New York and London, John Wiley, 2nd ed., 1957).

—: *Motion and time study: Design and measurement of work* (New York and London, John Wiley, 6th ed., 1969).

Benson, F.: "Further notes on the productivity of machines requiring attention at random intervals", in *Journal of the Royal Statistical Society*, Series B, Vol. XIV, 1952, pp. 200-210.

—: Cox, D. R.: "The productivity of machines requiring attention at random intervals", in *Journal of the Royal Statistical Society*, Series B, Vol. XIII, 1951, pp. 65-82.

Biel-Nisen, H. E.: "Universal maintenance standards", in *Journal of Methods-Time Measurement* (Fair Lawn, NJ), Vol. 7, Nos. 4 and 5, Nov. 1960-Feb. 1961.

Bowman, Edmond; Fetter, Robert: *Analysis for production management* (Homewood, Ill., Richard Irwin, 1961).

British Institute of Management: *Classification and coding: An introduction and review of classification and coding systems* (London, 1971).

British Standards Institution: *Glossary of terms used in work study* (London, 1969).

Buffa, E. S.: *Modern production management* (New York and London, John Wiley, 4th ed., 1973).

Burbidge, J. L.: *Principles of production control* (London, Macdonald and Evans, 3rd ed., 1971).

—: *Production planning* (London, Heinemann, 1971).

—: *The introduction of group technology* (London, Heinemann, 1974).

Bureau des temps élémentaires: *Vocabulaire technique concernant l'étude du travail* (Paris, Les Editions d'organisation, 1954).

—: "La préparation scientifique des décisions (recherche opérationnelle) appliquée à l'étude du travail", in *L'étude du travail* (Paris), Jan. 1960, pp. 7-24.

Carpentier, J.; Cazamian, P.: *Night work: Its effects on the health and welfare of the worker* (Geneva, ILO, 1977).

Garroll, P.: *How to chart data* (New York and London, McGraw-Hill, 1960).

Carson, G. B. (ed.) et al.: *Production handbook* (New York, Ronald Press, 3rd ed., 1972).

Cemach, H. P.: *Work study in the office* (Barking, UK, Applied Science Publishers, 4th ed., 1969).

de Chantal, R.: "Etude du travail et théorie des attentes", in *L'étude du travail*, June 1957, pp. 14-20.

Cox, D. R.: "Tables on operator efficiency in multi-machine operation", in *Journal of the Royal Statistical Society,* Series B, Vol. XV, 1953.

—: "A table for predicting the production from a group of machines under the care of one operative", in *Journal of the Royal Statistical Society,* Series B, Vol. XVI, 1954, pp. 285-287.

Crossan, R. M.; Nance, H. W.: *Master standard data: The economic approach to work measurement* (New York and London, McGraw-Hill, 2nd ed., 1972).

Currie, R. M.: *Simplified P.M.T.S.* (London, British Institute of Management, 1963).

—: *Financial incentives based on work measurement,* 2nd ed. revised J. E. Faraday (London, British Institute of Management, 1971).

—: *Work study,* 4th ed. revised J. E. Faraday (London, Pitman, 1977).

Edwards, G. A. B.: *Readings in group technology* (Brighton, UK, Machinery Publishing Co., 1971).

Evans, A. A.: *Hours of work in industrialised countries* (Geneva, ILO, 1975).

Grant, E. L.: *Statistical quality control* (New York and London, McGraw-Hill, 4th ed., 1972).

Heyde, Chris: *The sensible taskmaster* (Sydney, Heyde Dynamics, 1976).

Hunter, D.: *The diseases of occupations* (London, Hodder and Stoughton Educational, 6th ed., 1977).

International Labour Office (ILO): *Higher productivity in manufacturing industries,* Studies and reports, New series, No. 38 (Geneva, 3rd. impr., 1967).

—: *Accident prevention: A workers' education manual* (Geneva, 9th impr., 1978).

—: *Encyclopaedia of occupational health and safety,* 2 vols. (Geneva, 5th impr., 1976).

—: *Job evaluation,* Studies and reports, New series, No. 56 (Geneva, 9th impr., 1977).

—: *Payment by results,* Studies and reports, New series, No. 27 (Geneva, 14th impr., 1977).

—: *Protection of workers against noise and vibration in the working environment* (Geneva, 2nd impr., 1980).

—: *Management of working time in industrialised countries* (Geneva, 1978).

—: *New forms of work organisation, 1* (Geneva, 1979).

—: *New forms of work organisation, 2* (Geneva, 1979).

—: *Managing and developing new forms of work organisation,* edited by G. Kanawaty (Geneva, 2nd ed., 1981).

International Occupational Safety and Health Centre (CIS): *CIS Abstracts* (Geneva, ILO–CIS; published periodically).

Ishikawa, Kaoru: *Guide to quality control* (Tokyo, Asian Productivity Organisation, 1976).

Lehmann, J. T.: *La mesure des temps alloués* (Louvain, Librairie universitaire, 1965).

Lindholm, Rolf; Norstedt, Jan-Peder: *The Volvo report* (Stockholm, Swedish Employers' Confederation, 1975).

Mallick, R. W.; Gaudreau, A. T.: *Plant layout and practice* (New York and London, John Wiley, 1966).

Marić, D.: *Adapting working hours to modern needs* (Geneva, ILO, 1977).

Marriott, R.: *Incentive payment systems: A review of research and opinion* (London, Staples Press, 3rd ed., 1969).

Mary, J. A.: *L'expérience Guilliet* (Paris, Union des industries metallurgiques et minières (UIMM), 1975).

Maurice, M.: *Shift work: Economic advantages and social costs* (Geneva, ILO, 1975).

Mayer, Raymond E.: *Production and operations management* (New York and London, McGraw-Hill, 3rd ed., 1975).

Maynard, H. B.: *Production: An international appraisal of contemporary manufacturing systems and the changing role of the worker* (New York and London, McGraw-Hill, 1975).